AF361880

Detection, Control and Contamination of Mycotoxins

Detection, Control and Contamination of Mycotoxins

Editors

Chiara Cavaliere
Anna Laura Capriotti
Carmela Maria Montone
Andrea Cerrato

MDPI • Basel • Beijing • Wuhan • Barcelona • Belgrade • Manchester • Tokyo • Cluj • Tianjin

Editors

Chiara Cavaliere
Sapienza Università di Roma
Italy

Anna Laura Capriotti
Sapienza Università di Roma
Italy

Carmela Maria Montone
Sapienza Università di Roma
Italy

Andrea Cerrato
Sapienza Università di Roma
Italy

Editorial Office
MDPI
St. Alban-Anlage 66
4052 Basel, Switzerland

This is a reprint of articles from the Special Issue published online in the open access journal *Toxins* (ISSN 2072-6651) (available at: https://www.mdpi.com/journal/toxins/special_issues/detection_ mycotoxins).

For citation purposes, cite each article independently as indicated on the article page online and as indicated below:

LastName, A.A.; LastName, B.B.; LastName, C.C. Article Title. *Journal Name* **Year**, *Volume Number*, Page Range.

ISBN 978-3-0365-6350-3 (Hbk)
ISBN 978-3-0365-6351-0 (PDF)

Contents

Review

Recent Progress in Rapid Determination of Mycotoxins Based on Emerging Biorecognition Molecules: A Review

Yanru Wang [1], Cui Zhang [1], Jianlong Wang [1,*] and Dietmar Knopp [2,*]

[1] College of Food Science and Engineering, Northwest A&F University, Yangling, Xianyang 712100, China; yanruwang@nwafu.edu.cn (Y.W.); cuizhang@nwafu.edu.cn (C.Z.)
[2] Chair for Analytical Chemistry and Water Chemistry, Institute of Hydrochemistry, Technische Universitat München, Elisabeth-Winterhalter-Weg 6, D-81377 München, Germany
* Correspondence: wanglong79@nwsuaf.edu.cn (J.W.); dietmar.knopp@mytum.de (D.K.)

Abstract: Mycotoxins are secondary metabolites produced by fungal species, which pose significant risk to humans and livestock. The mycotoxins which are produced from *Aspergillus*, *Penicillium*, and *Fusarium* are considered most important and therefore regulated in food- and feedstuffs. Analyses are predominantly performed by official laboratory methods in centralized labs by expert technicians. There is an urgent demand for new low-cost, easy-to-use, and portable analytical devices for rapid on-site determination. Most significant advances were realized in the field bioanalytical techniques based on molecular recognition. This review aims to discuss recent progress in the generation of native biomolecules and new bioinspired materials towards mycotoxins for the development of reliable bioreceptor-based analytical methods. After brief presentation of basic knowledge regarding characteristics of most important mycotoxins, the generation, benefits, and limitations of present and emerging biorecognition molecules, such as polyclonal (pAb), monoclonal (mAb), recombinant antibodies (rAb), aptamers, short peptides, and molecularly imprinted polymers (MIPs), are discussed. Hereinafter, the use of binders in different areas of application, including sample preparation, microplate- and tube-based assays, lateral flow devices, and biosensors, is highlighted. Special focus, on a global scale, is placed on commercial availability of single receptor molecules, test-kits, and biosensor platforms using multiplexed bead-based suspension assays and planar biochip arrays. Future outlook is given with special emphasis on new challenges, such as increasing use of rAb based on synthetic and naïve antibody libraries to renounce animal immunization, multiple-analyte test-kits and high-throughput multiplexing, and determination of masked mycotoxins, including stereoisomeric degradation products.

Keywords: mycotoxins; antibodies; aptamers; short peptides; molecularly imprinted polymers; rapid tests; lateral flow assay; microplate assay; biosensor; multiplexing

Key Contribution: In this work, we summarize recent progress in the generation of native biomolecules and new bioinspired materials towards mycotoxins for the development of reliable bioreceptor-based analytical methods. Special focus, on a global scale, is placed on commercial availability of single reagents, test-kits, and biosensor platforms. Future outlook emphasizes new challenges.

Citation: Wang, Y.; Zhang, C.; Wang, J.; Knopp, D. Recent Progress in Rapid Determination of Mycotoxins Based on Emerging Biorecognition Molecules: A Review. *Toxins* **2022**, *14*, 73. https://doi.org/10.3390/toxins14020073

Received: 20 December 2021
Accepted: 15 January 2022
Published: 20 January 2022

Publisher's Note: MDPI stays neutral with regard to jurisdictional claims in published maps and institutional affiliations.

1. Introduction

Mycotoxins are secondary metabolites produced by different species of filamentous fungi, including *Aspergillus*, *Fusarium*, *Penicillium*, *Alternaria*, *Claviceps*, etc. [1–3]. They can be found in various food and feed, such as cereals, nuts, oilseeds, fruits, spices, coffee, wine, beer, and foods of animal origin, including dairy products, meat, and eggs [4–8]. Mycotoxin contamination in both food and feed commodities is considered to be inevitable due to the widespread occurrence of mycotoxin-producing fungi in the environment [9]. Although more than 400 mycotoxins with diverse structures have been identified, a limited number of compounds are considered a problem in food and feed safety. These include aflatoxins (AFs) [10,11], ochratoxin A (OTA) [12,13], fumonisins (FMs) [14], T-2/HT-2 toxins [15,16], deoxynivalenol (DON) [17,18], zearalenone (ZEN) [19,20], citrinin (CIT) [21], patulin (PAT) [22,23], and ergot alkaloids (EAs) [24,25] due to their significant prevalence in food and feed and severe health risks to humans and animals.

Among these mycotoxins, AFs have received the most attention due to their high toxicity. Aflatoxin B1 (AFB1) has been classified as Group 1 agent (potent human carcinogen) by the International Agency for Research on Cancer (IARC) of World Health Organization (WHO). It has long been associated with liver cancer, and more recent researches have exposed its negative role in nutrition outcomes and immune suppression effects [26]. OTA (IARC 1993) and fumonisin B1 (FB1) (IARC 2002) are suspect human carcinogens, which are classified as Group 2B agents (possibly carcinogenic in humans) [27]. The presence of other mycotoxins in diet has also been demonstrated to cause adverse and chronic health effects, such as gastrointestinal symptoms (DON) [28], endocrine-disrupting effects (ZEN) [29], growth retardation (DON, T-2 toxin) [30,31], nephrotoxicity (CIT) [32], and genotoxicity (DON, CIT, PAT) [33–35]. Furthermore, co-exposure of several mycotoxins to humans and animals through diet may cause additive or synergistic effects, which have been reported in studies using cell cultures and animals [36–38]. Table 1 lists the major mycotoxins and their main producing fungi species, affected food commodities, and toxic effects to humans and animals.

Owing to their poisonous character and widespread prevalence in food and feed products, maximum permitted levels (maximum residue limits, MRLs) for most toxic mycotoxins in multiple food and feed products have been set worldwide. The limit values differ among countries as well as to related commodities. Table 2 compares the maximum permitted levels of major mycotoxins in food as set by the European Union (EU), the United States (U.S.), and China. Among all the food commodities, infant foods have the lowest permitted levels for all mycotoxins.

Table 1. Summary of major mycotoxins and their characteristics.

Mycotoxins	Structure	Main Fungi Species	Commodities Affected	Toxic Effects
Aflatoxins: B_1, B_2, G_1, G_2, M_1*	AFB$_1$ 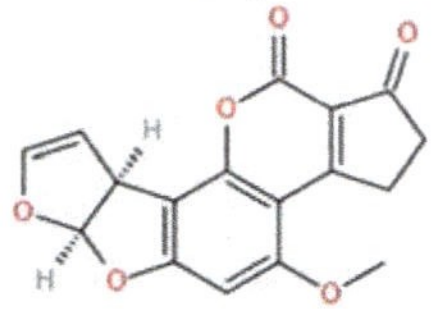	*Asp. flavus, Asp. parasiticus*	Nuts, spices, grains such as maize, rice, wheat, * milk and milk products, etc.	Carcinogenic, teratogenic, mutagenic, immunosuppressive [39]
DON	DON 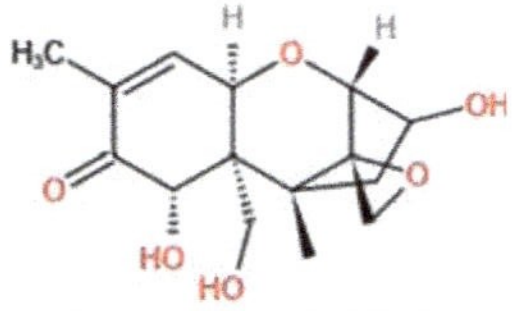	*F. graminearum, F. culmorum, F. cerealis*	Cereals, cereal products	Diarrhea, vomiting, anorexia, immune dysregulation [40]
ZEN	*trans*-ZEN 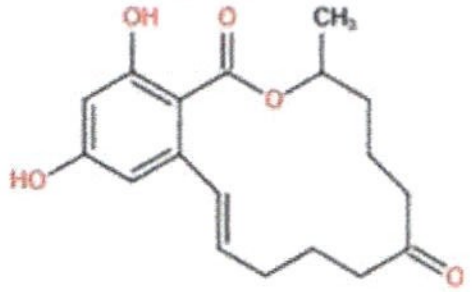 / *cis*-ZEN 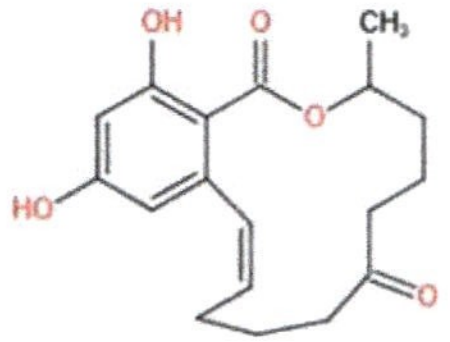	*F. graminearum* (*Gibberella zeae*), *F. culmorum, F. cerealis, F. equiseti, F. crookwellense, F. semitectum*	Cereals, cereal products, maize, rice, beer, etc.	Hyperoestrogenic, hepatotoxic, haematotoxic, immunotoxic, genotoxic [41]

Table 1. *Cont.*

Mycotoxins	Structure	Main Fungi Species	Commodities Affected	Toxic Effects
OTA	OTA 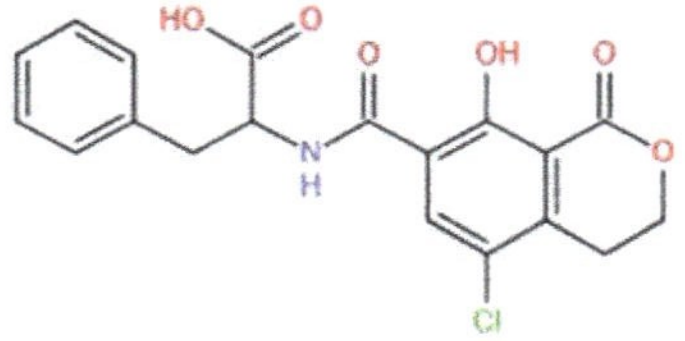	*Asp. ochraceus, Asp. niger, Asp. carbonarius, Asp. terreus, P. verrucosum, P. nordicum*	Cereals, wine, coffee, cocoa, beans, dried fruits, nuts, spices, cheese, etc.	Nephrotoxic, hepatotoxic, neurotoxic, teratogenic, immunotoxic [42]
Fumonisins: FB$_1$, FB$_2$, FB$_3$	FB$_1$ 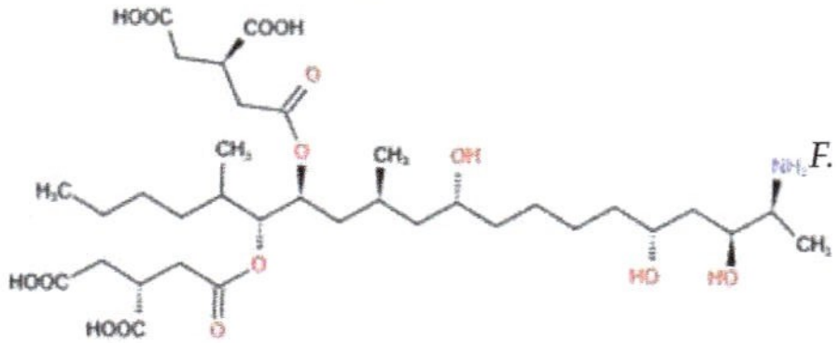	*F. verticillioides, F. proliferatum*	Mainly maize and maize-based products, sorghum, asparagus	Carcinogenic, cytotoxic, nephrotoxic, hepatotoxic [43,44]
T-2/HT-2 toxin	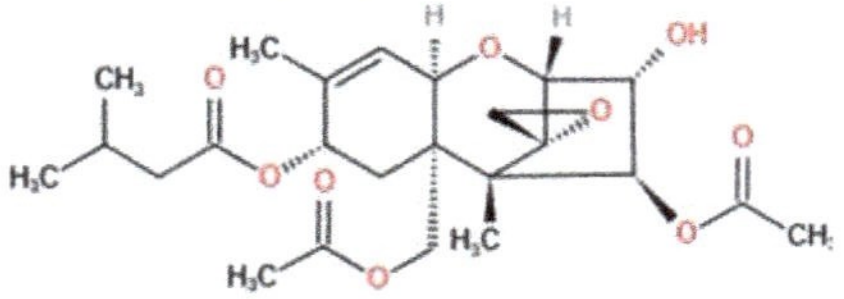 T-2 toxin 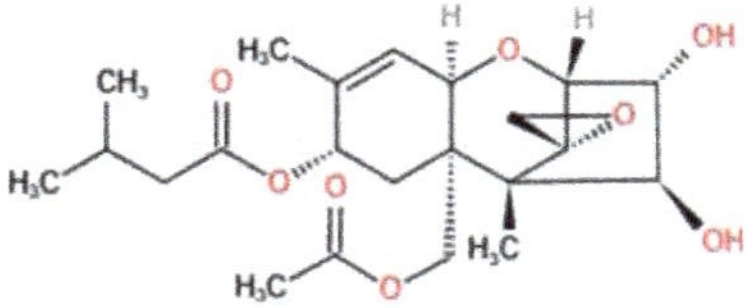 HT-2 toxin	*F. langsethiae, F. poae, F. sporotrichioides*	Wheat, rye, maize, soybeans	Growth retardation, myelotoxic, hemotoxic, necrotic lesions on contact sites [45]

Table 1. *Cont.*

Mycotoxins	Structure	Main Fungi Species	Commodities Affected	Toxic Effects
PAT	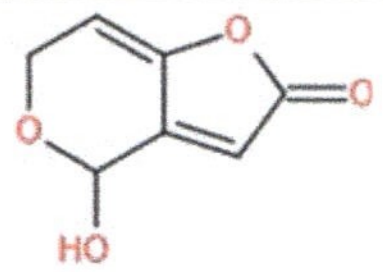	*P. expansum*	Fruits and vegetables	Nausea, vomiting and other gastro-intestinal symptoms, kidney damage [46]
CIT	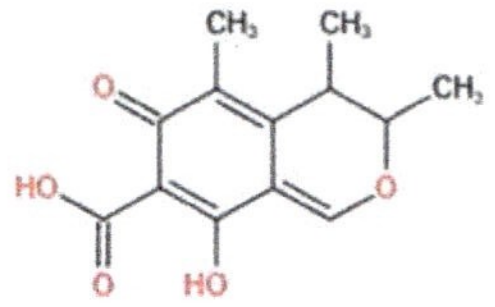	*P. citrinum, P. camemberti, Asp. terreus, Asp. niveus*	Fermented maize, cheese, corn, wheat, barley, red yeast rice, apples, brewed beer, cereal products	Nephrotoxic, may cause liver and kidney diseases, nervous system damage [47]

*AFM1 is only relevant to milk and milk products.

To address the legislation and ensure food safety, the development of analytical methods with high sensitivity and accuracy is of great demand. Current analytical methods include confirmatory methods and screening methods. The standardized methods for mycotoxin analysis are chromatographic methods, including thin-layer chromatography (TLC), gas-chromatography (GC) with electron capture detection (ECD) [48], flame ionization detection (FID) [49] or mass spectrometry (MS) [50], high performance liquid chromatography (HPLC) with ultraviolet detection (UV) [51], fluorescence detection (FLD) [52], and MS or tandem mass spectrometry (MS/MS) [53,54]. TLC was the predominant method in early days. Although it is still used by some laboratories, it has almost been replaced by HPLC and GC. The instrumental methods are usually used as the gold standard. Nevertheless, despite their accurate and precise determination, sophisticated instrumental methods have some limitations related to high cost, long detection time, and the requirement of skilled operator [55,56].

In response to these limitations, a couple of rapid methods with high sensitivity and specificity have been developed for the identification and quantification of mycotoxins [57–61]. Furthermore, researchers are still working on developing novel methods with improved sensitivity, specificity, robustness, time-saving, and cost-efficiency. Rapid methods are more preferred by analysts who need to know the results immediately (e.g., on-site screening of high numbers of samples) or in routine analysis in laboratories where the classical method is not available. Among all the rapid detection methods for mycotoxins, immunoassays have already found widespread use as screening methods, providing beneficial attributes, such as rapidness, simplicity, cost-efficiency, required sensitivity, and specificity [62–65].

The core principle of immunoassays is the molecular interaction between target and biorecognition element, i.e., the antibody. So far, antibodies have been regarded with no doubt as the gold-standard recognition element in immunoassays and biosensors. Polyclonal and monoclonal antibodies dominate the field. However, the development of molecular techniques for expression of complete antibodies or antibody fragments in different species and methods for production and screening of combinatorial libraries is challenging. It has opened a wide range of opportunities for the selection of rAbs and their engineering, i.e., production of tailored binders with predefined properties in different species, e.g., bacteria, yeast, and mammalian cells (Chinese hamster ovary, CHO cells). In addition, plants and crop species offer the necessary economy and scalability to enable extremely cost-effective and efficient production of antibodies (plant-based antibodies) [66–68]. Besides rAbs, other novel recognition elements are emerging in recent decades, including aptamers [69], short peptides [70], and molecularly imprinted polymers (MIPs) [71–73]. These reagents have the potential to overcome some of the disadvantages of conventional antibodies, e.g., stability and production issues. Considering the increasing number of emerging rapid methods for mycotoxin detection, it is important critically discuss the differences between the used biorecognition molecules and arising advantages and disadvantages of their application. Thus, in this review, we provide an overview of the current and emerging biorecognition molecules towards mycotoxins and discuss their strengths and weaknesses for mycotoxin monitoring (Figure 1). Furthermore, we also introduce the application of those recognition elements in various assay formats, e.g., microplate- and tube-based assays, lateral flow assays (LFA), immunoaffinity columns (IAC), and biosensors.

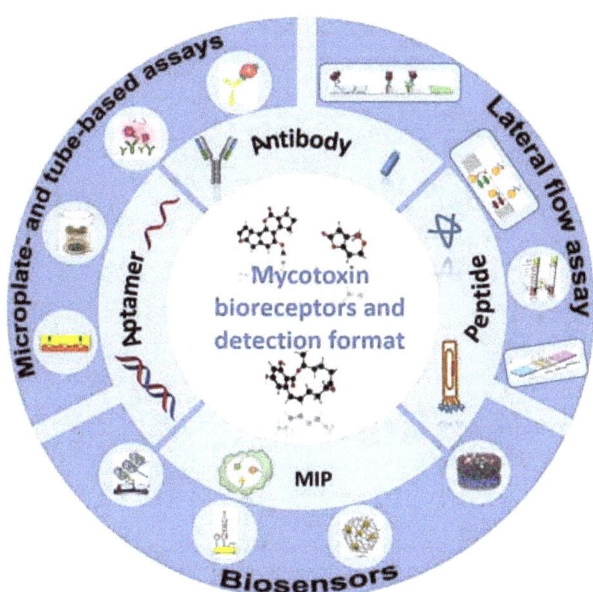

Figure 1. Schematic illustration of mycotoxins recognition elements and their application.

Table 2. Maximum permitted levels of mycotoxins in food according to regulations by China, European Union (EU) [1], and United States (U.S.).

Mycotoxins	Country	Maximum Permitted Level (µg/kg)
AFs	China	5–20 (0.5) *, (AFB1)
	EU	2–12 (0.1) *, (AFB1), 4–15, (sum of B1, B2, G1, G2),
	U.S.	20, (sum of B1, B2, G1, G2),
AFM1	China	0.5
	EU	0.05 (0.025) *
	U.S.	0.5
ZEN	China	60
	EU	50–400 (20) *
	U.S.	not set
OTA	China	2–10
	EU	2–80 (0.5) *
	U.S.	not set
DON	China	1000
	EU	500–1750, (200) *
	U.S.	1000
PAT	China	50
	EU	25–50, (10) *
	U.S.	50

Table 2. *Cont.*

Mycotoxins	Country	Maximum Permitted Level (µg/kg)
FMs	China	in preparation
	EU	800–4000, (200) *, (FB1, FB2)
	U.S.	2000–4000, (FB1, FB2, FB3)
T-2/HT-2	China	not set
	EU	in preparation [2]
CIT	EU	2000
EAs	EU	100–500, (20) *, (sum of 12 compounds)

[1] Regulations (EC) Nos. 2002/32/EC, 1881/2006, 2021/1399; [2] 2013/165/EU: Commission. Recommendation; * Number in brackets refers to infant food and young children.

2. Biorecognition Molecules

2.1. Antibodies

Among all the biorecognition molecules, antibodies are the most popular and widely applied due to their superiority in terms of affinity and specificity. There are mainly three types of antibodies, including pAb, mAb, and rAb. Affine polyclonal antibodies can be prepared in a relatively short period (around 10–12 weeks) at low cost. The first pAbs for mycotoxin detection were reported nearly 40 years ago [74,75]. In these publications, polyclonal antibodies were produced by simply collecting the serum of a New Zealand rabbit after several injections of antigens. PAbs are a mixture of antibodies towards different determinants of the antigen. Thus, they have disadvantages related to inconsistence among different antibodies of the same batch and between batches. Further, it is impossible to prepare pAbs with same characteristics using the identical reagents and immunization schedule but a different animal. This is almost a deal-breaker for long-term and higher sales commercial exploitation. However, pAbs have been and are still being widely applied in mycotoxin determination due to the benefits of ease of development, short production period, and relatively low cost [76–79].

In 1975, Köhler and Milstein invented the hybridoma cell technology, which allows the production of homogenous antibodies [80]. By hybridizing antibody-producing B-lymphocytes with myeloma cells, a hybridoma cell line can be selected and isolated. MAbs then can be produced by cultivation of hybridoma cells either in vivo or in vitro. Since the first mAbs described for AFs [81], AFM1 [82], OTA [83], DON [84], ZEN [85], T-2 toxin [86], and FMs [87], numerous mAbs have been developed and applied in both laboratory research and commercial assay products [88–92].

With the advancement of genetic engineering, the third generation of antibodies, named rAb technology, emerged [93]. Conventional IgG antibodies (MW 150 kDa) are composed of two identical heavy chains (50 kDa) and two identical light chains (25 kDa), which are linked together by disulfide bonds (Figure 2). It is a Y-shaped, multidomain protein with antigen-binding sites located on the complementarity determining regions (CDRs) of the variable domains of the heavy and light chains. Cloning and expression of the antibody variable domains in prokaryotic or eukaryotic systems can produce rAbs reproducibly and steadily. A wide variety of rAbs have been produced, including antigen binding fragment (Fab) [94], single-chain variable fragment (scFv) [95–97], and single-domain antibody (sdAb) [98,99]. For rAb development, antibody binding genes either from lymphocytes of the immunized animal or from hybridoma cells are cloned and displayed on phages [100,101], bacteria [102,103], yeast [104], or mammalian cells [105–107]. Acellular approaches use ribosome or mRNA display. Phage display is the most used technology for in-vitro rAb development. Antigen binding fragments can be enriched after 4–5 rounds of biopanning. The most powerful advantage of biopanning technique is that an antibody can be obtained with desired selectivity or affinity through optimization of panning conditions.

Compared with conventional antibodies, rAbs can be produced at a lower cost, with higher consistence and smaller size, and without the use of animals. Single-chain antibodies towards mycotoxins have been successfully expressed in bacteria and yeast [108–113]. However, Fab and scFv antibody fragments are usually suffer from instability and low production yield, which are the major limiting factors of this technology.

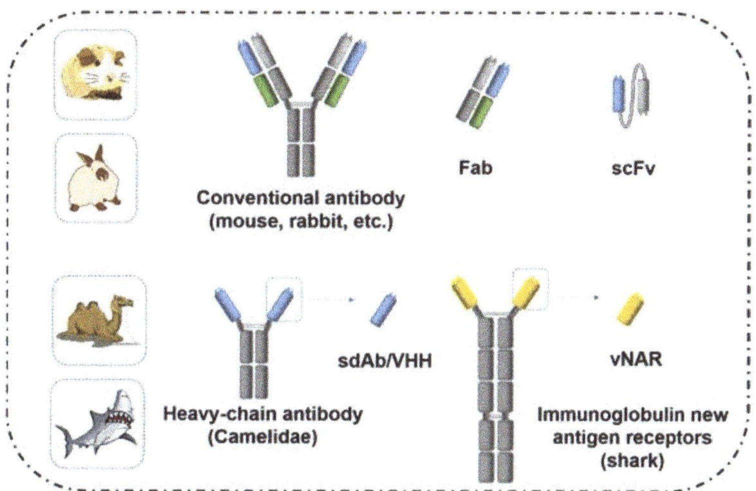

Figure 2. Schematic diagram of antibody structure. Abbreviations: Fab, antigen binding fragment; scFv, single-chain variable fragment; sdAb, single-domain antibody; VHH, variable domain of heavy chain of HCAb; vNAR, variable domain of new antigen receptors.

In the serum of Camelidae and cartilaginous fish is a considerable fraction of heavy-chain antibodies (HCAbs), which lack the light chains (Figure 2) [114,115]. While HCAbs of Camelidae lack the CH1 domain, that of cartilaginous fish, also called immunoglobulin new antigen receptors (IgNARs), have five constant domains. Thus, the variable domain of the heavy chain is linked directly to the hinge region in HCAbs. The antigen-binding fragment (~15 kDa) of the HCAbs, which constitutes only the variable domain of the heavy chain (VHH from camels and llamas; VNAR from sharks), is a single-domain antibody (sdAb), also called nanobody [116]. Compared with conventional antibodies and antibody fragments, including scFv and Fab, nanobodies have higher thermostability and solvent-resistance. The interloop disulfide bond in camelid VHH was considered to contribute strongly to its high stability and thermostability [117,118]. The presence of several amino acid substitutions in the framework region 2 provide VHH a more hydrophilic and soluble character. In most publications, the thermostability of nanobody was verified by testing its binding ability after treatment at extreme temperatures for various periods and comparing with pAb/mAb. The anti-idiotypic nanobody towards OTA developed by Zhang et al. [119] has enhanced thermostability compared to a mAb. The VHH retained more than 50% of its activity after being heated at 80 °C for 40 min, whereas the mAb lost most of its binding ability after 10 min incubation at the same temperature. Liu et al. [120] developed four different nanobodies against OTA. All nanobodies showed higher thermostability than mAb 6H8. Among them, Nb 32 is the most stable one, which could stand at 95 °C for 5 min without loss of its activity and retained 50% of its binding ability after incubation at 90 °C for 75 min. He et al. [121] developed a nanobody towards AFB1 and evaluated the solvent tolerance towards MeOH, DMSO, DMF, acetone, and acetonitrile. The data indicated the VHHs demonstrated higher resistance to MeOH than mAbs. Separate from mycotoxins, in a study with the herbicide parathion reported by Zhang et al., VHH9 could maintain nearly half of its binding activity under 40% of MeOH, DMSO, and acetonitrile [122]. Above all, nanobodies are superior biorecognition reagents compared with conventional antibodies, scFv and Fab fragments, which are less prone to loss of activity at high temperatures or in complex sample composition.

However, there are also some drawbacks in the development of nanobodies. First, camelid animals are not as easy to grow as small animals, such as mice, rabbits, or chicken. For that matter, using transgenic mice for immunization or panning of naïve or synthetic nanobody libraries might be an outcome [123]. Second, it is not easy to obtain a nanobody with high affinity, especially for small molecules. Up to now, mycotoxin-specific nanobodies were developed only towards AFs [121], OTA [120], 15-acetyl-deoxynivalenol [124], and tenuazonic acid [125]. The limited availability constitutes a clear shortage for the development of multi-mycotoxin assays. Third, due to its small size, the nanobody's random attachment to surfaces (e.g., polystyrene plate, nitrocellulose membrane, nanomaterials) can negatively impair its binding affinity [126,127]. The binding sites of the nanobody are more likely to be hindered sterically after immobilization compared with that of IgG.

Affinity and specificity are two important parameters for antigen binding probes. Preparation and designation of effective mycotoxin antigens that contain characteristic structure and could be exposed to the body is essential for successful isolation of specific and highly affine functional antibodies. Mycotoxins are small molecules (MW < 1000), which must be conjugated with a carrier protein in order to elicit an immune reaction. The structure of commonly used mycotoxin antigens and obtained antibody characteristics are summarized in Table 3. Mycotoxins have different functional groups, and therefore, a variety of coupling strategies were utilized. OTA, FB1, and CIT all have a carboxyl or amino group that can be activated and coupled to amino groups of carrier proteins to form stable amide linkages. AFs do not have an activatable group for direct conjugation with protein. Most established is the conjugation of AFB1 to a protein, such as keyhole limpet hemocyanin (KLH), bovine serum albumin (BSA), or ovalbumin (OVA), in the 1-position by means of a carboxymethyloxime (CMO) spacer. By far, most of all aflatoxin selective antibodies produced over the last decades have been generated by immunizations with this (commercially available) conjugate. The resulting antibodies all show similar selectivity. The affinity to the four major AFs usually follows the order AFB1 > AFG11 > AFB2 > AFG2 [128–132]. The immunization schedule and antibody screening techniques also have an important effect on the quality of resultant antibodies. By using a rapid cell fusion technique, Wu et al. [133] selected a cell line from 100,000 positive cell clones, which produced an mAb with similar recognition ability for AFB1, AFB2, AFG1, AFG2, AFM1, and AFM2. Devi et al. generated 10 hybridomas by immunizing AFB1-oxime-BSA to mice with an alternative immunization protocol [134]. One of them was highly specific to AFB1 because it only showed a weak cross-reaction with AFG1 (12%). High-affinity broad spectrum [135,136] and AFB1-specific [137–139] aflatoxin antibodies could also be generated using AFB2-conjugates that are less toxic than AFB1-conjugates. Synthesis of DON-protein conjugate was carried out mostly by converting the C3 hydroxyl group to carboxyl [84,140–142]. Similarly, T-2 antigen was prepared by esterification of the C3 hydroxyl group to obtain T-2-hemisuccinate (3-HS-T-2) [143–147].

Most of anti-ZEN antibodies reported were obtained by immunization with zearalenone-6′-carboxymethyloxime-protein conjugate [89,148–152]. Due to the protein binding position, those developed antibodies could not discriminate carbonyl and hydroxyl functional groups at position C6′ and thus usually showed high cross-reactivity with ZEN derivatives (including α-zearalenol, β-zearalenol, zearalanone, α-zearalanone, and β-zearalenone). To produce a specific antibody towards ZEN, Teshima et al. synthesized 5-aminozearalenone by a two-step approach and coupled it with protein at C-5 position of the compound [153]. The anti-ZEN mAb exhibited high specificity to ZEN, with weak cross-reactivity (<4%) to other analogs. Gao et al. coupled ZEN with cationic bovine serum albumin (cBSA) via a Mannich reaction [154]. By using this immunogen, specific anti-ZEN pAbs and mAbs were obtained, with cross-reactivity less than 7%. Sun et al. generated mAbs towards ZEN with a novel ZEN-BSA conjugate, which was prepared using 1,4-butanediol diglycidyl ether as a linker [155]. The selected antibody showed 53% cross-reactivity with zearalanone but weak cross-reactivity (<4%) with the other four analogs.

ZEN exists in two stereoisomeric forms: *trans-* and *cis*-zearalenone. *Trans*-ZEN is known naturally produced by *Fusarium* spp. and could isomerize to *cis*-ZEN photochemically, i.e., by UV light irradiation. The coexistence of *trans/cis-* ZEN has already been reported in edible oil [156], grains, and their products [157]. However, owing to the limited study of *cis*-ZEN, worldwide maximum levels for ZEN in food and feed are thus based on the *trans*-isomer. However, toxicological studies revealed an elevated estrogenic activity of *cis*-ZEN and/or its reductive metabolites α/β-*cis*-zearalenol compared to their respective *trans*-isomers [158]. To the best of our knowledge, there were neither studies performed to discriminate between both isomers based on bioanalytical methods nor reported stereoselective antibodies towards ZEN. At an earlier stage, we reported on first experimental evidence for an enzyme-generated chemiluminescence-induced *trans-cis* isomerization of chip-immobilized *trans*-ZEN in a microfluidic cell of a biosensor using a ZEA-mAb [159]. After that, the cross-reactivity of five commercially available anti-ZEN mAbs was tested with both isomers by competitive ELISA on microplates. Dependent on the source of the antibody, significantly reduced affinity of *cis*-ZEN was obtained (CR 12–72%) (data not published). This could be well explained by the fact that only *trans*-ZEN is commercially available for synthesis of the immunogen and generation of ZEN-antibodies. As consequence, the practical use of antibody-based assays, such as immunoassays, lateral flow assays, and immunoaffinity cartridges, may result in an underestimation of real ZEN content in samples that contain appreciable amount of *cis*-ZEN. Thus, suppliers of related assays are urgently requested to include the CR of *cis*-ZEN in the assay instructions for more reliable results interpretation. Moreover, the development of antibodies for stereospecific targeting of chiral haptens like ZEN is a special research challenge.

Table 3. Typical mycotoxin immunogens and obtained antibody characteristics.

Mycotoxin(s)	Immunogen Structure	Coupling Method	Antibody Type	Titer	IC$_{50}$	LOD	Reference
Total AFs	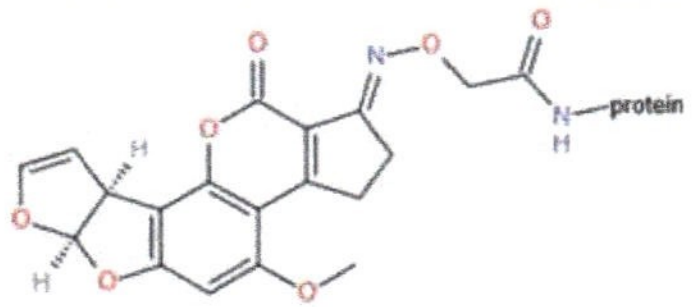 AFB1-oxime-BSA	Carbodiimide method	pAb	Higher than 1000	AFB1 1.8 ng/mL AFB2 16 ng/mL AFG1 20 ng/mL AFG2 320 ng/mL	AFB1 0.4 ng/mL	[160]
AFB1	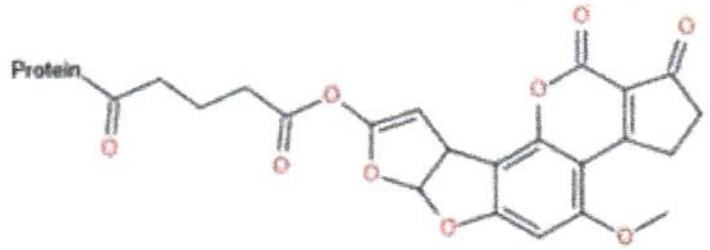 AFB2a-HG-BSA [1]	Mixed anhydride method	pAb	710–800	0.15 ng/assay	0.02 ng/assay	[137]
AFM1	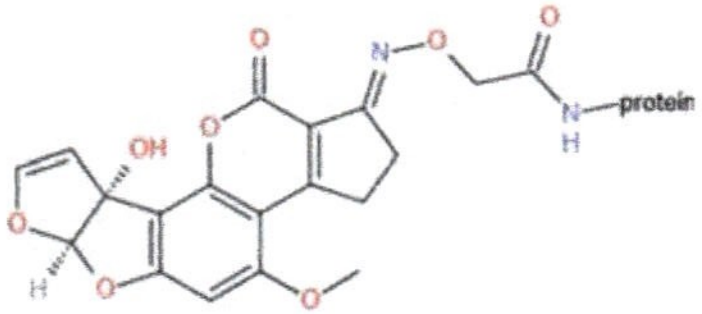 AFM1-BSA	Carbodiimide method	pAb and mAb	n.a.	25 ng/mL (mAb); 0.5 ng/mL (pAb)	n.a.	[82]
OTA	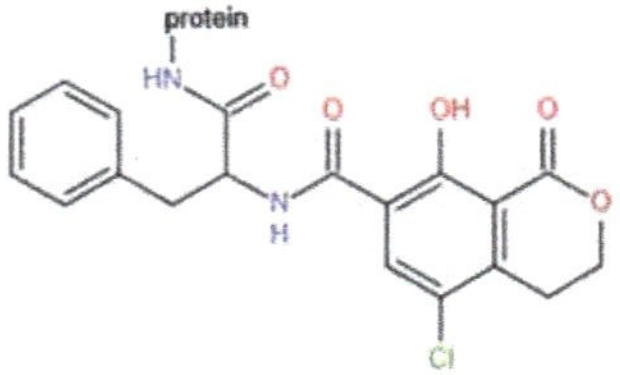 OTA-BSA	Carbodiimide method	pAb	n.a.	3 ng/mL	1 ng/mL	[161]

Table 3. *Cont.*

Mycotoxin(s)	Immunogen Structure	Coupling Method	Antibody Type	Titer	IC$_{50}$	LOD	Reference
DON	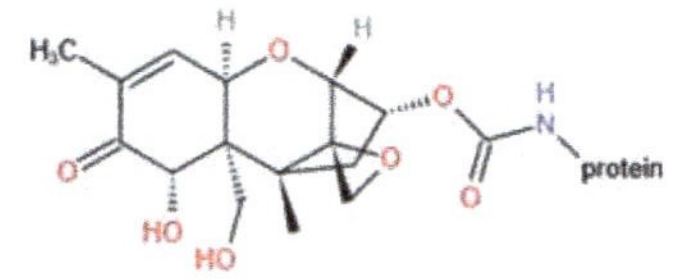 DON-BSA	N,N′-carbonyldiimidazole method	mAb	n.a.	9.84 ng/mL	n.a.	[162]
T-2 toxin	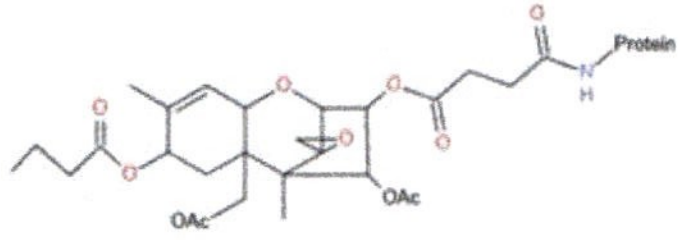 T-2-HS-BSA [2]	Carbodiimide method	pAb	303	3.5 ng/assay	1 ng/assay	[146]
ZEN	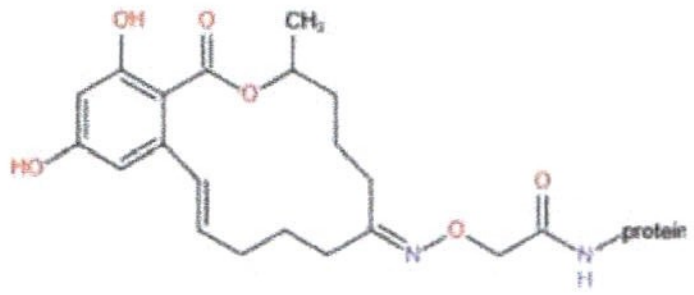 ZEN-oxime-BSA	Mixed anhydride procedure	pAb	5120	n.a.	0.5 ng/mL	[149]
	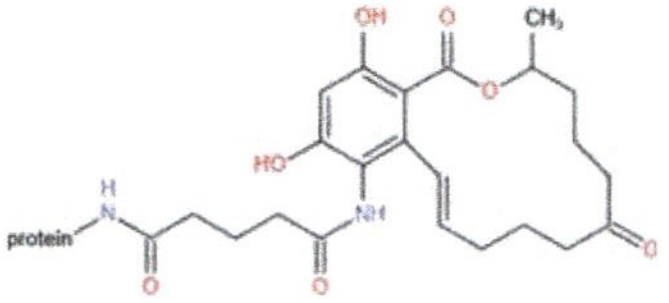 5-NH$_2$-ZEN-BSA	Glutaraldehyde method	mAb	520	11.2 ng/mL	0.3 ng/mL	[153]

Table 3. *Cont.*

Mycotoxin(s)	Immunogen Structure	Coupling Method	Antibody Type	Titer	IC$_{50}$	LOD	Reference
ZEN	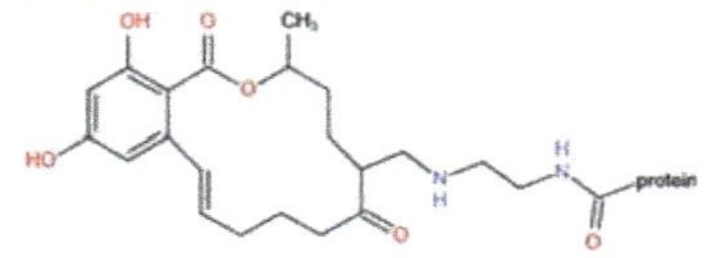	Mannich reaction	pAb and mAb	30,000	233.35 ng/mL (pAb); 55.72 ng/mL (mAb)	n.a.	[154]
	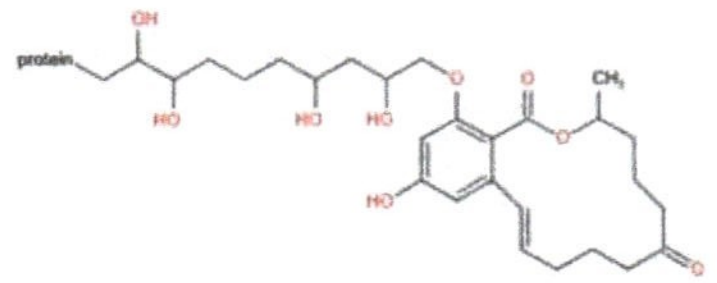	1,4-Butanediol diglycidyl ether method	mAb	1.024×10^6	1.115 ng/mL	n.a.	[155]
FB1	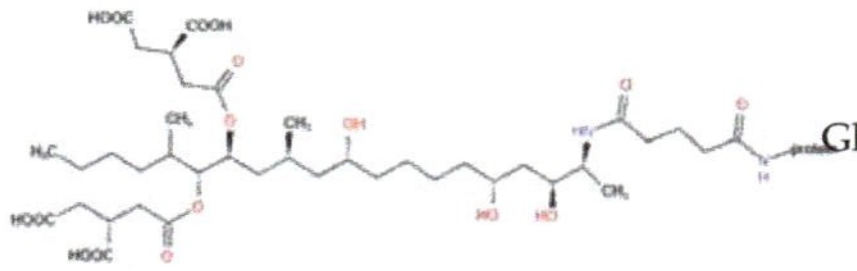 FB1-KLH	Glutaraldehyde method	pAb	10,000	0.45 ng/mL	0.1 ng/mL	[163]
CIT	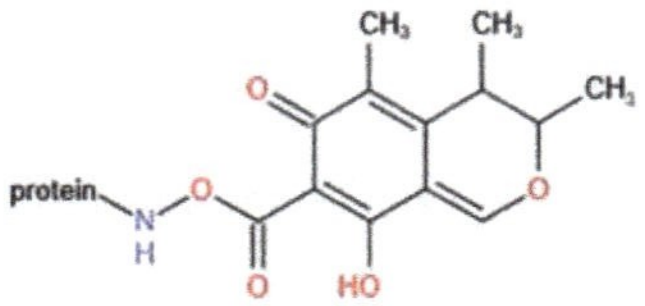 CIT-BSA	Activated ester method	mAb	32,000	0.28 ng/mL	0.01 ng/mL	[164]

Table 3. *Cont.*

Mycotoxin(s)	Immunogen Structure	Coupling Method	Antibody Type	Titer	IC$_{50}$	LOD	Reference
PAT	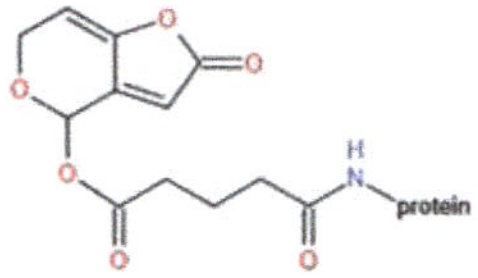 PAT-HG-BSA	Carbodiimide method	pAb	1100	n.a.	n.a.	[165]
	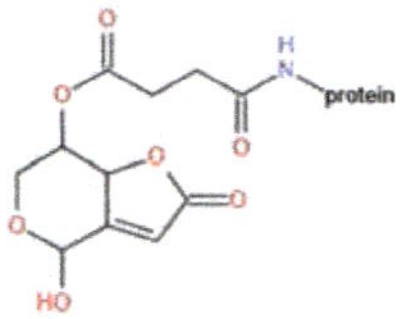 PAT-Ins-HS-BSA [3]	Carbodiimide method	pAb	n.a.	n.a.	10 ng/mL	[166]
	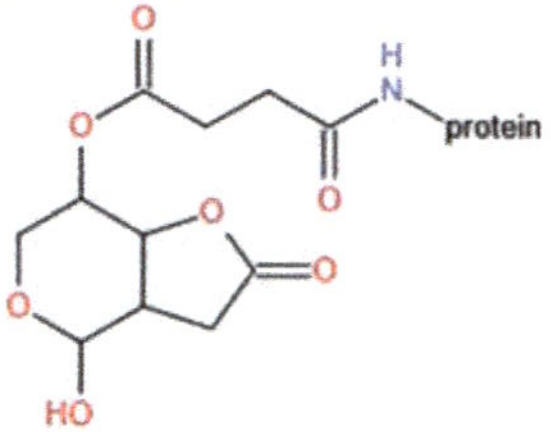 PAT-Sat-HS-BSA [4]	Carbodiimide method	pAb	n.a.	n.a.	10 ng/mL	[166]

n.a., data not available; [1] HG, hemiglutarate; [2] HS, hemisuccinate; [3] PAT-Ins-HS, 4-[(4-Hydroxy-2-oxo-2,6,7,7a-tetrahydro-4H-furo[3,2-c]pyran-7-yl)oxy]-4-oxobutanoic acid; [4] PAT-Sat-HS, 4-[(4-Hydroxy-2-oxohexahydro-4H-furo [3,2-c]pyran-7-yl)oxy]-4-oxobutanoic acid.

Unlike other toxins, PAT is highly unstable and can decompose during the course of protein conjugation. Furthermore, free PAT or bond-exposed epitope of immunogen are highly reactive with nucleophiles, which could bind with thiol and amino groups of proteins covalently and thus interfere with the generation of affine antibodies. For these reasons, high specific and affine anti-PAT antibodies are rarely reported. There are mainly two approaches to synthesize the immunogen. One is based on modification of the hydroxyl function [165,167]. By reaction with glutaric anhydride, PAT was converted to PAT hemiglutarate (PAT-HG) and then coupled with a carrier protein. However, no or slight competitive displacement by free PAT was observed with pAbs using PAT-HG as immunogen. The other approach is to synthesize a PAT derivative (PAT-SAT) from L-arabinose, which lacks the highly reactive C3-C4 double bond while maintaining the original skeleton of the toxin, and then conjugate it to protein [166,168]. PAbs that were produced using this antigen showed a high titer and inhibition effect by addition of free toxin. The competitive assay could detect PAT as low as 0.06 µg/L.

Even when using the same immunogen for antibody generation, in some cases, the obtained affinity of pAbs and mAbs is not identical. Taking OTA as an example, Reddy et al. developed pAbs by injecting OTA-BSA conjugate into a New Zealand rabbit [77]. The 50% inhibition binding (IC_{50}) of OTA was 5 ng/mL determined by indirect competitive ELISA (icELISA). Anti-OTA mAbs developed with the same antigen have higher affinity with IC_{50} around 0.3 ng/mL [169,170]. The single-domain antibody towards OTA developed by Liu et al. after immunization of an alpaca also had a good performance, with IC_{50} of 0.74 ng/mL and K_D value of 0.039 nM [120]. In most cases, scFv fragments derived from hybridoma cells have lower affinity than the parental mAb [171,172]. The scFv against AFB1 prepared by Min et al. retained 17 times less and anti-FMB1 scFv about 12-fold lower binding affinity than the parental mAbs [173]. However, one of the most powerful advantages of rAb development is that the affinity and selectivity of antibodies can be improved through in-vitro biopanning. Hu et al. [109], by using stringent panning conditions, isolated a scFv towards FMs with an 82-fold higher binding affinity than its parent mAb. There are also other factors that affect the quality of rAbs, such as the immune response of individual animals, immunization protocol, cell fusion technique, etc. Another benefit of rAb is the ease of gene modification, which could facilitate the directed evolution of antibodies [174]. Based on an anti-OTA nanobody, X. Wang et al. constructed a mutation library after identification of key amino acids of the antibody binding sites by homology modeling, molecular docking, and alanine scanning [175]. A mutant nanobody was then obtained by biopanning, which exhibited a K_D value of 52 nM, which is 1.4-fold and 1.36-fold lower than that of the original nanobody, respectively.

2.2. Aptamers

Aptamer ligands are short, single-stranded DNA or RNA sequences that adopt specific three-dimensional conformations and thus can bind the target specifically in a similar way to antibodies. Since they have been raised first in early 1990s [176,177], aptamers have attracted increasing attentions due to their advantages over antibodies in terms of robustness and cost-effectiveness. Antibodies are generally obtained from biological samples, while aptamers can be synthesized in vitro in a large quantity and at low cost once the sequence is determined [178]. Furthermore, aptamers exhibit higher stability under most environmental conditions and can resist chemical and physical denaturation without losing their binding activities. Aptamers are obtained by in-vitro screening of oligonucleotide libraries through systematic evolution of ligands by the exponential enrichment process, named SELEX. As shown in Figure 3, the SELEX technique starts with a large random oligonucleotide library (with 10^{14} to 10^{15} random sequences), and each oligonucleotide contains a random central region of 20 to 80 nucleotides, flanked by two fixed primer-binding regions on $3'$ and $5'$ ends. During selection, the targets are immobilized on a solid surface and incubated with the library. Then, free oligonucleotides are separated, and bound ones are eluted for enrichment by PCR and used for the next round of SELEX. After

7 to 30 rounds of SELEX, the enriched pool is cloned, sequenced, and characterized to select aptamers with desired properties. Aptamers recognize the targets by a combination of van der Waals forces, hydrogen bonding, electrostatic interaction, stacking interactions and shape complementarity, which is similar to antibody-antigen recognition [178].

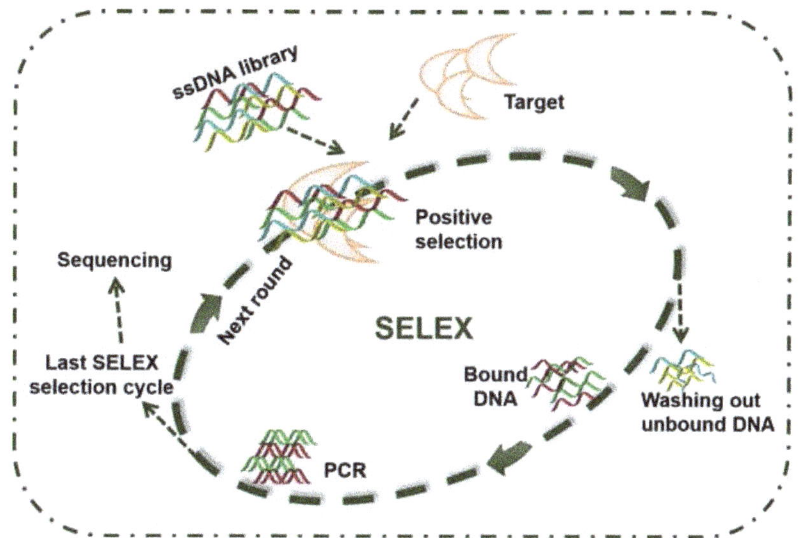

Figure 3. Selection of specific aptamers by SELEX technology.

Aptamers with high affinity are not always easy to obtain. The diversity of the initial oligonucleotide library is crucial to obtain high-affinity aptamers. More oligonucleotide sequences lead to higher chances to obtain useful target-specific aptamers. However, in fact, the starting library diversity is always limited due to synthesis technology and nucleotide preference. Besides, some sequences may get lost during PCR amplification, and the resultant sequences do not always have the desired binding affinities towards the target [179]. Another factor that impacts the aptamer's affinity is selection condition, including the amount of target, incubation condition, and the way to separate unbound oligonucleotides. Increasing selection pressure during SELEX can remove molecules with low affinity.

Over the past decade, aptamers towards a variety of mycotoxins have been developed. Cruz-Aguado and Penner prepared an aptamer towards OTA, which was the first aptamer identified for the detection of a mycotoxin [180]. The selected aptamer exhibited a dissociation constant in the nanomolar range and did not bind with other structurally similar chemicals. Since then, a number of aptamers have been developed and integrated in assays for the detection of AFB1 [181], M1 [182], FB1 [183], ZEN [184], DON [185,186], PAT [187], T-2 toxin [188], and EAs [189]. The sequences and affinity of those commonly used mycotoxin aptamers are summarized in Table 4. The dissociation constants, which indicate the affinity of aptamers, are from nanomolar to micromolar ranges.

Table 4. Sequences and dissociation constant (K_D) of commonly used mycotoxin aptamers.

Target	Sequence (5′-3′)	K_D	Reference
AFB1	GT TGG GCA CGT GTT GTC TCT CTG TGT CTC GTG CCC TTC GCT AGG CCC ACA	n.a. *	[181]
AFM1	ACT GCT AGA GAT TTT CCA CAT	n.a.	[190]
OTA	GAT CGG GTG TGG GTG GCG TAA AGG GAG CAT CGG ACA	0.2 μM	[180]
FB1	ATA CCA GCT TAT TCA ATT AAT CGC ATT ACC TTA TAC CAG CTT ATT CAA TTA CGT CTG CAC ATA CCA GCT TAT TCA ATT AGA TAG TAA GTG CAA TCT	100 ± 30 nM	[191]
ZEN	TCATCTATCTATGGTACATTACTATCTGTAATGTGATATG	41 ± 5 nM	[192]
DON	GCATCACTACAGTCATTACGCATCGTAGGGGGGATCGTTAAGGAAGTGCCCGGA GGCGGTATCGTGTGAAGTGCTGTCCC	n.a.	[185]
PAT	GGCCCGCCAACCCGCATCATCTACACTGATATTTTACCTT	21.83 ± 5.022 nM	[187]
T-2	GTATATCAAGCATCGCGTGTTTACACATGCGAGAGGTGAA	20.8 ± 3.1 nM	[188]
Ergot alkaloids	ACTCATCTGTGAAGAGAAGCAGCACAGAGGTCA GATGTCCGTCAGCCCCGATCGCCATCCAGGG ACTCCCCCCTATGCCTATGCGTGCTACCGTGAA	44 nM^2	[189]

* n.a., data not available.

Compared with pAbs and mAbs, aptamers may be superior in terms of stability, size, and production. However, the commercialization of aptamer-based methods for mycotoxin detection is not as fast as expected. The only report on commercial aptamer-based products was from NeoVentures Biotechnologies, Inc. (London, ON, Canada) for purification and determination of AFs and OTA [193]. Thus, the application of aptamers for rapid determination of mycotoxins in different matrices should be further studied.

2.3. Short Peptides

Molecular recognition by short peptides is a rapidly growing area of research. Peptides have regular structures and therefore can recognize functional groups on the targets through non-covalent interactions (e.g., electrostatic, hydrogen bonding, hydrophobic effects, and van der Waals forces). Advantages of peptide-based receptors are that they can be synthesized in vitro, easily modified and fused to other tags, and are less prone to activity loss under harsh conditions. Peptides with specific binding activity can be obtained by two different approaches, i.e., phage display and combinatorial synthesis. By phage display technology, peptides of a given length whose sequences are randomly generated are synthesized in vitro and expressed on the surface of bacteriophages. As shown in Figure 4, the phage library is incubated with the specific antigen immobilized on a microplate or magnetic beads. The unbound phages are washed away, and the bound phages are eluted and reinfected into bacteria for amplification. After several rounds of panning, peptides with high affinity to the antigen can be obtained. Theoretically, peptides with the ability to bind to a particular ligand can be selected if the library is large enough. In fact, with the limitation of library size and panning method, peptides with recognizing properties towards molecules, especially small molecules, are not easy to obtain. A number of phage-displayed peptides towards particular antibodies were developed and applied as mimotopes to replace the free toxins or their conjugates in immunoassays for mycotoxins [194–200]. However, obviously, there was no reported successful example of mycotoxin-specific peptides obtained by phage display method.

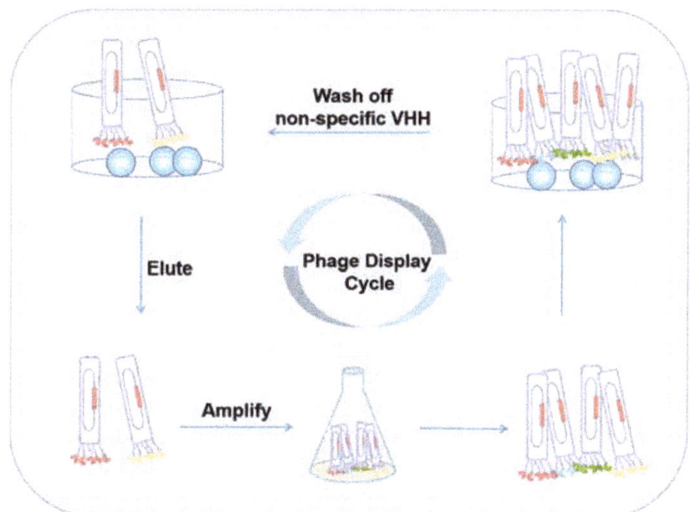

Figure 4. Specific peptide screening by phage display technology.

In contrast to phage display technology, which is based on panning of a large number of peptides that were randomly synthesized and displayed on phage particles, combinatorial peptides can be designed and synthesized on purpose based on the structure of mycotoxins. Tozzi et al. obtained tetrapeptides with binding properties towards AFs by a combinatorial approach, which was the first research report on peptides with binding ability towards mycotoxins [201]. The binding constants of selected peptides were in the range of only 8.3×10^3 M^{-1} to 12.0×10^3 M^{-1}, and the selectivity was similar with that shown by a commercial antibody. Molecular modeling software (e.g., SYBYL) can be applied to facilitate the design and synthesis of peptide sequences [202]. By using computational modeling, they designed two peptide ligands for OTA. Both of the peptides exhibited a binding strength to OTA with K_D value in the micromolar range, i.e., 11 to 15.7 µM, which are similar with those obtained by combinational chemistry [203,204]. With the advantages of easy availability and low cost, however, specific peptides have been developed only towards OTA and AFB1 [205]. The application of specific peptides as recognition elements in determination of other mycotoxins are not available yet.

2.4. Molecularly Imprinted Polymers (MIPs)

Other than the recognition elements mentioned above, molecularly imprinted polymers are not biological receptors but synthetic polymers. Thus, compared with other biorecognition elements, MIPs are more stable over varying conditions, such as temperature, pH value, and organic solvents, and easier to be produced at a relative lower cost and can be reused for several times [206]. MIPs are prepared by polymerization and crosslinking of functional monomers in the presence of the target molecule, called template, which is a catalyst and suitable porogen (Figure 5) [207]. After removing the original template, it results in a three-dimensional network that contains specific recognition cavities, which are complementary in shape and size with the target. These artificial materials thus can recognize a particular target molecule mimicking the biological activity of natural receptors.

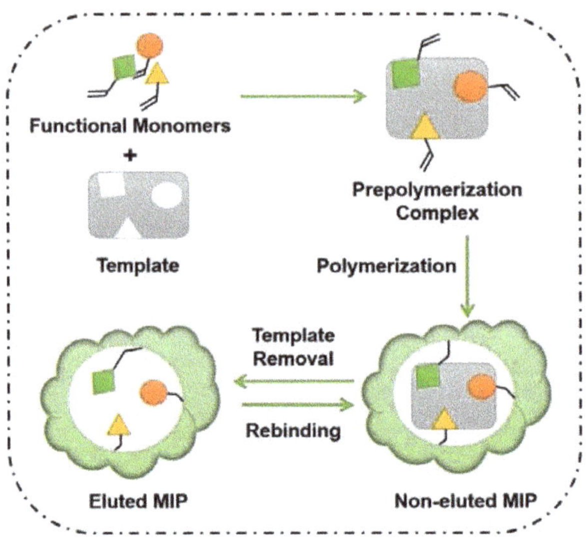

Figure 5. Synthesis procedure of specific MIP.

To obtain MIPs with high affinity and selectivity, two highly important factors should be considered, i.e., template molecule and monomer selection. Generally, the template molecule plays a vital role in the development of MIPs with high affinity. As for mycotoxins, the template is harmful and expensive; thus, it is not feasible for many laboratories to manipulate with hundreds of milligrams of mycotoxins. As an alternative, a template can be used that mimics the structure of target compound as best as possible [208,209]. Baggiani et al. developed the first MIPs that recognize OTA, using N-(4-chloro-1-hydroxy-2-naphthoylamido)-(L)–phenylalanine as a mimic template [210]. The dummy template is stable, less toxic, and easy to prepare. It was found that the carboxyl, phenolic hydroxyl, and peculiar substructures are critical structures for OTA imprinting.

The selection of optimal functional monomers is the primary step for the preparation of MIPs. However, the large library of monomers and the complexity of interactions among template and monomers make the selection a big challenge [211]. To facilitate the design of proper monomers, computational modeling method can be employed [212]. Sergeyeva et al. synthesized nanostructured polymeric membranes as a recognition element and developed an MIP-based fluorescent sensor for AFB1 determination [213]. The selection of functional monomers was performed from a virtual library using computational modeling. By using ethyl-2-oxocyclopentanecarboxylate as a dummy template, AFB1 MIP membrane of high selectivity was synthesized. In another research [214], also using computational method, molecular interactions between FB1 and different acrylic monomers were analyzed, and an appropriate monomer was selected for MIPs development. NanoMIPs were produced with high specificity and successfully applied in an MIP-based immunosorbent assay in the replacement of the primary antibody. Furthermore, the properties of the non-imprinted polymer (NIP, blank polymer), which is synthesized in parallel without addition of the template, is crucial for the evaluation of specific binding ability of the MIPs. Maier et al. developed an MIP-based SPE column for the enrichment of OTA from red wine, followed by HPLC quantification [215]. However, the MIP-based SPE did not reveal as much superior to NIP; i.e., the retention of OTA depended mainly on non-specific binding to the polymeric material other than specific retention by the imprinted binding sites. Baggiani et al. observed that if NIPs had no affinity toward a target molecule, the corresponding MIPs would display poor imprinting efficiency [216]. On the other hand, if the NIP has good binding behavior, the imprinted polymer will show enhanced binding ability. They concluded that the obtained results are valid for a wide variety of MIPs. Several researches have been published about blank polymers that performed good binding ability and selectivity [217,218]. Compared with MIPs, the synthesis of blank polymers avoids the use of template, which is more environmentally friendly and less expensive. Furthermore, the slow release of template during storage is also eliminated. There are other factors that influence the property of MIPs, including polymerization temperature [219], solvent [220], and polymerization procedure [221], etc. The use of MIPs as receptor is becoming more common in analysis area due to its inherent thermal and chemical stability, ease of preparation, and low cost.

Most common application of MIPs is solid-phase extraction, the so-called MISPE (Molecularly Imprinted Solid Phase Extraction), for purification of the toxins prior to further analysis, for example, chromatographic assay [222]. MIPs as sorbents have been developed for purification of AFs [223,224], ochratoxins [225], FMs [226], CIT [227], ZEN [208], T-2 toxin [228], PAT [229], metergoline [72,230], and alternariol [231]. Compared with other selective sorbents, such as immunoaffinity columns, MISPE has several advantages: (a) MIPs can bear a high number of binding sites, whereas biological acceptors only have one or two. Thus, the capacity of MISPE is usually higher than that of IACs [232]. Lucci et al. developed a clean-up method employing MIP as selective sorbent for the preconcentration of ZEN [233]. The column had a capacity of no less than 6.6 µg, whereas the ZearalaTest immunoaffinity column from VICAM was saturated when loading 1.6 µg of ZEN. (b) MIPs demonstrate very good thermal and chemical robustness, leading to repeatable usage without loss of activity. Taking OTA determination as an example, the MISPE could be reused for at least five times with wine [234] and 14 times with beer [235] after regeneration. (c) Most affinity sorbents are made of binding elements immobilized on a solid support, such as agarose. The development of MISPE is more convenient. Once the MIP for a target is obtained, the selective MISPE can be developed by simply packing a small amount of imprinted polymer into a cartridge. Furthermore, MIPs can also act as an adsorbent to remove and control mycotoxins in foodstuff, such as the decontamination of milk by removing AFs [236] or removing PAT from apple juice [237]. MIPs have also been employed in the development of sensors for mycotoxin analysis, which will be discussed in the following section.

3. Areas of Application of Recognition Elements for Detection of Mycotoxins

3.1. Sample Preparation

Sample purification and clean-up is usually required in chromatographic analysis of mycotoxins given the complex matrices and trace amounts of targets in food samples. This step is crucial to get clean and concentrated extracts, therefore improving assays' sensitivity to some extent. Owing to the high affinity and specificity of antibodies, immunoaffinity sorbents are powerful clean-up tools and are applicable in a wide range of food samples for single or multiple mycotoxin analysis [238–241]. Numerous immunoaffinity sorbents are commercially available worldwide (see Section 4) for the analysis of single or multiple mycotoxins. With similar properties, other molecular recognition elements have also been introduced as affinity sorbents, such as aptamer-based oligosorbents [242–246] and MISPE sorbents [232,247–249]. Sample purification technologies and their properties have been extensively discussed in previous publications [250–252] and therefore will not be covered in very much detail in this review. Rapid determination of mycotoxins should require only simple or no sample treatment; i.e., complicated clean-up steps should be avoided. There are also reports on immunoaffinity columns combined with immunoassays [253,254]. The most prevalent sample treatment in rapid analysis is liquid-liquid extraction (LLE). Mycotoxins are hydrophobic molecules, which are most dissolvable in organic solvents. Thus, they are usually extracted using polar solvents, such as methanol and acetonitrile. In conclusion, the presence of organic solvents in the extract requires relatively high stability and tolerance of the biorecognition molecule.

3.2. Microplate- and Tube-Based Assays

Microplates, also termed microtiter plates or multi-well plates, became essential tools in analytical chemistry. The commonly used microplate-based assay for mycotoxin analysis is the enzyme-linked immunosorbent assay (ELISA). There are two types of competitive ELISA formats used in mycotoxin determination, including direct (dcELISA) and indirect ELISA (icELISA). In direct ELISA, an unknown amount of mycotoxin in samples competes with analyte-enzyme conjugate for the coated anti-mycotoxin antibody, and the signal is then developed by adding the enzyme substrate. In indirect ELISA, analyte-protein conjugate (e.g., BSA, OVA, and KLH) is coated on the microplate, and competition for the limited amount of antibody takes place between the immobilized antigen and free analyte.

The anti-mycotoxin antibody (primary antibody) can be labeled with an enzyme directly, or a secondary antibody enzyme conjugate is added for color development. The most commonly used enzyme is horseradish peroxidase (HRP), which catalyzes the oxidation of TMB by hydrogen peroxide and results in a blue color. Alkaline phosphatase (AP) is more stable and sensitive compared with HRP but with a higher cost. In addition to enzymes, different types of reporters have been developed for signal enhancing, such as polyHRP [255], fluorophores [256,257], functionalized magnetic beads [258,259,259–263], and upconverting luminescent nanoparticles [264]. In comparison to enzymes, these signal transducers usually have higher stability, enhanced signal, and lower price. Glucose oxidase (GOx), which can convert glucose by utilizing molecular oxygen to gluconic acid and hydrogen peroxide (H_2O_2), has also been employed as a reporter [265]. In the presence of HRP, H_2O_2 was converted to hydroxyl radicals and induced tyramine-mediated AuNP aggregation, thereby resulting in a dramatic change in visible color and dynamic light scattering (DLS) intensity (D_H), which can be recorded with a DLS analyzer. By using this system, Zhan et al. developed a DLS-enhanced direct competitive ELISA for AFB1 detection in corn [266]. From Figure 6a, in the absence of AFB1, GOx-AFB1 was captured by anti-AFB1 mAb immobilized on the microplate, which could induce AuNPs aggregation in the presence of HRP and tyramine, with an intense D_H value. The LOD of the assay was 0.12 pg/mL, 153-fold lower than plasmonic ELISA and 385-fold lower than colorimetric dcELISA. QDs of variable size have different colors, thus facilitating the construction of microplate immunoassay for multiplex mycotoxin detection. Beloglazova et al. synthesized CdSe-based QDs with different emission spectrum and developed double-analyte multiplex assay (DAM) for simultaneous determination of ZEN and AFB1 [267]. In the DAM assay, two specific antibodies were immobilized in the same well of the microplate. Analytical signal was detected for both analytes by double scanning of the wells with different emission wavelength.

RAbs have been developed and applied in ELISA for AFB1 [268,269], OTA [270,271], ZEN [95,110], DON [171,272], FB1 [273,274], CIT [275], and T-2/HT-2 toxins [94]. One of the major advantages of the rAb-based ELISA is that rAbs-reporter fusions can be expressed directly based on genetic engineering, which eliminates the chemical synthesis of antibody-reporter or use of commercial secondary antibody. Various rAb-reporter fusions have been constructed, e.g., alkaline phosphatase (AP) [122], green fluorescent protein (GFP) [276], and HRP [277], which provide a valuable tool in the construction of microplate-based assays. Nanoluciferase (Nluc) is a novel luminescence tracer that offers excellent performance in immunoassays [278]. Wang's group isolated specific Nbs against *Alternaria* mycotoxin tenuazonic acid and fused with Nluc by genetic engineering technique [125]. Based on the bifunctional fusion, a two-step bioluminescent enzyme immunoassay was constructed. The IC_{50} value of the assay was 8.6 ng/mL, which is six-fold more sensitive than ELISA.

Aptamers can also be applied as bioreceptor in microplate-based assays, named as enzyme-linked aptamer sorbent assay (ELASA). In direct format, the aptamer is coated on the microplate, and competition occurs between free analyte and analyte-reporter conjugate. The immobilization strategy of aptamers on the plate is of great importance to maintain its high binding affinity. Attachment of biotinylated aptamer on streptavidin/avidin modified microplate is the most commonly used procedure [279]. In indirect competitive ELASA, antigens or short, complementary DNA strands are coated on the plate, followed by addition of biotinylated aptamer and samples containing an unknown amount of mycotoxin for competition. Then, streptavidin-modified enzyme and substrate is added for color development. The aptamer can also be functionalized directly with a reporter, such as HRP [280], thrombin [281], and fluorescein [282,283], etc., which reduces the detection period and sometimes increases the assay's sensitivity. By using single-stranded DNA-binding protein (SSB) as the competitive antigen and a specific aptamer as the bioreceptor, Xing et al. [284] constructed a novel green ELASA system for mycotoxin detection. As shown in Figure 6b, immobilized SSB and free targets compete for binding with the aptamer. SA-HRP was subsequently added for color development. This method was successfully applied for the analysis of AFB1, OTA, and ZEN in corn, with an LOD value of 112 ng/L, 319 ng/L, and 377 ng/L, respectively.

As stated already, only a few peptide receptors have been successfully designed and used as an alternative to antibodies in mycotoxin ELISA. By immobilization of anti-OTA peptide NFO4 on a microplate, Bazin et al. established a peptide-based dcELISA for OTA detection [204]. The assay could detect OTA up to 2 μg/L in red wine, which highlights the possibility of using a peptide as biorecognition element in immunoassays. On the other hand, peptides that serve as epitope mimics have been introduced as valuable substitutes for mycotoxin-protein conjugates in competitive immunoassays. Analyte-protein conjugates are usually involved in competitive immunoassays as competitive binders with the antibody. However, the synthesis of mycotoxin-protein conjugate can be difficult, time consuming, and even hazardous to users and the environment. Moreover, lot-to-lot variation and low conjugation efficiency make the synthesis of competing mycotoxin antigen one of the major challenges in developing immunoassays. Phage-displayed peptides (mimotopes) have been proposed as an alternative way to overcome these drawbacks [285,286]. Such mimotopes bind to the same antibody paratope as target toxin and thus can substitute hapten conjugates in an immunoassay. The ease of genetic engineering and low production cost make peptide mimotopes an attractive choice as antigen surrogates [287]. At present, a variety of peptide mimotopes have been identified and applied in the analysis of mycotoxins, including AFB1 [288], ZEN [289,290], OTA [291], FB1 [292], and DON [293]. Peltomaa et al. identified a ZEN-mimicking peptide by phage display and synthesized it with extended biotin sequence on C-terminus [294]. As can be seen in Figure 6c, anti-ZEN mAb was coated on the microplate, and peptide mimotope competed with free toxin in sample for limited antibody binding sites. Afterwards, streptavidin-conjugated upconversion nanoparticles were added to develop an upconversion luminescence signal for ZEN quantification. This dcULISA has an LOD of 20 pg/mL (63 pM) with high specificity towards ZEN.

By replacing primary antibody with MIPs, biomimetic or pseudo-ELISA have been proposed for mycotoxin determination. The attachment of MIPs on the microplate is a key step for the successful development of a biomimetic ELISA. Given the hydrophobicity property, coating of MIPs on the polystyrene microplate is rather complex. In one approach, polymers are grafted directly on the plate in the presence of template and form a molecularly imprinted film [295,296]. Chianella et al. developed a novel immobilization method by using MIP nanoparticles (nanoMIPs) [297]. As illustrated in Figure 6d, stable coating could be achieved by simply loading nanoMIPs into the microplate wells, followed by evaporation of the solution. This technique is simple and analogous to physical adsorption of antibody in ELISA [298]. By using this method, Munawar et al. proposed a nanoMIPs-based assay (MINA) for the determination of FB1 [214]. Competition between FB1 and HRP-FB1 conjugate for binding of immobilized nanoMIPs occurred, followed by colorimetric reaction with enzyme substrate. The optical density was then used for quantitative determination of FB1, which is analogous to ELISA. The assay was shown to be 22 times more sensitive compared to a mAb-based ELISA, with a limit of detection of 1.9 pM and a linear range of 10 pM–10 nM. The 53 maize samples were analyzed by MINA, and the results were good, correlating with those obtained using ELISA and HPLC [299]. In comparison with antibody-coated microplates, the major advantage of the MIP-coated plates is its high stability. They can be stored under high temperature for a prolonged time without affecting the sensitivity of the assay. This characteristic makes them applicable for cost-efficient, room-temperature storage and transportation.

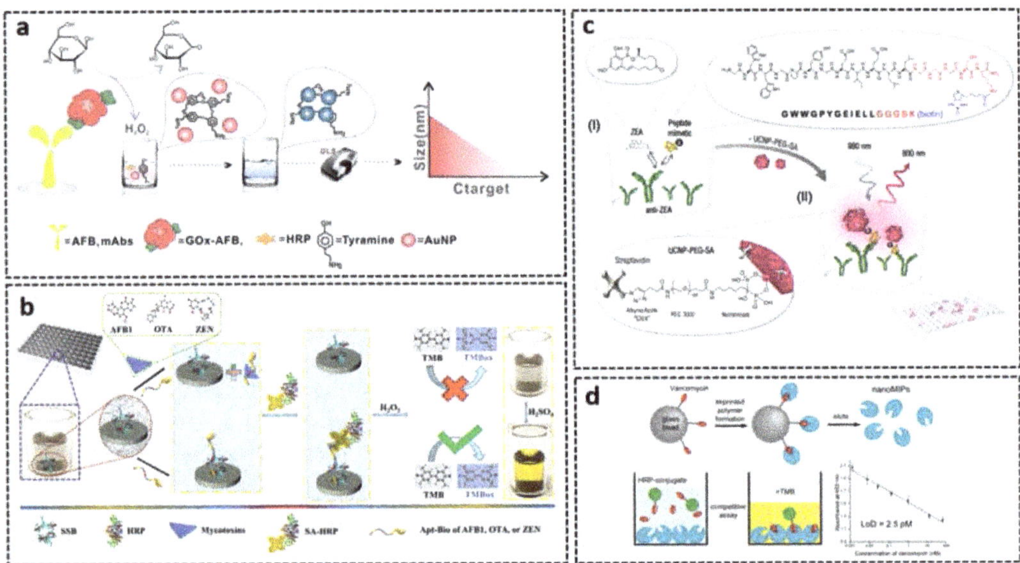

Figure 6. (**a**) Schematic illustration of DLS-dcELISA method combined with H$_2$O$_2$-mediated tyramine signal amplification system. (**b**) Scheme of green ELISA based on SSB-assisted aptamer. (**c**) Scheme of the competitive ULISA for the detection of ZEN. (**d**) Scheme of molecularly imprinted polymer nanoparticle-based assay for vancomycin determination. Reproduced with permission from [266,284,294,297].

3.3. Lateral Flow Assays

Lateral flow assay (LFA) is based on the movement of liquid sample along a strip of polymeric material, generally nitrocellulose (NC) membrane, for qualitative, semi-quantitative, and, to some extent, quantitative determination of analytes. Compared with microplate-based assays, LFA needs less detection period and is much easier to conduct; thus, it has been widely applied in point-of-care diagnostics and on-site monitoring. For the determination of small molecules like mycotoxins, LFA is based on a competitive format that is the same as microplate-based assays. Up to now, pAb- and mAb-based LFAs have been already developed for detecting mycotoxins, including AFB1 [300], AFM1 [301], DON [302], OTA [303], ZEN [304], FB1 [305], T-2 toxin [306], cyclopiazonic acid [307], and tenuazonic acid [308]. The assay is executed by adding small sample volume on the strip, allowing analytes of interest flow through the membrane. After a while, qualitative or semi-quantitative result is revealed by the appearance of a test line (T-line), and quantification can be realized by an optical reader. Given the benefits of strong red color, good stability, easy-of-synthesis, and low toxicity, gold nanoparticles are the most common labels in LFA [309]. To further enhance the color intensity of gold nanoparticles and facilitate the sensitivity of LFA, Xu et al. [310] synthesized polydopamine (PDA)-coated AuNPs as signal-amplification label for detection of ZEN in maize. PDA coating served as a linker of mAb and the nanoparticles. PDA-coated AuNPs was proven to be more stable and less easily aggregated with a stronger color brightness than that of AuNPs. From Figure 7a, in the absence of ZEN, a red band is observed due to the accumulation of Au@PDA-mAb on the T-line. Conversely, when there is an amount of ZEN in the sample, less or no recognition position is available to capture antigen on the T-line, which resulted in no line or a weaker line. The LOD of this assay is 7.4 pg/mL, which was 10-fold lower than that of AuNP-based LFA. In recent years, with the development of novel nanomaterials, extensive efforts have been devoted to increase the sensitivity of LFA for mycotoxin analysis by utilizing new labeled probes with stronger signal [311,312]. Therefore, different types of detection agents have been developed and applied in LFA for signal generation, including quantum dots (QDs) [306], upconverting nanoparticles (UCNPs) [313], magnetic nanoparticles [314], and near-infrared (NIR) fluorescent dye [315], etc. The analyte-antibody-probe complex continuously migrates by capillary action and competes for binding with the antigen deposited on the test line.

Due to the complexity of the co-occurrence of mycotoxins, the demands of simultaneous detection of multiple mycotoxins are increasing. To date, a number of multiplex LFAs for mycotoxins have been successfully developed with good performance, which could simultaneously detect up to six mycotoxins [315–321]. By using AuNPs and time-resolved fluorescent microspheres (TRFMs) as corresponding signal labels, two types of LFAs (AuNPs-LFA and TRFMs-LFA) were established by Z. Liu et al. for simultaneous detection of AFB1, ZEN, T-2, DON, and FB1 (Figure 7b) [322]. The visible LOD for the five mycotoxins were 10/2.5/1.0/10/0.5 µg/kg (AuNPs-LFA) and 2.5/0.5/0.5/2.5/0.5 µg/kg (TRFMs-LFA). By integration with a self-designed, smartphone-based, dual-mode device, quantification of the five mycotoxins was realized with LODs of 0.59/0.24/0.32/0.9/0.27 µg/kg and 0.42/0.10/0.05/0.75/0.04 µg/kg, respectively.

Although very few LFAs were introduced using rAbs as bioreceptors, the anti-idiotype nanobody (AIdnb) could serve as surrogate antigen on an immunochromatographic strip. Li's group reported a time-resolved fluorescence lateral flow assay based on two anti-idiotypic nanobodies for simultaneous detection of AFB1 and ZEN in maize products [323]. As can be seen in Figure 7c, AIdnb were coated on NC membrane as test lines for AFB1 and ZEN, respectively. Anti-AFB1 mAb and anti-ZEN mAb were conjugated with Eu/Tb (III)-nanospheres as detector. For negative samples, the probes were captured by AIdnb immobilized on T-lines. For positive samples, target toxins in the samples reacted with mAb-probe, resulting in less or no probe captured on the T-lines. The intensity of T-line and C-line (control line) was measured by a homemade portable fluorescence spectrophotometer. A linear relationship between T/C values and logarithm of concentration of AFB1 and ZEN was constructed and applied for quantification.

In addition, peptide mimotopes, with the function of binding to the corresponding antibody, can also be used as antigen mimetics in LFA for mycotoxin analysis [324]. Yan et al. applied phage-displayed peptide and peptide-MBP (myelin basic peptide) fusion onto the T-line as the mimetic antigen. CdSe/ZnS QDs and QD-nanobeads with excellent optical property were conjugated with corresponding mAb as a signal reporter for rapid and simultaneous detection of FB1, ZEN, and OTA [325]. Under optimal conditions, the peptide-MBP-based LFA could detect 0.25 ng/mL FB1, 3.0 ng/mL ZEN, and 0.5 ng/mL OTA visually within 10 min.

Given the nature of nucleotide, aptamer can be hybridized with complementary DNA, and once the targets are present, the hybridization is deconstructed. Based on this property, aptamer-based LFAs have been designed for AFB1, ZEN, and OTA [326–330]. Wu et al. developed an aptamer-based lateral flow test strip for ZEN detection based on the competitive combination of aptamer with toxin and its complementary DNA (DNA 1) on the test line [329]. In this format, 3′-thiol- and poly A-modified OTA aptamer was synthesized and labeled with AuNPs. As shown in Figure 7d, in the absence of ZEN, AuNPs-Apt hybridizes with DNA 1 that is labeled with streptavidin and biotin-modified complementary DNA 2 that is immobilized on the T-line, causing a visible red line. If ZEN is present in the sample, AuNPs-Apt would bind with toxins, resulting in no line or a red line with weaker intensity. The more target analytes in the sample, the weaker intensity of the T-line. The control zone is loaded with biotin-modified polyT, which would hybridize with the polyA tail of the aptamer regardless of the presence of ZEN. The strip could detect

ZEN in a range of 5–200 ng/mL, and the visual LOD was 20 ng/mL. Since a minimum of 500 s for hybridizing DNAs on microarray is usually required, it is difficult to obtain a strong and valid signal on the NC membrane by hybridization within 10 min [331]. To address this problem, Shim et al. developed an aptamer-based dipstick assay for AFB1 determination. In this approach, the biotin-modified aptamer was first incubated with sample solution and Cy5 dye-modified complementary DNA probe, which could assure adequate time for DNA hybridization [181]. Streptavidin and anti-Cy5 antibody were immobilized on test and control zone, respectively. The assay could be finished within 30 min with an LOD of 0.1 ng/mL for AFB1 in buffer.

Most of the reported aptamer-based strip assays are based on the competition binding of free toxin and complementary DNA with aptamers. One key factor that affects the reaction is the length of complementary DNA. If the length of complementary DNA is the same with aptamer, the latter would rather hybridize with complementary DNA than combine with the target. Taking advantage of the high affinity of aptamer-complementary strand and the binding efficiency of aptamer-target, Zhu et al. designed a dual-competitive LFA for AFB1 determination [332]. In this assay, AFB1-BSA was deposited on the test zone and competed for binding to aptamer with free toxin. In the presence of AFB1, the Cy5-labeled aptamer combines with toxin, which could not be captured by the immobilized antigen, resulting in a decrease in fluorescence signal on T-line. When the aptamer-AFB1 complex arrived at control zone, the aptamer hybridized with complementary DNA and dissociated with the toxin at the same time, owing to the higher affinity of hybridization. As a result, the higher the concentration of AFB1, the higher the intensity of the signal on the C-line. To increase the validity of the strip, the ST/SC ratio was employed for quantification of AFB1. The assay achieved an LOD of 0.1 ng/mL and a linear range of 0.1–1000 ng/mL.

Besides the benefits mentioned in the second section, aptamers can also show superiority over antibodies in the application on LFAs for mycotoxin determination. On one hand, aptamers can be synthesized with biotin, thiol, or fluorescent molecules for single-site conjugation, which facilitate quantitative analysis. Secondly, given the single-stranded DNA property, aptamer LFAs can be designed based on hybridization with complementary DNA, which eliminates the use of antigen. However, the limitation of this assay is hybridization deficiency, making it difficult to obtain a strong and reliable line on both test and control zones.

3.4. Biosensors

Biosensors are portable bioanalytical devices that incorporate biological recognition elements for binding of target molecules and a signal transducer to convert the biorecognition event into a measurable signal [333]. Based on various bioinspired recognition elements, mycotoxin sensors can be categorized into immunosensors [334], aptasensors [335,336], peptide-based sensors [205], and MIPs-based sensors [337,338]. Compared with other analytical methods mentioned above, it is much easier to realize real-time monitoring of reaction changes dynamically using biosensors, with output of the results in digital formats. Not only can the detection period be shorter, but the sensitivity, simplicity, robustness, and reusability can also be improved, making it possible to develop low-cost high-throughput screening methods for mycotoxins. A variety of transducers have been explored for mycotoxin sensor development. The electrochemical, optical, mass sensitive, calorimetric, and magnetic transducers stand out as the most important sensing platforms [339–341]. There is an increasing tendency for development of electrochemical immuno- and aptasensors [342–344].

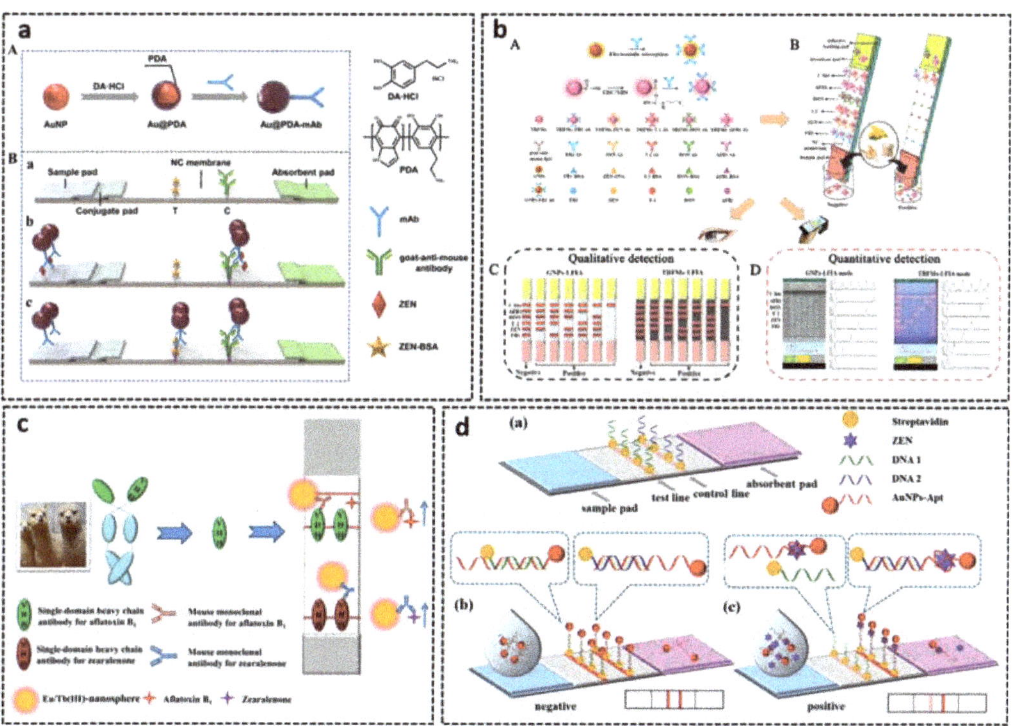

Figure 7. (**a**) Scheme of polydopamine-coated gold nanoparticles-based lateral flow immunoassay for ZEN detection. (**b**) Schematic diagram of smartphone-based GNPs and TRFMs-LFIAs for multiplex mycotoxins detection. (**c**) Schematic illustration of anti-idiotypic nanobody-based TRFICA for AFB1 and ZEN. (**d**) Scheme of aptamer-based lateral flow test strip for ZEN.Reproduced with permission from [310,322,323,329].

PAbs and mAbs are most commonly applied in the fabrication of mycotoxin immunosensors. J. Tang's group developed an impedimetric immunosensor for OTA determination in red wine based on OTA-specific pAbs [345]. In this platform (Figure 8a), OTA-BSA was immobilized on the electrode. This conjugate competes with free OTA for graphene oxide nanosheets labeled anti-OTA pAb. This immunosensor exhibited an LOD of 0.055 pg/mL, with a working range between 0.1 pg/mL to 30 ng/mL. Zong et al. [346] developed a chemiluminescence immunosensor for AFB1, OTA, and CIT, using specific mAbs and glass-slide-immobilized antigens. A signal-on photoelectrochemical immunoassay for AFB1 based on enzymatic product-etching MnO_2 nanosheets for dissociation of carbon dots was developed [347]. Under optimal conditions, the photocurrent increased with the increasing target AFB1 within a dynamic working range from 0.01 to 20 ng/mL with an LOD of 2.1 pg/mL. X. Tang et al. immobilized diacetoxyscirpenol-OVA on a microtiter strip and constructed a pressure-dependent immunosensor by labeling secondary antibody with Au@PtNP [348]. The mentioned immunosensors are based on indirect competitive immunoreactions, which require conjugating free toxin to a carrier protein or a signal probe as competitor. Peptide mimotopes, with the ability to bind to the same antibody paratope as the antigen, have been integrated into biosensor fabrication as a promising surrogate [293]. Hou et al. demonstrated an electrochemical immunosensor using phage-displayed peptide mimotope as the competing antigen for the detection of OTA [291]. In this case, the anti-OTA mAb was immobilized on a PEG-modified electrode. After competitive reaction of OTA and OTA-mimotope for binding to the anti-OTA mAb, the HRP-conjugated anti-M13 bacteriophage antibody was added to the sensor, and the quantification was realized by square-wave voltammetry measurement. The peptide-based immunosensor showed high selectivity and sensitivity, allowing the detection of OTA as low as 2.04 fg/mL in a linear range of 7.17–548.76 fg/mL. Label-free electrochemical biosensors have also been reported. For fabrication of an electrochemical immunosensor, Jiang et al. synthesized MoS_2-thionin composites to modify the glassy carbon electrode (GCE), followed by coating of Pt-conjugated anti-ZEN mAb [349]. Square wave voltammetry (SWV) measurement was conducted to determine the concentration of ZEN. The peak current deceased with the increase of ZEN concentration. A linear range from 0.01 to 50 ng/mL and an LOD of 0.005 ng/mL was achieved by the electrochemical sensing platform. Recently, a label-free photoelectrochemical immunosensor based on antibody-immobilized photocatalyst g-C3N4/Au/WO3 was developed, which allowed the detection of AFB1 in the range form 1.0 pg/mL to 100 ng/mL [350].

Recombinant antibodies have great potential in biosensing systems. Z. Tang et al. developed a competitive FRET-based immunosensor by using QDs-labeled OTA and QDs-labeled nanobody as energy donor and acceptor, respectively [351]. Compared with traditional antibodies, the small size of Nb decreases the FRET distance between two QDs, making it more suitable for a sensitive FRET-based assay. The Nb-FRET immunosensor could detect OTA as low as 5 pg/mL within 5 min. A voltammetric immunosensor was constructed for AFB1 detection by X. Liu et al. [352]. The anti-AFB1 nanobody was coated on the surface of AuNPs/WS$_2$/MWCNTs nanocomposites serving as recognition element and AFB1-streptavidin conjugate as competitor. This assay displayed a linear range from 0.5 pg/mL to 10 ng/mL with an LOD of 68 fg/mL. By using an anti-DON Fab fragment as recognition element, Romanazzo et al. [113] developed an Enzyme-Linked-Immunomagnetic-Electrochemical (ELIME) assay for DON detection in food samples. The sensor achieved a working range from 100 ng/mL to 4500 ng/mL and an EC50 of 380 ng/mL.

Specific peptides are also able to be applied in biosensors. Based on the crystal structure a AFB1-specific antibody, B. Liu et al. constructed a peptides library specific to AFB1 by using molecular docking and amino mutation [205]. The peptide P24 with highest affinity with AFB1 was selected and employed in an electrochemical immunosensor with signal enhancement of porous AuNPs. As shown in Figure 8b, P24 was immobilized on the surface of porous AuNPs/GCE electrode as a recognition element. The electrical current was detected by differential pulse voltammetry method for quantification of AFB1. The LOD of the assay was 9.4×10^{-4} µg/L with a linear range from 0.01 µg/L to 20 µg/L.

In the past decade, numerous aptasensors towards several mycotoxins have been introduced, including OTA [353], AFB1 [354], AFM1 [355], PAT [356], FB1 [357], ZEN, and T-2 toxin [358]. Based on their chemical nature, aptamers are more effective and robust under extreme pH and temperature conditions, making them attractive as reliable recognition elements in biosensors. Mycotoxin aptasensors mainly depend on the interactions between an aptamer and target toxin and its complementary strand or a signal probe, taking advantage of the unique property of nucleic acids, including configurational or conformational modifications under the formation of aptamer-target complex. Various detection modes have been applied, which can be mainly categorized into optical and electrochemical sensors. Optical methods, such as colorimetry [359], fluorescence [360], luminescence [361], FRET [362], and surface-enhanced Raman spectroscopy-based aptasensors [363], benefit from easy generation and provide high sensitivity. Based on self-assembly of rolling circle amplification (RCA), Hao et al. developed a fluorescent DNA hydrogel aptasensor for OTA [364]. As illustrated in Figure 8c, the OTA aptamer was first hybridized with the primer. In the presence of OTA, the aptamer tends to bind with the target, leading to the dissociation of primer. Free primer would combine with the padlock probe, which would initiate the RCA reaction, resulting in a formation of fluorescent DNA hydrogel. On the contrary, in the absence of OTA, no DNA hydrogel can be produced. The LOD of this aptasensor was 0.01 ng/mL, with a linear range from 0.05 to 100 ng/mL. In a similar approach, Abnous et al. designed a colorimetric aptasensor for AFM1 in milk based on the combination of CRISPR-Cas12a, RCA, and catalytic activity of gold nanoparticles [365]. The sensing method achieved an LOD of 0.05 ng/L, with a detection range from 0.2 to 300 ng/L. In comparison, electrochemical aptasensors are more cost-effective and feasible for on-site application owing to more simple instrumentation and fewer reagents [366–369]. This sensing mode mainly depends on the detection of changes of electric current occurring on electrode surface produced by recognition reaction.

Despite the binding affinity of aptamers, several factors should be considered to construct an aptasensor with high sensitivity. One is the aptamer's immobilization strategy. The fabrication of an electrochemical aptasensor requires the immobilization of aptamer on an electrode, which could remarkably affect the binding activity of aptamers. To increase the immobilization efficiency, various efforts have been made to modify the sensing platform. For example, carbon quantum dots/octahedral Cu_2O nanocomposite has been used to modify the glass carbon electrode and combine with aptamer through amino-carboxylic interaction [370]. The sensing platform allowed the detection of AFB1 with an LOD of 0.9 ± 0.04 ag/mL and a dynamic range from 3 ag/mL to 1.9 µg/mL. In addition, chitosan-functionalized acetylene black and multiwalled carbon nanotubes (CS@AB-MWCNTs) nanocomposite, with large specific surface area, good conductivity, and film-forming property, has also been proved to improve the immobilization of aptamer on electrode, thus increasing the detection sensitivity [371]. The other factor affecting the performance of an aptasensor is the signal amplification method. Many functional nanomaterials with outstanding physicochemical properties provide a powerful tool to improve the sensitivity of the developed electrochemical sensors. By using upconversion nanoparticles-doped Bi_2S_3 nanorods as photoactive materials, Gao et al. constructed a near-infrared light-induced photofuel cell-based aptasensor, allowing the detection for AFB1 in the range of 0.01–100 ng/mL, with an LOD of 7.9 pg/mL [372]. DNA amplification methods, including polymerase chain reaction (PCR) [373,374], hybridization chain reaction (HCR) [375,376], rolling circle amplification (RCA) [377,378], strand displacement amplification (SDA) [379,380], toehold-mediated strand displacement amplification (TMSD) [381], catalytic hairpin assembly (CHA) [382], and DNA machines, have also been applied in aptasensor construction to enhance the sensitivity [383]. Taking advantage of HCR, DNA walkers, and the properties of MoO_x nanomaterials, Wang and coworkers demonstrated an aptasensor for determination of OTA [384]. The sensitivity was greatly improved, with a detection limit as low as 3.3 fg/mL.

MIPs have received extensive attention for electrochemical sensors construction due to their unique advantages, such as high intrinsic stability and ease of preparation [385]. MIP-based electrochemical sensors have been utilized to detect mycotoxins, including AFB1 [386], OTA [387–389], DON [390], ZEN [391], FB1 [392,393], CIT [394], PAT [395–397], and T-2 toxin [398]. To obtain an ideal MIP-based electrochemical sensor with high sensitivity, the fabrication of the MIP on the electrode surface as well as the electrode modification strategy must be considered. Numerous methods have been utilized in the fabrication of MIPs, including electropolymerization, bulk polymerization, and precipitation polymerization, and among them, electropolymerization is a convenient way to prepare MIP membranes on the surface of the electrode given its rapid preparation, easy control of film thickness, and improved cohesiveness. Selvam et al. constructed an MIP-based disposable sensor for PAT [399]. In this strategy (Figure 8d), SeS_2-loaded Co MOF was synthesized via a tangible hydrothermal technology and loaded on a screen-printed electrode surface to improve the conductivity and stability. Then, Au@PANI (gold polyaniline) nanocomposite was prepared and loaded on the MOF screen-printed electrode to achieve higher sensitivity. Finally, the MIP sensor was fabricated on the Au@PANI/SeS2@Co MOF-modified screen-printed electrode platform via electropolymerization. An electron-blocking layer was formed when PAT was captured by the imprinted cavities, which caused a decrease in the electrochemical signal. This sensor possessed excellent performance, with an LOD of 0.66 pM for PAT and a logarithmic linear range from 0.001 to 100 nM. By using a similar approach, Huang et al. constructed an MIP-based electrochemical sensing platform for PAT determination by electropolymerization [395]. The combination of thionine, PtNP, and nitrogen-doped graphene (NGE) was used to modify the glassy carbon electrode to enhance the electric signal. The LOD of the fabricated sensor was 0.001 ng/mL in the PAT concentration range of 0.002–2 ng/mL. In another study, an MIP sensor for DON detection was developed by preparation of an MIP membrane on COOH-MWCNTs-modified electrode surface via electropolymerization [390]. The sensor displayed effective surface area, good conductivity, high selectivity, and a good response towards DON, with an LOD of 0.07 μM in wheat flour samples.

To summarize, electrochemical biosensors are the most prominent among mycotoxin sensors owing to their sensitivity, low cost, and miniaturization. The quantification for mycotoxins is based on the interaction between analytes and recognition elements, which is transformed to electrical signals using amperometric, potentiometric, conductimetric, and impedimetric measurements.

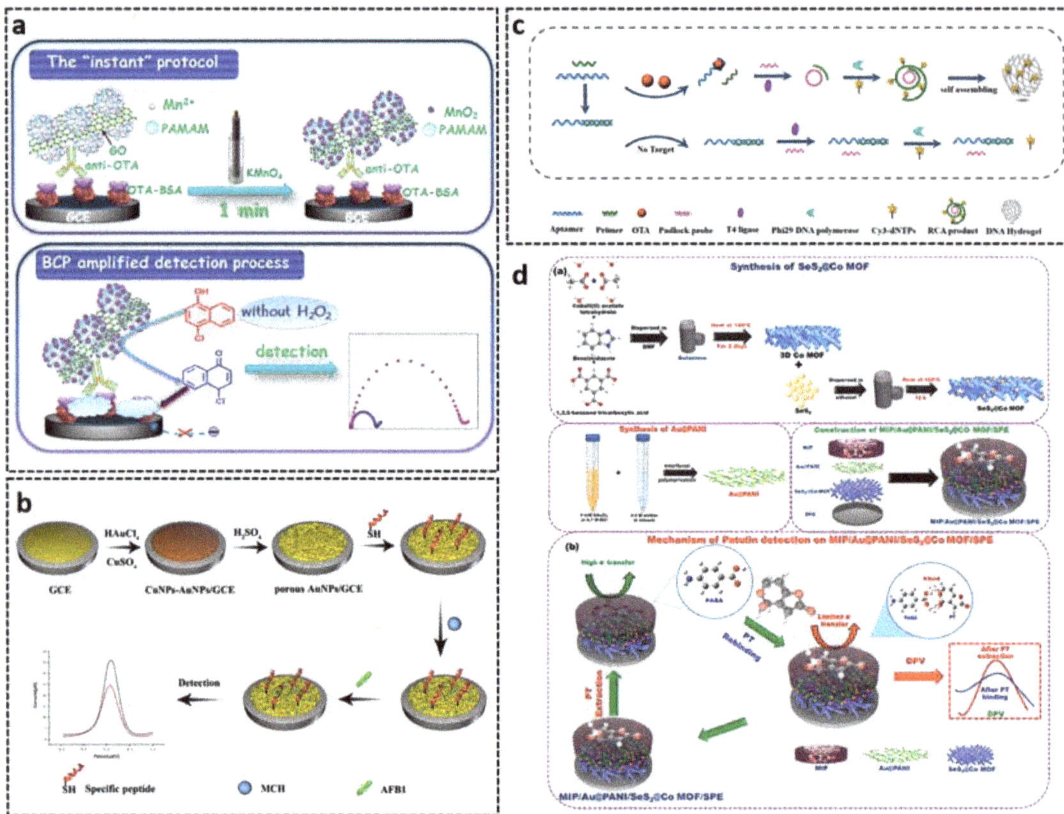

Figure 8. (**a**) Scheme of the amplified impedimetric immunosensor for OTA detection. (**b**) Scheme of electrochemical immunosensor for AFB1 detection based on specific peptide. (**c**) Scheme of fluorescent DNA hydrogel aptasensor for the detection of OTA. (**d**) Scheme of SeS$_2$-loaded Co MOF with Au@PANI-comprised electroanalytical MIP-based sensor for PAT. Reproduced with permission from [205,345,364,399].

4. Commercial Biorecognition Elements, Test Kits, and Analysis Systems for Mycotoxins Detection

Up to date, there is a high number of commercial recognition elements and test kits for mycotoxin analysis available on the market, and most of them are conventional antibodies and antibody-based test kits, including pAbs and mAbs, ELISA test kits, lateral flow assays, and immunoaffinity columns. In this section, we give a general overview on currently available commercial products and important worldwide suppliers. The collection does not claim to be complete. Besides direct marketing by manufacturers, distribution occurs mainly by regional retailers or specialized dot-com companies. The latter organize a direct contact between the product manufacturer and the customer; i.e., they act as an agent only [400].

4.1. Mycotoxin Antibodies

Mouse monoclonal mycotoxin antibodies, i.e., isotype IgG, are predominant on the market. However, polyclonal ones are still offered. To date, alternative biorecognition elements, with the exception of rAbs, are not commercially available as single products, i.e., not being part of a test kit (e.g., MIP-solid phase extraction columns). Some rAbs for mycotoxins were offered recently by Creative Biolabs (www.creative-biolabs.com accessed on 10 December 2021). Beside bioequivalent reagents (full-size IgG) with the identical primary sequence (for ZEN, OTA), scFv and Fab (for ZEN, OTA, AFM1) and also VHH single-domain antibodies (for 15-AcDON) are obtainable. Unfortunately, web-based product information documents are not complete. For example, data for cross-reactivity with metabolites and other mycotoxins are missing. Furthermore, if disposable, application notes should be made available for download to interested users.

As listed in Table 5, mAb and/or pAb are commercially available towards common mycotoxins. There are only very few suppliers of antibodies against PAT, T-2/HT-2, and EAs. A mAb-based ELISA kit against CIT can be obtained from Creative Diagnostics (www.creative-diagnostics.com accessed on 17 January 2022) (not listed in Table 5). Generally, important information, such as used immunogen, host species, purity, reactivity, advice for reconstitution and storage, recommended use, etc., are outlined in the disclosed data sheet. These details are essential to create your own immunoassay.

Table 5. Commercial products for bioanalytical determination of mycotoxins.

Mycotoxin	U.S.: Company/location/website *						
	Eurofins Abraxis Warminster, PA USA abraxis.eurofins-technologies.com	Neogen Lansing, MI U.S. www.neogen.com	Beacon Analytical Systems, Inc. Saco, ME USA www.beaconkits.com	Envirologix Inc. Portland, Main USA www.envirologix.com	Charm Sciences, Inc. Lawrence, MA USA www.charm.com	Vicam Milford, MA USA www.vicam.com	Romer Labs, Inc Newark, DE USA www.romerlabs.com
AFB1	n.a.	n.a.	n.a.	n.a.	n.a.	IAC	ELISA, LFD
AFB2	n.a.	n.a.	n.a.	n.a.	n.a.	IAC	n.a.
AFG1	n.a.	n.a.	n.a.	n.a.	n.a.	n.a.	n.a.
AFG2	n.a.	n.a.	n.a.	n.a.	n.a.	n.a.	n.a.
Total AFs	ELISA	ELISA, IAC, LFD	ELISA (plate, tube)	LFD	LFD	IAC, LFD	ELISA, LFD
AFM1	ELISA	ELISA, LFD	ELISA	n.a.	LFD	IAC, LFD	ELISA
OTA	ELISA (OTA, B, C)	ELISA, LFD	n.a.	LFD	n.a.	IAC	ELISA, LFD
ZEN	IAC, LFD	n.a.	ELISA (plate, tube)	LFD	LFD	IAC, LFD	ELISA, LFD
DON	ELISA	ELISA, IAC, LFD	ELISA	LFD	LFD	IAC, LFD	ELISA, LFD
FB1	n.a.	n.a.	n.a.	n.a.	n.a.	n.a.	LFD
Total FMs	n.a.	ELISA, LFD	ELISA	LFD	LFD	IAC, LFD	ELISA, LFD
T-2	n.a.	n.a.	ELISA	n.a.	n.a.	IAC, LFD	ELISA
T-2/HT-2	n.a.	ELISA, LFD	ELISA	LFD	LFD	IAC, LFD	n.a.
PAT	n.a.	n.a.	n.a.	n.a.	n.a.	n.a.	n.a.
CIT	n.a.	n.a.	ELISA	n.a.	n.a.	IAC	n.a.
Ergot alkaloids	n.a.	LFD	n.a.	n.a.	n.a.	n.a.	n.a.

Table 5. *Cont.*

Mycotoxin	Europe: Company/location/website						
	Biosense Laboratories AS Bergen Norway www.biosense.com	R-Biopharm Darmstadt Germany r-biopharm.com	Abcam, Inc. Cambridge U.K. www.abcam.com (abcam.cn abcam.jp)	antibodies-online GmbH Aachen# Germany www.antikoerper-online.de	Agrisera AG Umea Sweden www.agrisera.com	Biomol Hamburg Germany www.biomol.com	Aokin AG Berlin Germany www.aokin.com
AFB1	n.a.	ELISA	mAb	pAb, mAb, ELISA	ELISA	mAb, ELISA, LFD	mAb
AFB2	n.a.	n.a.	n.a.	n.a.	n.a.	mAb	n.a.
AFG1	n.a.	n.a.	n.a.	n.a.	n.a.	mAb	n.a.
AFG2	n.a.	n.a.	mAb	n.a.	n.a.	mAb	n.a.
Total AFs	ELISA	ELISA, LFD, IAC	n.a.	mAb, ELISA	ELISA	mAb, ELISA, LFD	IAC
AFM1	n.a.	ELISA	n.a.	ELISA	pAb	mAb, ELISA, LFD	mAb, IAC
OTA	n.a.	ELISA, IAC	pAb, mAb	ELISA	pAb	mAb, ELISA, LFD	mAb, IAC
ZEN	ELISA	ELISA, LFD, IAC	mAb	pAb, mAb, ELISA	ELISA	mAb, ELISA, LFD	mAb, IAC
DON	ELISA	ELISA, LFD, IAC	pAb	pAb, mAb, ELISA	pAb	mAb, ELISA, LFD	mAb, IAC
FB1	n.a.	ELISA, LFD, IAC	mAb	ELISA	n.a.	mAb, ELISA, LFD	mAb, IAC
Total FMs	ELISA	n.a.	n.a.	n.a.	n.a.	mAb	n.a.
T-2	ELISA	ELISA, IAC	n.a.	ELISA	n.a.	ELISA, LFD	mAb
T-2/HT-2	n.a.	ELISA, LFD, IAC	n.a.	n.a.	n.a.	n.a.	mAb, IAC
PAT	n.a.	MISPE	n.a.	pAb	pAb	n.a.	n.a.
CIT	n.a.	ELISA, IAC	n.a.	n.a.	n.a.	n.a.	n.a.
Ergot alkaloids	n.a.	n.a.	n.a.	n.a.	n.a.	mAb	mAb

Mycotoxin	China: Company/location/website						
	Cusabio Technology Co., Ltd. Wuhan cusabio.cn	Lvdu Bio-sciences 6 Technology Co., Ltd. Binzhou, Shandong lvdu.net	Jiangsu Suwei Microbiological Research Co., Ltd. Wuxi jssuwei.com	Beijing WDWK Biotechnology Co., Ltd. Beijing wdwkbio.com	Nankai Biotech Co. Ltd. Hangzhou nkbiotech.com	Beijing KWINBON Biotechnology Co., Ltd. Beijing kwinbon.com	Shandong Meizheng Bio-Tech Co., Ltd. Rizhao, Shandong meizhengbio.com
AFB1	ELISA	mAb, IAC, LFD, ELISA	ELISA, LFD, IAC	ELISA, LFD	LFD	ELISA, LFD, IAC	ELISA, LFD, IAC
AFB2	n.a.	n.a.	n.a.	n.a.	n.a.	n.a.	n.a.
AFG1	n.a.	n.a.	n.a.	n.a.	n.a.	n.a.	n.a.
AFG2	n.a.	n.a.	n.a.	n.a.	n.a.	n.a.	n.a.
Total AFs	ELISA	IAC, LFD	ELISA, IAC	n.a.	n.a.	ELISA	ELISA, LFD, IAC
AFM1	ELISA	mAb, IAC, LFD, ELISA	ELISA, LFD, IAC	ELISA, LFD	LFD	ELISA, LFD	ELISA, LFD, IAC
OTA	ELISA	mAb, IAC, LFD, ELISA	ELISA, LFD, IAC	ELISA, LFD	LFD	ELISA, IAC	ELISA
ZEN	ELISA	mAb, IAC, LFD, ELISA	ELISA, LFD, IAC	ELISA, LFD	LFD	ELISA, LFD, IAC	ELISA, LFD
DON	ELISA	mAb, IAC, LFD, ELISA	ELISA, LFD, IAC	ELISA, LFD	LFD	ELISA, LFD, IAC	ELISA, LFD, IAC
FB1	ELISA	mAb, IAC, ELISA	n.a.	n.a.	n.a.	n.a.	n.a.
Total FMs	n.a.	LFD, ELISA	ELISA	ELISA, LFD	LFD	ELISA, LFD	ELISA, LFD
T-2	n.a.	mAb, IAC, LFD, ELISA	ELISA	ELISA, LFD	LFD	ELISA	ELISA
T-2/HT-2	n.a.	n.a.	n.a.	n.a.	n.a.	n.a.	n.a.
PAT	n.a.	mAb	n.a.	n.a.	n.a.	n.a.	n.a.

Mycotoxin	China: Company/location/website						
	Cusabio Technology Co., Ltd. Wuhan cusabio.cn	Lvdu Bio-sciences 6 Technology Co., Ltd. Binzhou, Shandong lvdu.net	Jiangsu Suwei Microbiological Research Co., Ltd. Wuxi jssuwei.com	Beijing WDWK Biotechnology Co., Ltd. Beijing wdwkbio.com	Nankai Biotech Co. Ltd. Hangzhou nkbiotech.com	Beijing KWINBON Biotechnology Co., Ltd. Beijing kwinbon.com	Shandong Meizheng Bio-Tech Co., Ltd. Rizhao, Shandong meizhengbio.com
CIT	n.a.	IAC	n.a.	n.a.	n.a.	n.a.	n.a.
Ergot alkaloids	n.a.	n.a.	n.a.	n.a.	n.a.	n.a.	n.a.

n.a., information not available. Used abbreviations: mAb, monoclonal antibody; pAb, polyclonal antibody; ELISA, enzyme-linked immunosorbent assay; LFD, lateral flow device; IAC, immunoaffinity chromatography. * All the websites were accessed on 10 December 2021.

4.2. Microplate or Tube ELISA Test Kits

All commercial test kits are based on mAbs and pAbs. Products with use of alternative biorecognition elements are unknown. ELISA is commonly used in mycotoxin detection with the advantages of high throughput, sensitivity, and accuracy. Both direct and indirect immunoassay formats have been involved in commercial ELISA kits. The microplate is precoated with antibody (direct format) or mycotoxin-protein conjugate (indirect format) and blocked with protein. In direct format, mycotoxin standards or samples are then added to each well with conjugated mycotoxin-enzyme. In indirect format, standard solution or samples are added together with specific antibody and enzyme-conjugated secondary antibody. To achieve a fast detection, the incubation period (15–20 min) is usually less than that reported in research articles (30–60 min). The whole detection can be finished mostly within 30 min. Various signal readout techniques (e.g., colorimetry, fluorescence, chemiluminescence) in ELISAs have been reported. Aokin rapid analysis systems (www.aokin.com accessed on 10 December 2021) commercialized a hand-held and portable instrument (Aokin mycontrol analyzer FP) based on highly sensitive, patented kinetic fluorescence polarization technology, together with a set of detection kits (mycontrol kits, incl. SPE cleanup columns) for most of regulated mycotoxins to allow rapid and quantitative determination in food and feedstuffs on-site. Altogether, most commercial ELISA kits predominantly use colorimetry (note: In Table 5, the signal technique is not specified for individual ELISA kits.). The mycotoxin ELISA kits can be applied to food (e.g., milk), agricultural commodities (e.g., wheat, rice, maize, etc.), and feed products after a simple extraction procedure. Generally, the applications are accurately specified in the instruction manuals.

4.3. Lateral Flow Assays

Given the benefits of low costs, user-friendliness, rapidness (usually <15 min), little interferences, portability, long shelf life, and operation by nonspecialized personnel on-site, lateral flow assays have attracted considerable interest in food-safety area. The goal is to accommodate all the reagents required for a quantifiable test on a simple membrane, strip, or capillary. It should be possible to place a volume of sample on the carrier or dip it into the liquid sample (e.g., sample extract) and to determine the presence of target analyte from resulting depth of color or length of colored band. Depending on the kind of label, evaluation can be done by naked eye (visual inspection), or spots can be read out by an electronic device, e.g., a smartphone. Labeling of antibodies by nanogold, which leads to red lined for colorimetric evaluation, is dominant. Most commercially available test strips for mycotoxins are encased individually in a plastic container. With few exceptions (PAT, CIT, EAs), related tests are offered for all regulated mycotoxins (Table 5). Generally, these tests are used for qualitative and semi-quantitative determinations, especially for sample screening regarding compliance/exceedance of limit values and maximum residue limits (MRLs). So far, quantitative tests are less widely available. Analogous to ELISA formats, different test configurations are possible. Owing to the small size of the targets, common principle of LFDs for mycotoxins is the indirect competitive immunoassay. The LFD is a combination of thin-layer chromatography, i.e., diffusion over a distinct distance on a membrane, and detection of a specifically labeled immune-reactant. The main elements are a sample pad (for addition of liquid), conjugate pad (with adsorbed, labeled target antibody probe, i.e., the primary antibody), nitrocellulose membrane with fixed test line (T-line, with adsorbed analyte protein conjugate) and control line (C-line, with adsorbed secondary antibody specific for the primary antibody; this line serves as an indispensable confirmation that the test worked correctly), and absorption pad (for absorption of diffused liquid), all of which are generally fixed on a plastic carrier. Briefly, after the addition of sample extract to the strip, any mycotoxin present in the sample will bind to the labeled antibody probe and migrate together with unbound antibody along the membrane caused by capillary forces. The mycotoxin-protein conjugate on the T-line competes with free toxin in the sample extract for binding to the labeled antibody. If a visible line appears on the test zone, the concentration of mycotoxin is less than the cut-off level, which is a negative sample. Conversely, a positive sample results in no visible line on the test zone. Quantification of the target can be realized by a commercial strip reader, based on the intensity of T-line or T/C signal ratio. With the aim to obtain even more sensitive assays, new labels are used, for example, fluorescent nanoparticles, such as quantum dots (Shandong Lvdu Bio-Sciences & Technology Co., LTD., Binzhou, China) and Lanthanide chelates (Beijing WDWK Biotechnology Co., LTD., Beijing, China). So far, rAbs-, aptamers-, and MIP-based LFDs are not commercialized.

4.4. Immunoaffinity Columns

Immunoaffinity chromatography (IAC; sometimes also termed immunoextraction, IE) is a special kind of solid-phase extraction (SPE) and combines immunological and traditional methods. It has been widely used as mycotoxin clean-up prior to TLC, HPLC, GC, or LC-MSMS. It is focused on the time- and cost-saving removal of sample interferences and the selective preconcentration of the trace target(s) in front of the chromatographic analysis. The technique uses cartridges or columns made from plastic material or glass and filled with anti-mycotoxin-loaded sorbent (termed immunoaffinity support or immunosorbent). The method can be performed off-line or on-line depending on the available support material. High-performance supports, which are needed for on-line techniques, must be rigid and robust, mechanically stable, and perfusive. The common principle, after addition of sample extract, is to start with a washing step to remove unbound and weakly bound sample constituents. This is followed by desorption of specifically trapped target analyte(s) with a suitable eluent and, finally, injection into the chromatographic system. Except PAT and EAs, IAC columns are offered for all regulated mycotoxins (Table 5). Users should perform the procedures as described in test instructions of suppliers. Generally, the tests are offered for single use. Increasingly, IAC columns for the simultaneous enrichment of multiple mycotoxins are coming to the market. The current examples for mention include Myco6in1+ columns from VICAM (www.vicam.com accessed on 10 December 2021) and 11 + Myco MS-Prep® from R-Biopharm (https://r-biopharm.com accessed on 10 December 2021) for simultaneous determination of AFs, OTA, FMs, DON, ZEN, and T-2/HT-2 in combination with LC-MS/MS. (note: Columns from VICAM also can detect nivalenol). To the best of our knowledge, commercial affinity columns for sample preparation based on rAbs and aptamers are not on the market yet. In contrast, MIP-based columns are already available. The company AFFINISEP (www.affinisep.com accessed on 10 December 2021) offers a set of cartridges (AFFINIMIP® SPE) designed for the analysis of one specific family of mycotoxins (AFs, FMs, DON, ZEN, OTA, PAT) and a multi-mycotoxin cartridge for simultaneous extraction of mentioned families plus T-2/HT-2 in combination with LC-MS/MS. Further, an MIP clean-up column (EASIMIP™ Patulin) is available from R-Biopharm.

4.5. Multiplexed Analysis Platforms

The ability to test multiple mycotoxins simultaneously, termed multiplexing, has several advantages over traditional single-analyte testing and has gained increasing interest over the past decade. As presented previously, new rapid tests were developed to address this challenge. Furthermore, there is an increasing need to provide cost-efficient, rapid, and fully automated methods for routine analytical laboratories. One option is to integrate a set of single devices on a modular platform and control the complete analysis by suitable software. Examples, such as Cobas Analyzer from Roche Diagnostics (www.roche.com accessed on 10 December 2021) and ADVIA Centauer XP Immunoassay System from Siemens Healthineers (www.siemens-healthineers.com accessed on 10 December 2021), can be found in high-throughput laboratories, e.g., clinical in-vitro diagnostics and pharma screening. Related investments are only affordable by big players in the food-safety testing market.

Therefore, another direction of research is focused on the development of continuous devices, e.g., flow injection analysis (FIA), and new miniaturized platforms for multiplexed high-throughput analysis. The latter can be separated into two technologies. The first is particle-based methods (bead-based arrays), which use a set of differently labeled micro- or nanoparticles (also termed microspheres or beads) as carriers for positioning of detection reagents, and tests are performed in suspension (suspension arrays) [401]. The second is planar biochips (also termed microarrays, lab-on-the-chip), with use of site-specific positioning of the detection reagents on a microchip, encased in a microfluidic cassette [402]. The strengths of suspension arrays are high array density, high-throughput capacity, and opportunity for configuration of a multiplex assay on demand (customizable, i.e., distinct sets of microspheres). Both types of assays are strongly dependent on the availability of special reagents and materials, i.e., appropriately functionalized carriers (particles or microchips), dispenser (arrayer for spotting of chips), biochip reader or scanner for microarrays, and flow cytometer for suspension arrays, including evaluation software. Because of the complexity of food samples, the maximum number of cycles of determination/regeneration (operating life) with one and the same biorecognition surface is limited, and the trend goes in the direction of single-use materials (beads and biochips). The latter is also caused by the steadily more efficient and cost-effective production (Table 6).

Table 6. Providers of multiplexed immunochemical analyses systems (biochips/beads).

Provider	Principle	Internet Address *
Luminex Corporation, Austin, TX, USA	suspension assay	www.luminexcorp.com
Becton Dickinson Biosciences, Franklin Lakes, NJ, USA	suspension assay	www.bdbiosciences.com
Quanterix Corp., Lexington, MA, USA	suspension assay	www.Quanterix.com
Merck Millipore, Burlington, MA, USA	suspension assay	www.merckmillipore.com
Bio-Rad-Laboratories, Hercules, CA, USA	suspension assay	www.bio-rad.com
SAFIA Technologies GmbH, Berlin, Germany	suspension assay	www.safia.tech
Foss GmbH, Hamburg, Germany	suspension assay	www.fossanalytics.com
Unisensor Seraing, Belgium	suspension assay	www.unisensor.be
Randox-Laboratories, Crumlin, UK	planar array (biochip)	www.randoxfood.com
GWK Präzisionstechnik GmbH, München, Germany	planar array (biochip)	www.gwk-munich.com

* All websites were last accessed on 10 December 2021.

The number of commercial biosensors that can detect the interaction of receptors with their targets in a preferably label-free manner and on time is steadily growing (Table 7). Important detection techniques are surface plasmon resonance (SPR) [403,404], quartz crystal microbalance (QCM) [405,406], microcantilever arrays [407], biolayer interferometry (BLI) [408], and electroswitchable biosurfaces (ESB) [409,410].

Table 7. Providers of biosensors that are based on different detection principles.

Provider	Principle [1]	Internet Address *
GE Healthcare, Chicago, IL, USA	SPR	www.biacore.com/lifesciences
Biolin Scientific, Gothenburg, Sweden	QCM	www.biolinscientific.com/qsense
Micromotive GmbH, Mainz Germany	Microcantilever array	www.micromotive.de
2bind GmbH, Regensburg, Germany	Bio-layer-interferometrie	www.2bind.com
Dynamic Biosensors GmbH, München, Germany [2]	ESB	www.dynamic-biosensors.com

[1] SPR, surface plasmon resonance; QCM, quartz crystal microbalance; ESB, electro-switchable biosurfaces. [2] Antibodies and microchips are available from Technical University Munich. * All websites were last accessed on 10 December 2021.

5. Conclusions and Future Perspective

Rapid determination methods based on biorecognition elements have been presented as promising tools for monitoring of mycotoxin contamination in food samples, which is a powerful supplement to highly sophisticated instrumental methods. In this review, the basic characteristics and application potential of commonly used recognition elements, including traditional pAbs and mAbs, and upcoming new receptors, such as rAbs, aptamers, peptides, as well as MIPs, were presented. Tremendous efforts have been dedicated over the last decade to develop receptors with further enhanced specificity, binding affinity, stability, and lower cost via improved antigen design, optimized screening strategies, expression or synthesis methods, and integration of new computational modeling approaches [411–413]. Consequentially, massive progress in the application of new receptors in various analytical formats, including microplate- or tube-based assays, lateral flow assays, solid-phase affinity support materials, and biosensors, have followed mainly in the academic sector. The product market on the global scale, however, clearly lags behind in bringing the new reagents, test kits, and technologies to the customers.

Disregarding some limitations, such as long production period and high costs associated with conventionally used pAbs and mAbs, they still dominate the field, both as available purified biorecognition reagents, receptors used in test kits, and affine binders of sample clean-up materials. The market for these reagents and tests is a competitive one, with global trading. It can be quite difficult and cumbersome to the customer to identify the original source of the antibody, and it often happens that, e.g., a mAb-producing cell clone was licensed to several companies for marketing either as purified antibody reagent or being part of a test-kit, branded or distributed under different names. However, there is a continuous demand for more cost-effective mycotoxin receptors with customized selectivity, affinity, and stability, i.e., engineered to fit the needs of the final application. From the current point of view, rAbs, especially the nanobodies, could be the most promising solution owing to progress of related technology. Furthermore, the availability and use of synthetic and naïve antibody libraries, which are rather large nowadays, could lead to the renouncing of obsolescent animal immunization. It is worth mentioning, for example, the EU directive 2010/63/EU on the protection of animals used for scientific purposes [414]. Creditably, Creative-Biolabs first made rAbs for OTA, ZEN, AFM1, and 15-AcDON commercially available.

The number of publications on the determination of mycotoxins based on various receptors was established and is presented in Figure 9. As can be seen, researches based on antibodies for mycotoxins detection are predominant. Aptamers are capable of recognizing targets with similar or even higher affinity compared to antibodies, with appealing characteristics in the aspects of production, stability, and signal labeling. However, it is still a challenge to obtain aptamers towards small molecules with desired characteristics. Peptide receptors can be obtained by phage-display technique or combinatorial synthesis, which are less prone to denaturation under high temperature and organic solvent. Nevertheless, mycotoxin-specific peptides and their applications are rarely reported except for OTA and AFB1. Yet, the application of peptide mimotopes as antigen substitutes might be a promising aspect in the future. Among all the receptors, MIPs are the most easily obtained, with increased thermal and mechanical stability. As biomimetic recognition materials, MIPs have attractive features mainly in the application to sample preparation. However, their affinity and specificity are generally lower compared to the other binding receptors. In future research, a variety of limitations, including but not limit to template leakage, non-specific binding, and low affinity, should be addressed.

Up to date, detection methods towards a variety of fungi metabolites with high toxicity and widespread occurrence have been extensively studied. In addition to common mycotoxins, for which the maximum permitted levels in food and feed products are regulated, those without a regulation also pose great harmfulness to humans and livestock, e.g., ergot alkaloid, citreoviridin, and sterigmatocystin, etc. However, studies on toxicology, risk assessment, and detection methods towards those emerging mycotoxins are still limited.

Thus, there is an ongoing demand for the development of recognition elements, assays, and rapid test-kits for new mycotoxins, e.g., NX-toxins, degradation products (e.g., cis-ZEN), as well as unregulated but important mycotoxins (e.g., the enniatins).

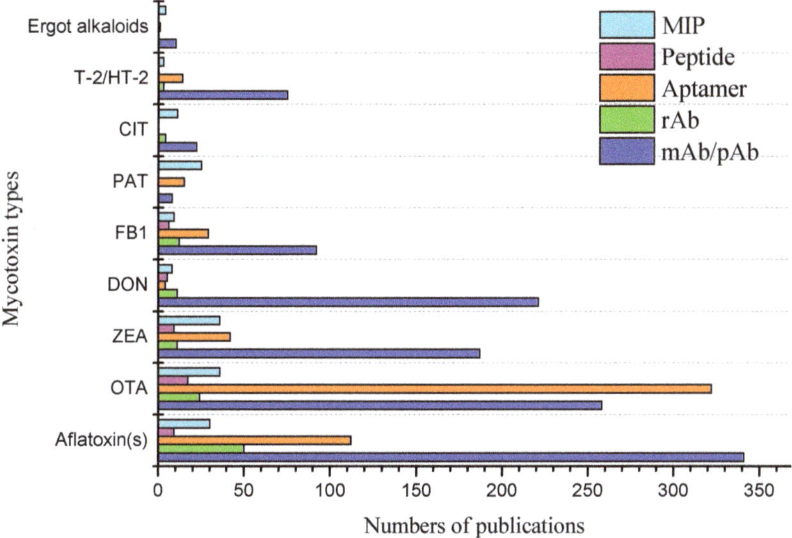

Figure 9. Numbers of publications on mycotoxins determination based on various recognition elements. Data were obtained in Web of Science until 3 December 2021.

To summarize, there is still a great deal of room for and challenges associated with advancements in recognition elements-related assays and biosensor development, which will move their application from laboratory to market. Especially, it can be expected that novel, customized recognition elements, such as nanobodies, aptamers, and MIPs, and rapid test kits based on these receptors might be increasingly available to users in the future. We suppose biosensors that allow label-free, multi-analyte determination will be found mainly in food and feed laboratories. In addition, new, cost-effective, and portable devices for rapid on-site analysis will be increasingly available. Finally, the crucial factors for the selection of the most appropriate method can be seen in regulatory issues, the objective of the analytical determination, sample type, facility of the laboratory, and, not to be overlooked, the experience of the analytical staff.

Author Contributions: Conceptualization, Y.W. and D.K.; writing—original draft preparation, Y.W. and C.Z.; writing—review and editing, D.K.; supervision, D.K. and J.W.; funding acquisition, Y.W. All authors have read and agreed to the published version of the manuscript.

Funding: This research was funded by National Natural Science Foundation of China (31501560) and China Postdoctoral Science Foundation funded project (2019M653767).

Institutional Review Board Statement: Not applicable.

Informed Consent Statement: Not applicable.

Data Availability Statement: Not applicable.

Conflicts of Interest: The authors declare no conflict of interest.

References

1. Agriopoulou, S.; Stamatelopoulou, E.; Varzakas, T. Advances in Occurrence, Importance, and Mycotoxin Control Strategies: Prevention and Detoxification in Foods. *Foods* **2020**, *9*, 137. [CrossRef]
2. Rai, A.; Das, M.; Tripathi, A. Occurrence and toxicity of a fusarium mycotoxin, zearalenone. *Crit. Rev. Food Sci. Nutr.* **2020**, *60*, 2710–2729. [CrossRef]
3. Marin, S.; Ramos, A.J.; Cano-Sancho, G.; Sanchis, V. Mycotoxins: Occurrence, toxicology, and exposure assessment. *Food Chem. Toxicol.* **2013**, *60*, 218–237. [CrossRef]
4. Pleadin, J.; Frece, J.; Markov, K. Mycotoxins in food and feed. In *Advances in Food and Nutrition Research*; Academic Press Inc.: Cambridge, MA, USA, 2019; Volume 89, pp. 297–345. ISBN 9780128171714.
5. Al-Jaal, B.; Salama, S.; Al-Qasmi, N.; Jaganjac, M. Mycotoxin contamination of food and feed in the Gulf Cooperation Council countries and its detection. *Toxicon* **2019**, *171*, 43–50. [CrossRef]
6. Misihairabgwi, J.M.; Ezekiel, C.N.; Sulyok, M.; Shephard, G.S.; Krska, R. Mycotoxin contamination of foods in Southern Africa: A 10-year review (2007–2016). *Crit. Rev. Food Sci. Nutr.* **2019**, *59*, 43–58. [CrossRef] [PubMed]
7. Gruber-Dorninger, C.; Jenkins, T.; Schatzmayr, G. Global mycotoxin occurrence in feed: A ten-year survey. *Toxins* **2019**, *11*, 375. [CrossRef] [PubMed]
8. Rodrigues, I.; Naehrer, K. A three-year survey on the worldwide occurrence of mycotoxins in feedstuffs and feed. *Toxins* **2012**, *4*, 663–675. [CrossRef]
9. Williams, J.; Phillips, T.; Jolly, P.E.; Stiles, J.; Jolly, C.; Aggarwal, D. Human aflatoxicosis in developing countries: A review of toxicology, exposure, potential health consequences, and interventions. *Am. J. Clin. Nutr.* **2004**, *80*, 22–1106. [CrossRef] [PubMed]
10. Ismail, A.; Gonçalves, B.L.; de Neeff, D.V.; Ponzilacqua, B.; Coppa, C.F.S.C.; Hintzsche, H.; Sajid, M.; Cruz, A.G.; Corassin, C.H.; Oliveira, C.A.F. Aflatoxin in foodstuffs: Occurrence and recent advances in decontamination. *Food Res. Int.* **2018**, *113*, 74–85. [CrossRef] [PubMed]
11. Mousavi Khaneghah, A.; Moosavi, M.; Omar, S.S.; Oliveira, C.A.F.; Karimi-Dehkordi, M.; Fakhri, Y.; Huseyn, E.; Nematollahi, A.; Farahani, M.; Sant'Ana, A.S. The prevalence and concentration of aflatoxin M1 among different types of cheeses: A global systematic review, meta-analysis, and meta-regression. *Food Control* **2021**, *125*, 107960. [CrossRef]
12. Jørgensen, K. Occurrence of ochratoxin A in commodities and processed food - A review of EU occurrence data. *Food Addit. Contam.* **2005**, *22*, 26–30. [CrossRef] [PubMed]
13. Ponsone, M.L.; Chiotta, M.L.; Palazzini, J.M.; Combina, M.; Chulze, S. Control of ochratoxin a production in grapes. *Toxins* **2012**, *4*, 364–372. [CrossRef] [PubMed]
14. Scott, P.M. Recent research on fumonisins: A review. *Food Addit. Contam. Part A* **2012**, *29*, 242–248. [CrossRef] [PubMed]
15. Schwake-Anduschus, C.; Langenkämper, G.; Unbehend, G.; Dietrich, R.; Märtlbauer, E.; Münzing, K. Occurrence of Fusarium T-2 and HT-2 toxins in oats from cultivar studies in Germany and degradation of the toxins during grain cleaning treatment and food processing. *Food Addit. Contam. Part A* **2010**, *27*, 1253–1260. [CrossRef]
16. Kiš, M.; Vulić, A.; Kudumija, N.; Šarkanj, B.; Jaki Tkalec, V.; Aladić, K.; Škrivanko, M.; Furmeg, S.; Pleadin, J. A Two-Year Occurrence of Fusarium T-2 and HT-2 Toxin in Croatian Cereals Relative of the Regional Weather. *Toxins* **2021**, *13*, 39. [CrossRef] [PubMed]
17. Karami-Osboo, R.; Mirabolfathy, M.; Aliakbari, F. Natural deoxynivalenol contamination of corn produced in Golestan and Moqan areas in Iran. *J. Agric. Sci. Technol.* **2010**, *12*, 233–239.
18. Wang, L.; Liao, Y.; Peng, Z.; Chen, L.; Zhang, W.; Nüssler, A.K.; Shi, S.; Liu, L.; Yang, W. Food raw materials and food production occurrences of deoxynivalenol in different regions. *Trends Food Sci. Technol.* **2019**, *83*, 41–52. [CrossRef]
19. Mahato, D.K.; Devi, S.; Pandhi, S.; Sharma, B.; Maurya, K.K.; Mishra, S.; Dhawan, K.; Selvakumar, R.; Kamle, M.; Mishra, A.K.; et al. Occurrence, impact on agriculture, human health, and management strategies of zearalenone in food and feed: A review. *Toxins* **2021**, *13*, 92. [CrossRef]
20. Ropejko, K.; Twarużek, M. Zearalenone and Its Metabolites-General Overview, Occurrence, and Toxicity. *Toxins* **2021**, *13*, 35. [CrossRef]
21. Silva, L.J.G.; Pereira, A.M.P.T.; Pena, A.; Lino, C.M. Citrinin in foods and supplements: A review of occurrence and analytical methodologies. *Foods* **2021**, *10*, 14. [CrossRef]
22. Saleh, I.; Goktepe, I. The characteristics, occurrence, and toxicological effects of patulin. *Food Chem. Toxicol.* **2019**, *129*, 301–311. [CrossRef]
23. Mahato, D.K.; Kamle, M.; Sharma, B.; Pandhi, S.; Devi, S.; Dhawan, K.; Selvakumar, R.; Mishra, D.; Kumar, A.; Arora, S.; et al. Patulin in food: A mycotoxin concern for human health and its management strategies. *Toxicon* **2021**, *198*, 12–23. [CrossRef]
24. Agriopoulou, S. Ergot alkaloids mycotoxins in cereals and cereal-derived food products: Characteristics, toxicity, prevalence, and control strategies. *Agronomy* **2021**, *11*, 931. [CrossRef]
25. Krska, R.; Crews, C. Significance, chemistry and determination of ergot alkaloids: A review. *Food Addit. Contam. Part A Chem. Anal. Control. Expo. Risk Assess.* **2008**, *25*, 722–731. [CrossRef]
26. Shephard, G.S. Impact of mycotoxins on human health in developing countries. *Food Addit. Contam. Part A Chem. Anal. Control. Expo. Risk Assess.* **2008**, *25*, 146–151. [CrossRef] [PubMed]
27. Anfossi, L.; Giovannoli, C.; Baggiani, C. Mycotoxin detection. *Curr. Opin. Biotechnol.* **2016**, *37*, 120–126. [CrossRef] [PubMed]

28. Liao, Y.; Peng, Z.; Chen, L.; Nüssler, A.K.; Liu, L.; Yang, W. Deoxynivalenol, gut microbiota and immunotoxicity: A potential approach? *Food Chem. Toxicol.* **2018**, *112*, 342–354. [CrossRef] [PubMed]

29. Frizzell, C.; Ndossi, D.; Verhaegen, S.; Dahl, E.; Eriksen, G.; Sørlie, M.; Ropstad, E.; Muller, M.; Elliott, C.T.; Connolly, L. Endocrine disrupting effects of zearalenone, alpha- and beta-zearalenol at the level of nuclear receptor binding and steroidogenesis. *Toxicol. Lett.* **2011**, *206*, 210–217. [CrossRef] [PubMed]

30. Amuzie, C.J.; Pestka, J.J. Suppression of insulin-like growth factor acid-labile subunit expression-A novel mechanism for deoxynivalenol-induced growth retardation. *Toxicol. Sci.* **2009**, *113*, 412–421. [CrossRef] [PubMed]

31. Liu, X.; Guo, P.; Liu, A.; Wu, Q.; Xue, X.; Dai, M.; Hao, H.; Qu, W.; Xie, S.; Wang, X.; et al. Nitric oxide (NO)-mediated mitochondrial damage plays a critical role in T-2 toxin-induced apoptosis and growth hormone deficiency in rat anterior pituitary GH3 cells. *Food Chem. Toxicol.* **2017**, *102*, 11–23. [CrossRef]

32. Degen, G.H.; Ali, N.; Gundert-Remy, U. Preliminary data on citrinin kinetics in humans and their use to estimate citrinin exposure based on biomarkers. *Toxicol. Lett.* **2018**, *282*, 43–48. [CrossRef]

33. Song, E.; Xia, X.; Su, C.; Dong, W.; Xian, Y.; Wang, W.; Song, Y. Hepatotoxicity and genotoxicity of patulin in mice, and its modulation by green tea polyphenols administration. *Food Chem. Toxicol.* **2014**, *71*, 122–127. [CrossRef]

34. Zhu, X.; Zeng, Z.; Chen, Y.; Li, R.; Tang, X.; Zhu, X.; Huo, J.; Liu, Y.; Zhang, L.; Chen, J. Genotoxicity of three mycotoxin contaminants of rice: 28-day multi-endpoint assessment in rats. *Mutat. Res. Genet. Toxicol. Environ. Mutagen.* **2021**, *867*. [CrossRef]

35. Rašić, D.; Želježić, D.; Kopjar, N.; Kifer, D.; Šegvić Klarić, M.; Peraica, M. DNA damage in rat kidneys and liver upon subchronic exposure to single and combined ochratoxin A and citrinin. *World Mycotoxin J.* **2019**, *12*, 163–172. [CrossRef]

36. Zhou, H.; George, S.; Hay, C.; Lee, J.; Qian, H.; Sun, X. Individual and combined effects of Aflatoxin B1, Deoxynivalenol and Zearalenone on HepG2 and RAW 264.7 cell lines. *Food Chem. Toxicol.* **2017**, *103*, 18–27. [CrossRef]

37. Wan, L.Y.M.; Turner, P.C.; El-Nezami, H. Individual and combined cytotoxic effects of Fusarium toxins (deoxynivalenol, nivalenol, zearalenone and fumonisins B1) on swine jejunal epithelial cells. *Food Chem. Toxicol.* **2013**, *57*, 276–283. [CrossRef] [PubMed]

38. Sun, L.H.; Lei, M.Y.; Zhang, N.Y.; Gao, X.; Li, C.; Krumm, C.S.; Qi, D.S. Individual and combined cytotoxic effects of aflatoxin B1, zearalenone, deoxynivalenol and fumonisin B1 on BRL 3A rat liver cells. *Toxicon* **2015**, *95*, 6–12. [CrossRef] [PubMed]

39. Rushing, B.R.; Selim, M.I. Aflatoxin B1: A review on metabolism, toxicity, occurrence in food, occupational exposure, and detoxification methods. *Food Chem. Toxicol.* **2019**, *124*, 81–100. [CrossRef]

40. Liu, D.; Ge, L.; Wang, Q.; Su, J.; Chen, X.; Wang, C.; Huang, K. Low-level contamination of deoxynivalenol: A threat from environmental toxins to porcine epidemic diarrhea virus infection. *Environ. Int.* **2020**, *143*, 105949. [CrossRef]

41. Kuiper-Goodman, T.; Scott, P.M.; Watanabe, H. Risk assessment of the mycotoxin zearalenone. *Regul. Toxicol. Pharmacol.* **1987**, *7*, 253–306. [CrossRef]

42. Tao, Y.; Xie, S.; Xu, F.; Liu, A.; Wang, Y.; Chen, D.; Pan, Y.; Huang, L.; Peng, D.; Wang, X.; et al. Ochratoxin A: Toxicity, oxidative stress and metabolism. *Food Chem. Toxicol.* **2018**, *112*, 320–331. [CrossRef] [PubMed]

43. Ekwomadu, T.I.; Akinola, S.A.; Mwanza, M. Fusarium mycotoxins, their metabolites (Free, emerging, and masked), food safety concerns, and health impacts. *Int. J. Environ. Res. Public Health* **2021**, *18*, 11741. [CrossRef]

44. Li, Q.; Yuan, Q.; Wang, T.; Zhan, Y.; Yang, L.; Fan, Y.; Lei, H.; Su, J. Fumonisin B1 Inhibits Cell Proliferation and Decreases Barrier Function of Swine Umbilical Vein Endothelial Cells. *Toxins* **2021**, *13*, 863. [CrossRef]

45. Nathanail, A.V.; Varga, E.; Meng-Reiterer, J.; Bueschl, C.; Michlmayr, H.; Malachova, A.; Fruhmann, P.; Jestoi, M.; Peltonen, K.; Adam, G.; et al. Metabolism of the Fusarium Mycotoxins T-2 Toxin and HT-2 Toxin in Wheat. *J. Agric. Food Chem.* **2015**, *63*, 7862–7872. [CrossRef] [PubMed]

46. Speijers, G.J.A.; Franken, M.A.M.; van Leeuwen, F.X.R. Subacute toxicity study of patulin in the rat: Effects on the kidney and the gastro-intestinal tract. *Food Chem. Toxicol.* **1988**, *26*, 23–30. [CrossRef]

47. de Oliveira Filho, J.W.G.; Islam, M.T.; Ali, E.S.; Uddin, S.J.; de Oliveira Santos, J.V.; de Alencar, M.V.O.B.; Júnior, A.L.G.; Paz, M.F.C.J.; de Brito, M.d.R.M.; e Sousa, J.M.d.C.; et al. A comprehensive review on biological properties of citrinin. *Food Chem. Toxicol.* **2017**, *110*, 130–141. [CrossRef] [PubMed]

48. Kong, W.; Zhang, X.; Shen, H.; Ou-Yang, Z.; Yang, M. Validation of a gas chromatography-electron capture detection of T-2 and HT-2 toxins in Chinese herbal medicines and related products after immunoaffinity column clean-up and pre-column derivatization. *Food Chem.* **2012**, *132*, 574–581. [CrossRef]

49. Schothorst, R.C.; Jekel, A.A. Determination of trichothecenes in wheat by capillary gas chromatography with flame ionisation detection. *Food Chem.* **2001**, *73*, 111–117. [CrossRef]

50. Tanaka, T.; Yoneda, A.; Inoue, S.; Sugiura, Y.; Ueno, Y. Simultaneous determination of trichothecene mycotoxins and zearalenone in cereals by gas chromatography-mass spectrometry. *J. Chromatogr. A* **2000**, *882*, 23–28. [CrossRef]

51. Hussain, S.; Asi, M.R.; Iqbal, M.; Khalid, N.; Wajih-Ul-Hassan, S.; Ariño, A. Patulin mycotoxin in mango and orange fruits, juices, pulps, and jams marketed in Pakistan. *Toxins* **2020**, *12*, 52. [CrossRef]

52. Moez, E.; Noel, D.; Brice, S.; Benjamin, G.; Pascaline, A.; Didier, M. Aptamer assisted ultrafiltration cleanup with high performance liquid chromatography-fluorescence detector for the determination of OTA in green coffee. *Food Chem.* **2020**, *310*, 125851. [CrossRef]

53. Tang, Y.; Mu, L.; Cheng, J.; Du, Z.; Yang, Y. Determination of Multi-Class Mycotoxins in Apples and Tomatoes by Combined Use of QuEChERS Method and Ultra-High-Performance Liquid Chromatography Tandem Mass Spectrometry. *Food Anal. Methods* **2020**, *13*, 1381–1390. [CrossRef]

54. Pantano, L.; La Scala, L.; Olibrio, F.; Galluzzo, F.G.; Bongiorno, C.; Buscemi, M.D.; Macaluso, A.; Vella, A. Quechers LC-MS/MS screening method for mycotoxin detection in cereal products and spices. *Int. J. Environ. Res. Public Health* **2021**, *18*, 3774. [CrossRef] [PubMed]

55. Janik, E.; Niemcewicz, M.; Podogrocki, M.; Ceremuga, M.; Gorniak, L.; Stela, M.; Bijak, M. The existing methods and novel approaches in mycotoxins' detection. *Molecules* **2021**, *26*, 3981. [CrossRef]

56. Iqbal, S.Z. Mycotoxins in food, recent development in food analysis and future challenges; a review. *Curr. Opin. Food Sci.* **2021**, *42*, 237–247. [CrossRef]

57. Tittlemier, S.A.; Brunkhorst, J.; Cramer, B.; DeRosa, M.C.; Lattanzio, V.M.T.; Malone, R.; Maragos, C.; Stranska, M.; Sumarah, M.W. Developments in mycotoxin analysis: An update for 2019-2020. *World Mycotoxin J.* **2021**, *14*, 3–26. [CrossRef]

58. Chen, X.; Wu, H.; Tang, X.; Zhang, Z.; Li, P. Recent Advances in Electrochemical Sensors for Mycotoxin Detection in Food. *Electroanalysis* **2021**, *33*, 1–11. [CrossRef]

59. Hou, Y.; Jia, B.; Sheng, P.; Liao, X.; Shi, L.; Fang, L.; Zhou, L.; Kong, W. Aptasensors for mycotoxins in foods: Recent advances and future trends. *Compr. Rev. Food Sci. Food Saf.* **2021**, in press. [CrossRef] [PubMed]

60. Jia, M.; Liao, X.; Fang, L.; Jia, B.; Liu, M.; Li, D.; Zhou, L.; Kong, W. Recent advances on immunosensors for mycotoxins in foods and other commodities. *TrAC - Trends Anal. Chem.* **2021**, *136*, 116193. [CrossRef]

61. Nolan, P.; Auer, S.; Spehar, A.; Elliott, C.T.; Campbell, K. Current trends in rapid tests for mycotoxins. *Food Addit. Contam. Part A Chem. Anal. Control. Expo. Risk Assess.* **2019**, *36*, 800–814. [CrossRef]

62. Peltomaa, R.; Benito-Peña, E.; Moreno-Bondi, M.C. Bioinspired recognition elements for mycotoxin sensors. *Anal. Bioanal. Chem.* **2018**, *410*, 747–771. [CrossRef]

63. Li, R.; Wen, Y.; Wang, F.; He, P. Recent advances in immunoassays and biosensors for mycotoxins detection in feedstuffs and foods. *J. Anim. Sci. Biotechnol.* **2021**, *12*, 108. [CrossRef]

64. Singh, J.; Mehta, A. Rapid and sensitive detection of mycotoxins by advanced and emerging analytical methods: A review. *Food Sci. Nutr.* **2020**, *8*, 2183–2204. [CrossRef] [PubMed]

65. He, Q.; Xu, Y. Antibody Developments and Immunoassays for Mycotoxins. *Curr. Org. Chem.* **2018**, *21*, 2622–2631. [CrossRef]

66. Tschofen, M.; Knopp, D.; Hood, E.; Stöger, E. Plant Molecular Farming: Much More than Medicines. *Annu. Rev. Anal. Chem.* **2016**, *9*, 271–294. [CrossRef]

67. Melnik, S.; Neumann, A.C.; Karongo, R.; Dirndorfer, S.; Stübler, M.; Ibl, V.; Niessner, R.; Knopp, D.; Stöger, E. Cloning and plant-based production of antibody MC10E7 for a lateral flow immunoassay to detect [4-arginine]microcystin in freshwater. *Plant Biotechnol. J.* **2018**, *16*, 27–38. [CrossRef] [PubMed]

68. Neumann, A.C.; Melnik, S.; Niessner, R.; Stöger, E.; Knopp, D. Microcystin-LR enrichment from freshwater by a recombinant plant-derived antibody using Sol-Gel-Glass immunoextraction. *Anal. Sci.* **2019**, *35*, 207–214. [CrossRef]

69. Ning, Y.; Hu, J.; Lu, F. Aptamers used for biosensors and targeted therapy. *Biomed. Pharmacother.* **2020**, *132*, 110902. [CrossRef]

70. Thyparambil, A.; Bazin, I.; Guiseppi-Elie, A. Molecular Modeling and Simulation Tools in the Development of Peptide-Based Biosensors for Mycotoxin Detection: Example of Ochratoxin. *Toxins* **2017**, *9*, 395. [CrossRef]

71. Malik, M.I.; Shaikh, H.; Mustafa, G.; Bhanger, M.I. Recent Applications of Molecularly Imprinted Polymers in Analytical Chemistry. *Sep. Purif. Rev.* **2019**, *48*, 179–219. [CrossRef]

72. De Middeleer, G.; Dubruel, P.; De Saeger, S. Molecularly imprinted polymers immobilized on 3D printed scaffolds as novel solid phase extraction sorbent for metergoline. *Anal. Chim. Acta* **2017**, *986*, 57–70. [CrossRef] [PubMed]

73. Hatamabadi, D.; Mostafiz, B.; Beirami, A.D.; Banan, K.; Moghaddam, N.S.T.; Afsharara, H.; Keçili, R.; Ghorbani-Bidkorbeh, F. Are molecularly imprinted polymers (Mips) beneficial in detection and determination of mycotoxins in cereal samples? *Iran. J. Pharm. Res.* **2020**, *19*, 1–18. [CrossRef] [PubMed]

74. Langone, J.J.; Vunakis, H. Van Aflatoxin B: Specific Antibodies and Their Use in Radioimmunoassay. *J. Natl. Cancer Inst.* **1976**, *56*, 591–595. [CrossRef] [PubMed]

75. Aalund, O.; Brunfeldt, K.; Hald, B.; Krogh, P.; Poulsen, K. A Radioimmunoassay for Ochratoxin A: A Preliminary Investigation. *Acta Pathol. Microbiol. Scand. Sect. C Immunol.* **1975**, *83 C*, 390–392. [CrossRef]

76. Bonel, L.; Vidal, J.C.; Duato, P.; Castillo, J.R. Ochratoxin A nanostructured electrochemical immunosensors based on polyclonal antibodies and gold nanoparticles coupled to the antigen. *Anal. Methods* **2010**, *2*, 335. [CrossRef]

77. Thirumala-Devi, K.; Mayo, M.A.; Reddy, G.; Reddy, S.V.; Delfosse, P.; Reddy, D.V.R. Production of polyclonal antibodies against Ochratoxin A and its detection in chilies by ELISA. *J. Agric. Food Chem.* **2000**, *48*, 5079–5082. [CrossRef]

78. Wang, J.J.; Liu, B.H.; Hsu, Y.T.; Yu, F.Y. Sensitive competitive direct enzyme-linked immunosorbent assay and gold nanoparticle immunochromatographic strip for detecting aflatoxin M1 in milk. *Food Control* **2011**, *22*, 964–969. [CrossRef]

79. Liu, B.H.; Hsu, Y.T.; Lu, C.C.; Yu, F.Y. Detecting aflatoxin B1 in foods and feeds by using sensitive rapid enzyme-linked immunosorbent assay and gold nanoparticle immunochromatographic strip. *Food Control* **2013**, *30*, 184–189. [CrossRef]

80. Köhler, G.; Milstein, C. Continuous cultures of fused cells secreting antibody of predefined specificity. *Nature* **1975**, *256*, 495–497. [CrossRef] [PubMed]

81. Groopman, J.D.; Trudel, L.J.; Donahue, P.R.; Marshak-Rothstein, A.; Wogan, G.N. High-affinity monoclonal antibodies for aflatoxins and their application to solid-phase immunoassays. *Proc. Natl. Acad. Sci. USA* **1984**, *81*, 7728–7731. [CrossRef]

82. Woychik, N.A.; Hinsdill, R.D.; Chu, F.S. Production and characterization of monoclonal antibodies against aflatoxin M1. *Appl. Environ. Microbiol.* **1984**, *48*, 1096–1099. [CrossRef]

83. Rousseau, D.M.; Candlish, A.A.G.; Slegers, G.A.; Van Peteghem, C.H.; Stimson, W.H.; Smith, J.E. Detection of ochratoxin A in porcine kidneys by a monoclonal antibody-based radioimmunoassay. *Appl. Environ. Microbiol.* **1987**, *53*, 514–518. [CrossRef]

84. Casale, W.L.; Pestka, J.J.; Patrick Hart, L. Enzyme-Linked Immunosorbent Assay Employing Monoclonal Antibody Specific for Deoxynivalenol (Vomitoxin) and Several Analogues. *J. Agric. Food Chem.* **1988**, *36*, 663–668. [CrossRef]

85. Dixon, D.E.; Warner, R.L.; Hart, L.P.; Pestka, J.J.; Ram, B.P. Hybridoma cell line production of a specific monoclonal antibody to the mycotoxins zearalenone and.alpha.-zearalenol. *J. Agric. Food Chem.* **1987**, *35*, 122–126. [CrossRef]

86. Hunter, K.W.; Brimfield, A.A.; Miller, M.; Finkelman, F.D.; Chu, S.F. Preparation and characterization of monoclonal antibodies to the trichothecene mycotoxin T-2. *Appl. Environ. Microbiol.* **1985**, *49*, 168–172. [CrossRef] [PubMed]

87. Azcona-Olivera, J.I.; Abouzied, M.M.; Plattner, R.D.; Pestka, J.J. Production of monoclonal antibodies to the mycotoxins fumonisins B1, B2, and B3. *J. Agric. Food Chem.* **1992**, *40*, 531–534. [CrossRef]

88. Peng, D.; Yang, B.; Pan, Y.; Wang, Y.; Chen, D.; Liu, Z.; Yang, W.; Tao, Y.; Yuan, Z. Development and validation of a sensitive monoclonal antibody-based indirect competitive enzyme-linked immunosorbent assay for the determination of the aflatoxin M1 levels in milk. *Toxicon* **2016**, *113*, 18–24. [CrossRef]

89. Zhang, X.; Eremin, S.A.; Wen, K.; Yu, X.; Li, C.; Ke, Y.; Jiang, H.; Shen, J.; Wang, Z. Fluorescence Polarization Immunoassay Based on a New Monoclonal Antibody for the Detection of the Zearalenone Class of Mycotoxins in Maize. *J. Agric. Food Chem.* **2017**, *65*, 2240–2247. [CrossRef] [PubMed]

90. Liu, J.W.; Lu, C.C.; Liu, B.H.; Yu, F.Y. Development of novel monoclonal antibodies-based ultrasensitive enzyme-linked immunosorbent assay and rapid immunochromatographic strip for aflatoxin B1 detection. *Food Control* **2016**, *59*, 700–707. [CrossRef]

91. Zhang, X.; Wen, K.; Wang, Z.; Jiang, H.; Beier, R.C.; Shen, J. An ultra-sensitive monoclonal antibody-based fluorescent microsphere immunochromatographic test strip assay for detecting aflatoxin M1 in milk. *Food Control* **2016**, *60*, 588–595. [CrossRef]

92. Li, W.; Powers, S.; Dai, S.Y. Using commercial immunoassay kits for mycotoxins: "Joys and sorrows"? *World Mycotoxin J.* **2014**, *7*, 417–430. [CrossRef]

93. Peltomaa, R.; Barderas, R.; Benito-Peña, E.; Moreno-Bondi, M.C. Recombinant antibodies and their use for food immunoanalysis. *Anal. Bioanal. Chem.* **2021**, *414*, 193–217. [CrossRef] [PubMed]

94. Arola, H.O.; Tullila, A.; Kiljunen, H.; Campbell, K.; Siitari, H.; Nevanen, T.K. Specific Noncompetitive Immunoassay for HT-2 Mycotoxin Detection. *Anal. Chem.* **2016**, *88*, 2446–2452. [CrossRef]

95. Sompunga, P.; Pruksametanan, N.; Rangnoi, K.; Choowongkomon, K.; Yamabhai, M. Generation of human and rabbit recombinant antibodies for the detection of Zearalenone by phage display antibody technology. *Talanta* **2019**, *201*, 397–405. [CrossRef]

96. Hu, Z.Q.; Li, H.P.; Liu, J.L.; Xue, S.; Gong, A.D.; Zhang, J.B.; Liao, Y.C. Production of a phage-displayed mouse ScFv antibody against fumonisin B1 and molecular docking analysis of their interactions. *Biotechnol. Bioprocess Eng.* **2016**, *21*, 134–143. [CrossRef]

97. Ren, W.; Xu, Y.; Huang, Z.; Li, Y.; Tu, Z.; Zou, L.; He, Q.; Fu, J.; Liu, S.; Hammock, B.D. Single-chain variable fragment antibody-based immunochromatographic strip for rapid detection of fumonisin B1 in maize samples. *Food Chem.* **2020**, *319*, 126546. [CrossRef]

98. Cai, C.; Zhang, Q.; Nidiaye, S.; Yan, H.; Zhang, W.; Tang, X.; Li, P. Development of a specific anti-idiotypic nanobody for monitoring aflatoxin M1 in milk and dairy products. *Microchem. J.* **2021**, *167*, 106326. [CrossRef]

99. Wang, F.; Li, Z.F.; Bin Wan, D.; Vasylieva, N.; Shen, Y.D.; Xu, Z.L.; Yang, J.Y.; Gettemans, J.; Wang, H.; Hammock, B.D.; et al. Enhanced Non-Toxic Immunodetection of Alternaria Mycotoxin Tenuazonic Acid Based on Ferritin-Displayed Anti-Idiotypic Nanobody-Nanoluciferase Multimers. *J. Agric. Food Chem.* **2021**, *69*, 4911–4917. [CrossRef] [PubMed]

100. Unkauf, T.; Miethe, S.; Führer, V.; Schirrmann, T.; Frenzel, A.; Hust, M. Generation of recombinant antibodies against toxins and viruses by phage display for diagnostics and therapy. In *Advances in Experimental Medicine and Biology*; Springer: New York, NY, USA, 2016; Volume 917, pp. 55–76.

101. Brichta, J.; Hnilova, M.; Viskovic, T. Generation of hapten-specificrecombinantantibodies: Antibody phage display technology: A review. *Vet. Med.* **2005**, *50*, 231–252. [CrossRef]

102. Ståhl, S.; Uhlén, M. Bacterial surface display: Trends and progress. *Trends Biotechnol.* **1997**, *15*, 185–192. [CrossRef]

103. Löfblom, J. Bacterial display in combinatorial protein engineering. *Biotechnol. J.* **2011**, *6*, 1115–1129. [CrossRef] [PubMed]

104. Min, W.-K.; Kim, S.-G.; Seo, J.-H. Affinity maturation of single-chain variable fragment specific for aflatoxin B1 using yeast surface display. *Food Chem.* **2015**, *188*, 604–611. [CrossRef]

105. Davies, S.L.; James, D.C. Engineering Mammalian Cells for Recombinant Monoclonal Antibody Production. In *D. Production of recombinant monoclonal antibodies in mammalian cells*; Al-Rubeai, M., Ed.; Springer: New York, NY, USA, 2009; pp. 153–173.

106. Frenzel, A.; Hust, M.; Schirrmann, T. Expression of recombinant antibodies. *Front. Immunol.* **2013**, *4*, 217. [CrossRef]

107. Schirrmann, T.; Al-Halabi, L.; Dübel, S.; Hust, M. Production systems for recombinant antibodies. *Front. Biosci.* **2008**, *13*, 4576–4594. [CrossRef]

108. Li, X.; Li, P.; Lei, J.; Zhang, Q.; Zhang, W.; Li, C. A simple strategy to obtain ultra-sensitive single-chain fragment variable antibodies for aflatoxin detection. *RSC Adv.* **2013**, *3*, 22367–22372. [CrossRef]

109. Hu, Z.Q.; Li, H.P.; Wu, P.; Li, Y.B.; Zhou, Z.Q.; Zhang, J.B.; Liu, J.L.; Liao, Y.C. An affinity improved single-chain antibody from phage display of a library derived from monoclonal antibodies detects fumonisins by immunoassay. *Anal. Chim. Acta* **2015**, *867*, 74–82. [CrossRef]

110. Chang, H.-J.; Choi, S.-W.; Chun, H.S. Expression of functional single-chain variable domain fragment antibody (scFv) against mycotoxin zearalenone in Pichia pastoris. *Biotechnol. Lett.* **2008**, *30*, 1801–1806. [CrossRef]

111. Chalyan, T.; Potrich, C.; Schreuder, E.; Falke, F.; Pasquardini, L.; Pederzolli, C.; Heideman, R.; Pavesi, L. AFM1 Detection in Milk by Fab' Functionalized Si3N4 Asymmetric Mach-Zehnder Interferometric Biosensors. *Toxins* **2019**, *11*, 409. [CrossRef]

112. Edupuganti, S.R.; Edupuganti, O.P.; Hearty, S.; O'Kennedy, R. A highly stable, sensitive, regenerable and rapid immunoassay for detecting aflatoxin B1 in corn incorporating covalent AFB1 immobilization and a recombinant Fab antibody. *Talanta* **2013**, *115*, 329–335. [CrossRef] [PubMed]

113. Romanazzo, D.; Ricci, F.; Volpe, G.; Elliott, C.T.; Vesco, S.; Kroeger, K.; Moscone, D.; Stroka, J.; Van Egmond, H.; Vehniäinen, M.; et al. Development of a recombinant Fab-fragment based electrochemical immunosensor for deoxynivalenol detection in food samples. *Biosens. Bioelectron.* **2010**, *25*, 2615–2621. [CrossRef] [PubMed]

114. Hamers-Casterman, C.; Atarhouch, T.; Muyldermans, S.; Robinson, G.; Hammers, C.; Songa, E.B.; Bendahman, N.; Hammers, R. Naturally occurring antibodies devoid of light chains. *Nature* **1993**, *363*, 446–448. [CrossRef]

115. Greenberg, A.S.; Avila, D.; Hughes, M.; Hughes, A.; McKinney, E.C.; Flajnik, M.F. A new antigen receptor gene family that undergoes rearrangement and extensive somatic diversification in sharks. *Nature* **1995**, *374*, 168–173. [CrossRef]

116. Muyldermans, S.; Baral, T.N.; Retamozzo, V.C.; De Baetselier, P.; De Genst, E.; Kinne, J.; Leonhardt, H.; Magez, S.; Nguyen, V.K.; Revets, H.; et al. Camelid immunoglobulins and nanobody technology. *Vet. Immunol. Immunop.* **2009**, *128*, 178–183. [CrossRef] [PubMed]

117. Turner, K.B.; Zabetakis, D.; Goldman, E.R.; Anderson, G.P. Enhanced stabilization of a stable single domain antibody for SEB toxin by random mutagenesis and stringent selection. *Protein Eng. Des. Sel.* **2014**, *27*, 89–95. [CrossRef]

118. Govaert, J.; Pellis, M.; Deschacht, N.; Vincke, C.; Conrath, K.; Muyldermans, S.; Saerens, D. Dual beneficial effect of interloop disulfide bond for single domain antibody fragments. *J. Biol. Chem.* **2012**, *287*, 1970–1979. [CrossRef]

119. Zhang, C.; Zhang, Q.; Tang, X.; Zhang, W.; Li, P. Development of an anti-idiotypic VHH antibody and toxin-free enzyme immunoassay for ochratoxin A in cereals. *Toxins* **2019**, *11*, 280. [CrossRef] [PubMed]

120. Liu, X.; Tang, Z.; Duan, Z.; He, Z.; Shu, M.; Wang, X.; Gee, S.J.; Hammock, B.D.; Xu, Y. Nanobody-based enzyme immunoassay for ochratoxin A in cereal with high resistance to matrix interference. *Talanta* **2017**, *164*, 154–158. [CrossRef]

121. He, T.; Wang, Y.; Li, P.; Zhang, Q.; Lei, J.; Zhang, Z.; Ding, X.; Zhou, H.; Zhang, W. Nanobody-based enzyme immunoassay for aflatoxin in agro-products with high tolerance to cosolvent methanol. *Anal. Chem.* **2014**, *86*, 8873–8880. [CrossRef]

122. Zhang, Y.; Xu, Z.; Wang, F.; Cai, J.; Dong, J.; Zhang, J.; Si, R.; Wang, C.; Wang, Y.; Shen, Y.; et al. Isolation of Bactrian Camel Single Domain Antibody for Parathion and Development of One-Step dc-FEIA Method Using VHH-Alkaline Phosphatase Fusion Protein. *Anal. Chem.* **2018**, *90*, 12886–12892. [CrossRef]

123. Muyldermans, S. Applications of Nanobodies. *Annu. Rev. Anim. Biosci.* **2021**, *9*, 401–421. [CrossRef] [PubMed]

124. Doyle, P.J.; Arbabi-Ghahroudi, M.; Gaudette, N.; Furzer, G.; Savard, M.E.; Gleddie, S.; McLean, M.D.; Mackenzie, C.R.; Hall, J.C. Cloning, expression, and characterization of a single-domain antibody fragment with affinity for 15-acetyl-deoxynivalenol. *Mol. Immunol.* **2008**, *45*, 3703–3713. [CrossRef]

125. Wang, F.; Li, Z.F.; Yang, Y.Y.; Bin Wan, D.; Vasylieva, N.; Zhang, Y.Q.; Cai, J.; Wang, H.; Shen, Y.D.; Xu, Z.L.; et al. Chemiluminescent Enzyme Immunoassay and Bioluminescent Enzyme Immunoassay for Tenuazonic Acid Mycotoxin by Exploitation of Nanobody and Nanobody-Nanoluciferase Fusion. *Anal. Chem.* **2020**, *92*, 11935–11942. [CrossRef] [PubMed]

126. Shriver-Lake, L.C.; Goldman, E.R.; Dean, S.N.; Liu, J.L.; Davis, T.M.; Anderson, G.P. Lipid-tagged single domain antibodies for improved enzyme-linked immunosorbent assays. *J. Immunol. Methods* **2020**, *481–482*, 112790. [CrossRef]

127. Anderson, G.P.; Liu, J.L.; Shriver-Lake, L.C.; Zabetakis, D.; Sugiharto, V.A.; Chen, H.W.; Lee, C.R.; Defang, G.N.; Wu, S.J.L.; Venkateswaran, N.; et al. Oriented Immobilization of Single-Domain Antibodies Using SpyTag/SpyCatcher Yields Improved Limits of Detection. *Anal. Chem.* **2019**, *91*, 9424–9429. [CrossRef] [PubMed]

128. Zhang, D.; Li, P.; Zhang, Q.; Zhang, W.; Huang, Y.; Ding, X.; Jiang, J. Production of ultrasensitive generic monoclonal antibodies against major aflatoxins using a modified two-step screening procedure. *Anal. Chim. Acta.* **2009**, *636*, 63–69. [CrossRef]

129. Soukhtanloo, M.; Talebian, E.; Golchin, M.; Mohammadi, M.; Amirheidari, B. Production and Characterization of Monoclonal Antibodies against Aflatoxin B$_1$. *J. Immunoass. Immunochem.* **2014**, *35*, 335–343. [CrossRef] [PubMed]

130. Ertekin, Ö.; Pirinçci, Ş.Ş.; Öztürk, S. Monoclonal IgA Antibodies for Aflatoxin Immunoassays. *Toxins* **2016**, *8*, 148. [CrossRef]

131. Lee, N.A.; Wang, S.; Allan, R.D.; Kennedy, I.R. A rapid aflatoxin B1 ELISA: Development and validation with reduced matrix effects for peanuts, corn, pistachio, and Soybeans. *J. Agric. Food Chem.* **2004**, *52*, 2746–2755. [CrossRef]

132. Kolosova, A.Y.; Shim, W.-B.; Yang, Z.-Y.; Eremin, S.A.; Chung, D.-H. Direct competitive ELISA based on a monoclonal antibody for detection of aflatoxin B1. Stabilization of ELISA kit components and application to grain samples. *Anal. Bioanal. Chem.* **2006**, *384*, 286–294. [CrossRef]

133. Wu, Y.; Yu, B.; Cui, P.; Yu, T.; Shi, G.; Shen, Z. Development of a quantum dot–based lateral flow immunoassay with high reaction consistency to total aflatoxins in botanical materials. *Anal. Bioanal. Chem.* **2021**, *413*, 1629–1637. [CrossRef]

134. Devi, K.T.; Mayo, M.A.; Reddy, K.L.N.; Delfosse, P.; Reddy, G.; Reddy, S.V.; Reddy, D.V.R. Production and characterization of monoclonal antibodies for aflatoxin B1. *Lett. Appl. Microbiol.* **1999**, *29*, 284–288. [CrossRef]

135. Cervino, C.; Knopp, D.; Weller, M.; Niessner, R. Novel Aflatoxin Derivatives and Protein Conjugates. *Molecules* **2007**, *12*, 641–653. [CrossRef]

136. Cervino, C.; Weber, E.; Knopp, D.; Niessner, R. Comparison of hybridoma screening methods for the efficient detection of high-affinity hapten-specific monoclonal antibodies. *J. Immunol. Methods* **2008**, *329*, 184–193. [CrossRef]

137. LAU, H.P.; GAUR, P.K.; CHU, F.S. Preparation and characterization of aflatoxin B2a-hemiglutarate and its use for the production of antibody against aflatoxin B1. *J. Food Saf.* **1980**, *3*, 1–13. [CrossRef]

138. Zhang, D.; Li, P.; Yang, Y.; Zhang, Q.; Zhang, W.; Xiao, Z.; Ding, X. A high selective immunochromatographic assay for rapid detection of aflatoxin B1. *Talanta* **2011**, *85*, 736–742. [CrossRef]

139. Gaur, P.K.; Lau, H.P.; Pestka, J.J.; Chu, F.S. Production and characterization of aflatoxin B(2a) antiserum. *Appl. Environ. Microbiol.* **1981**, *41*, 478–482. [CrossRef]

140. Han, L.; Li, Y.T.; Jiang, J.Q.; Li, R.F.; Fan, G.Y.; Lv, J.M.; Zhou, Y.; Zhang, W.J.; Wang, Z.L. Development of a direct competitive ELISA kit for detecting deoxynivalenol contamination in wheat. *Molecules* **2020**, *25*, 50. [CrossRef]

141. Kong, D.; Wu, X.; Li, Y.; Liu, L.; Song, S.; Zheng, Q.; Kuang, H.; Xu, C. Ultrasensitive and eco-friendly immunoassays based monoclonal antibody for detection of deoxynivalenol in cereal and feed samples. *Food Chem.* **2019**, *270*, 130–137. [CrossRef]

142. Lee, H.M.; Song, S.O.; Cha, S.H.; Wee, S.B.; Bischoff, K.; Park, S.W.; Son, S.W.; Kang, H.G.; Cho, M.H. Development of a monoclonal antibody against deoxynivalenol for magnetic nanoparticle-based extraction and an enzyme-linked immunosorbent assay. *J. Vet. Sci.* **2013**, *14*, 143–150. [CrossRef]

143. Peng, D.; Chang, F.; Wang, Y.; Chen, D.; Liu, Z.; Zhou, X.; Feng, L.; Yuan, Z. Development of a sensitive monoclonal-based enzyme-linked immunosorbent assay for monitoring T-2 toxin in food and feed. *Food Addit. Contam. Part A* **2016**, *33*, 683–692. [CrossRef]

144. Wang, Y.; Zhang, L.; Peng, D.; Xie, S.; Chen, D.; Pan, Y.; Tao, Y.; Yuan, Z. Construction of electrochemical immunosensor based on gold-nanoparticles/carbon nanotubes/chitosan for sensitive determination of T-2 toxin in feed and swine meat. *Int. J. Mol. Sci.* **2018**, *19*, 3895. [CrossRef]

145. Xu, H.; Dong, Y.; Guo, J.; Jiang, X.; Liu, J.; Xu, S.; Wang, H. Monoclonal antibody production and the development of an indirect competitive enzyme-linked immunosorbent assay for screening T-2 toxin in milk. *Toxicon* **2018**, *156*, 1–6. [CrossRef]

146. Chu, F.S.; Grossman, S.; Wei, R.D.; Mirocha, C.J. Production of antibody against T-2 toxin. *Appl. Environ. Microbiol.* **1979**, *37*, 104–108. [CrossRef]

147. Li, Y.; Luo, X.; Yang, S.; Cao, X.; Wang, Z.; Shi, W.; Zhang, S. High specific monoclonal antibody production and development of an ELISA method for monitoring T-2 toxin in rice. *J. Agric. Food Chem.* **2014**, *62*, 1492–1497. [CrossRef]

148. Cha, S.H.; Kim, S.H.; Bischoff, K.; Kim, H.J.; Son, S.W.; Kang, H.G. Production of a highly group-specific monoclonal antibody against zearalenone and its application in an enzyme-linked immunosorbent assay. *J. Vet. Sci.* **2012**, *13*, 119–125. [CrossRef]

149. Liu, M.T.; Ram, B.P.; Hart, L.P.; Pestka, J.J. Indirect enzyme-linked immunosorbent assay for the mycotoxin zearalenone. *Appl. Environ. Microbiol.* **1985**, *50*, 332–336. [CrossRef]

150. Pestka, J.J.; Liu, M.-T.; Knudson, B.K.; Hogberg, M.G. Immunization of Swine for Production of Antibody Against Zearalenone. *J. Food Prot.* **1985**, *48*, 953–957. [CrossRef]

151. Thongrussamee, T.; Kuzmina, N.S.; Shim, W.B.; Jiratpong, T.; Eremin, S.A.; Intrasook, J.; Chung, D.H. Monoclonal-based enzyme-linked immunosorbent assay for the detection of zearalenone in cereals. *Food Addit. Contam. Part A Chem. Anal. Control. Expo. Risk Assess.* **2008**, *25*, 997–1006. [CrossRef]

152. Thouvenot, D.; Morfin, R.F. Radioimmunoassay for zearalenone and zearalanol in human serum: Production, properties, and use of porcine antibodies. *Appl. Environ. Microbiol.* **1983**, *45*, 16–23. [CrossRef]

153. Teshima, R.; Kawase, M.; Tanaka, T.; Sawada, J.I.; Ikebuchi, H.; Terao, T.; Hirai, K.; Sato, M.; Ichinoe, M. Production and Characterization of a Specific Monoclonal Antibody against Mycotoxin Zearalenone. *J. Agric. Food Chem.* **1990**, *38*, 1618–1622. [CrossRef]

154. Gao, Y.; Yang, M.; Peng, C.; Li, X.; Cai, R.; Qi, Y. Preparation of highly specific anti-zearalenone antibodies by using the cationic protein conjugate and development of an indirect competitive enzyme-linked immunosorbent assay. *Analyst* **2012**, *137*, 229–236. [CrossRef]

155. Sun, Y.; Hu, X.; Zhang, Y.; Yang, J.; Wang, F.; Wang, Y.; Deng, R.; Zhang, G. Development of an immunochromatographic strip test for the rapid detection of zearalenone in corn. *J. Agric. Food Chem.* **2014**, *62*, 11116–11121. [CrossRef]

156. Köppen, R.; Riedel, J.; Proske, M.; Drzymala, S.; Rasenko, T.; Durmaz, V.; Weber, M.; Koch, M. Photochemical trans -/ cis -Isomerization and quantitation of zearalenone in edible oils. *J. Agric. Food Chem.* **2012**, *60*, 11733–11740. [CrossRef]

157. Brezina, U.; Kersten, S.; Valenta, H.; Sperfeld, P.; Riedel, J.; Dänicke, S. UV-induced cis-trans isomerization of zearalenone in contaminated maize. *Mycotoxin Res.* **2013**, *29*, 221–227. [CrossRef]

158. Gadzała-Kopciuch, R.; Kuźniewska, A.; Buszewski, B. Analytical approaches and preparation of biological, food and environmental samples for analyses of zearalenone and its metabolites. *Rev. Anal. Chem.* **2020**, *39*, 157–167. [CrossRef]

159. Oswald, S.; Niessner, R.; Knopp, D. First experimental evidence for an enzyme-generated chemiluminescence-induced trans-cis isomerization of chip-immobilized zearalenone in a microfluidic cell of a biosensor. *Luminesence* **2014**, *29*, 23–24.

160. Chu, F.S.; Ueno, I. Production of antibody against aflatoxin B1. *Appl. Environ. Microbiol.* **1977**, *33*, 1125. [CrossRef]

161. Chu, F.S.; Chang, F.C.C.; Hinsdill, R.D. Production of antibody against ochratoxin A. *Appl. Environ. Microbiol.* **1976**, *31*, 831–835. [CrossRef]

162. Han, L.; Li, Y.; Jiang, J.; Li, R.; Fan, G.; Lei, Z.; Wang, H.; Wang, Z.; Zhang, W. Preparation and characterisation of monoclonal antibodies against deoxynivalenol. *Ital. J. Anim. Sci.* **2020**, *19*, 560–568. [CrossRef]

163. Yu, F.-Y.; Chu, F.S. Production and Characterization of Antibodies against Fumonisin B1. *J. Food Prot.* **1996**, *59*, 992–997. [CrossRef]

164. Kalayu Yirga, S.; Ling, S.; Yang, Y.; Yuan, J.; Wang, S. The Preparation and Identification of a Monoclonal Antibody against Citrinin and the Development of Detection via Indirect Competitive ELISA. *Toxins* **2017**, *9*, 110. [CrossRef]

165. Mhadhbi, H.; Benrejeb, S.; Martel, A. Studies on the affinity chromatography purification of anti-patulin polyclonal antibodies by enzyme linked immunosorbent assay and electrophoresis. *Food Addit. Contam.* **2005**, *22*, 1243–1251. [CrossRef] [PubMed]

166. De Champdoré, M.; Bazzicalupo, P.; De Napoli, L.; Montesarchio, D.; Di Fabio, G.; Cocozza, I.; Parracino, A.; Rossi, M.; D'Auria, S. A new competitive fluorescence assay for the detection of patulin toxin. *Anal. Chem.* **2007**, *79*, 751–757. [CrossRef]

167. McElroy, L.J.; Weiss, C.M. The production of polyclonal antibodies against the mycotoxin derivative patulin hemiglutarate. *Can. J. Microbiol.* **1993**, *39*, 861–863. [CrossRef]

168. Pennacchio, A.; Varriale, A.; Esposito, M.G.; Staiano, M.; D'Auria, S. A near-infrared fluorescence assay method to detect patulin in food. *Anal. Biochem.* **2015**, *481*, 55–59. [CrossRef]

169. Liu, B.H.; Tsao, Z.J.; Wang, J.J.; Yu, F.Y. Development of a monoclonal antibody against ochratoxin A and its application in enzyme-linked immunosorbent assay and gold nanoparticle immunochromatographic strip. *Anal. Chem.* **2008**, *80*, 7029–7035. [CrossRef]

170. Cho, Y.J.; Lee, D.H.; Kim, D.O.; Min, W.K.; Bong, K.T.; Lee, G.G.; Seo, J.H. Production of a monoclonal antibody against ochratoxin A and its application to immunochromatographic assay. *J. Agric. Food Chem.* **2005**, *53*, 8447–8451. [CrossRef]

171. Maragos, C.M.; Li, L.; Chen, D. Production and characterization of a single chain variable fragment (scFv) against the mycotoxin deoxynivalenol. *Food Agric. Immunol.* **2012**, *23*, 51–67. [CrossRef]

172. Min, W.-K.; Cho, Y.-J.; Park, J.-B.; Bae, Y.-H.; Kim, E.-J.; Park, K.; Park, Y.-C.; Seo, J.-H. Production and characterization of monoclonal antibody and its recombinant single chain variable fragment specific for a food-born mycotoxin, fumonisin B1. *Bioprocess Biosyst. Eng.* **2010**, *33*, 109–115. [CrossRef]

173. Min, W.K.; Kweon, D.H.; Park, K.; Park, Y.C.; Seo, J.H. Characterisation of monoclonal antibody against aflatoxin B1 produced in hybridoma 2C12 and its single-chain variable fragment expressed in recombinant Escherichia coli. *Food Chem.* **2011**, *126*, 1316–1323. [CrossRef]

174. Wang, R.; Gu, X.; Zhuang, Z.; Zhong, Y.; Yang, H.; Wang, S. Screening and Molecular Evolution of a Single Chain Variable Fragment Antibody (scFv) against Citreoviridin Toxin. *J. Agric. Food Chem.* **2016**, *64*, 7640–7648. [CrossRef]

175. Wang, X.; Chen, Q.; Sun, Z.; Wang, Y.; Su, B.; Zhang, C.; Cao, H.; Liu, X. Nanobody affinity improvement: Directed evolution of the anti-ochratoxin A single domain antibody. *Int. J. Biol. Macromol.* **2020**, *151*, 312–321. [CrossRef]

176. Ellington, A.D.; Szostak, J.W. In vitro selection of RNA molecules that bind specific ligands. *Nature* **1990**, *346*, 818–822. [CrossRef] [PubMed]

177. Tuerk, C.; Gold, L. Systematic evolution of ligands by exponential enrichment: RNA ligands to bacteriophage T4 DNA polymerase. *Science* **1990**, *249*, 505–510. [CrossRef]

178. Ruscito, A.; Smith, M.; Goudreau, D.N.; Derosa, M.C. Current status and future prospects for aptamer-based mycotoxin detection. *J. AOAC Int.* **2016**, *99*, 865–877. [CrossRef]

179. Hasegawa, H.; Savory, N.; Abe, K.; Ikebukuro, K. Methods for Improving Aptamer Binding Affinity. *Molecules* **2016**, *21*, 421. [CrossRef]

180. Cruz-Aguado, J.A.; Penner, G. Determination of Ochratoxin A with a DNA Aptamer. *J. Agric. Food Chem.* **2008**, *56*, 10456–10461. [CrossRef] [PubMed]

181. Shim, W.B.; Kim, M.J.; Mun, H.; Kim, M.G. An aptamer-based dipstick assay for the rapid and simple detection of aflatoxin B1. *Biosens. Bioelectron.* **2014**, *62*, 288–294. [CrossRef]

182. Malhotra, S.; Pandey, A.K.; Rajput, Y.S.; Sharma, R. Selection of aptamers for aflatoxin M1 and their characterization. *J. Mol. Recognit.* **2014**, *27*, 493–500. [CrossRef]

183. Yue, S.; Jie, X.; Wei, L.; Bin, C.; Wang, D.; Yi, Y.; Lin, Q.; Li, J.; Zheng, T. Simultaneous detection of ochratoxin a and fumonisin b1 in cereal samples using an aptamer-photonic crystal encoded suspension array. *Anal. Chem.* **2014**, *86*, 11797–11802. [CrossRef]

184. Chryseis Le, L.; Cruz-Aguado, J.; Penner, G.A. DNA Ligands for Aflatoxins and Zearalenone. World Patent Application Application PCT/CA2010/001292, 20 August 2010.

185. Ong, C.C.; Siva Sangu, S.; Illias, N.M.; Chandra Bose Gopinath, S.; Saheed, M.S.M. Iron nanoflorets on 3D-graphene-nickel: A 'Dandelion' nanostructure for selective deoxynivalenol detection. *Biosens. Bioelectron.* **2020**, *154*, 112088. [CrossRef] [PubMed]

186. Wen, X.; Huang, Q.; Nie, D.; Zhao, X.; Cao, H.; Wu, W.; Han, Z. A multifunctional n-doped cu–mofs (N–cu–mof) nanomaterial-driven electrochemical aptasensor for sensitive detection of deoxynivalenol. *Molecules* **2021**, *26*, 2243. [CrossRef] [PubMed]

187. Wu, S.; Duan, N.; Zhang, W.; Zhao, S.; Wang, Z. Screening and development of DNA aptamers as capture probes for colorimetric detection of patulin. *Anal. Biochem.* **2016**, *508*, 58–64. [CrossRef]

188. Chen, X.; Huang, Y.; Duan, N.; Wu, S.; Xia, Y.; Ma, X.; Zhu, C.; Jiang, Y.; Wang, Z. Screening and identification of DNA aptamers against T-2 toxin assisted by graphene oxide. *J. Agric. Food Chem.* **2014**, *62*, 10368–10374. [CrossRef] [PubMed]

189. Rouah-Martin, E.; Mehta, J.; van Dorst, B.; de Saeger, S.; Dubruel, P.; Maes, B.U.W.; Lemiere, F.; Goormaghtigh, E.; Daems, D.; Herrebout, W.; et al. Aptamer-based molecular recognition of lysergamine, metergoline and small ergot alkaloids. *Int. J. Mol. Sci.* **2012**, *13*, 17138–17159. [CrossRef]

190. Nguyen, B.H.; Tran, L.D.; Do, Q.P.; Le Nguyen, H.; Tran, N.H.; Nguyen, P.X. Label-free detection of aflatoxin M1 with electro-chemical Fe 3O4/polyaniline-based aptasensor. *Mater. Sci. Eng. C* **2013**, *33*, 2229–2234. [CrossRef]

191. McKeague, M.; Bradley, C.R.; De Girolamo, A.; Visconti, A.; Miller, J.D.; Derosa, M.C. Screening and initial binding assessment of fumonisin b(1) aptamers. *Int. J. Mol. Sci.* **2010**, *11*, 4864–4881. [CrossRef]

192. Chen, X.; Huang, Y.; Duan, N.; Wu, S.; Ma, X.; Xia, Y.; Zhu, C.; Jiang, Y.; Wang, Z. Selection and identification of ssDNA aptamers recognizing zearalenone. *Anal. Bioanal. Chem.* **2013**, *405*, 6573–6581. [CrossRef]

193. Penner, G. Commercialization of an aptamer-based diagnostic test. *IVD Technol.* **2012**, *18*, 31–37. Available online: https://www.researchgate.net/publication/291178318 (accessed on 12 October 2021).

194. Wang, Y.; Wang, H.; Li, P.; Zhang, Q.; Kim, H.J.; Gee, S.J.; Hammock, B.D. Phage-displayed peptide that mimics aflatoxins and its application in immunoassay. *J. Agric. Food Chem.* **2013**, *61*, 2426–2433. [CrossRef]

195. He, Q.; Xu, Y.; Zhang, C.; Li, Y.; Huang, Z. Phage-borne peptidomimetics as immunochemical reagent in dot-immunoassay for mycotoxin zearalenone. *Food Control* **2014**, *39*, 56–61. [CrossRef]

196. Lai, W.; Fung, D.Y.C.; Yang, X.; Renrong, L.; Xiong, Y. Development of a colloidal gold strip for rapid detection of ochratoxin A with mimotope peptide. *Food Control* **2009**, *20*, 791–795. [CrossRef]

197. He, Q.-H.; Xu, Y.; Huang, Y.-H.; Liu, R.-R.; Huang, Z.-B.; Li, Y.-P. Phage-displayed peptides that mimic zearalenone and its application in immunoassay. *Food Chem.* **2011**, *126*, 1312–1315. [CrossRef]

198. Yuan, Q.; Pestka, J.J.; Hespenheide, B.M.; Kuhn, L.A.; Linz, J.E.; Hart, L.P. Identification of mimotope peptides which bind to the mycotoxin deoxynivalenol-specific monoclonal antibody. *Appl. Environ. Microbiol.* **1999**, *65*, 3279–3286. [CrossRef] [PubMed]

199. Liu, R.; Yu, Z.; He, Q.; Xu, Y. An immunoassay for ochratoxin A without the mycotoxin. *Food Control* **2007**, *18*, 872–877. [CrossRef]

200. Thirumala-Devi, K.; Miller, J.S.; Reddy, G.; Reddy, D.V.R.; Mayo, M.A. Phage-displayed peptides that mimic aflatoxin B1 in serological reactivity. *J. Appl. Microbiol.* **2001**, *90*, 330–336. [CrossRef]

201. Tozzi, C.; Anfossi, L.; Baggiani, C.; Giovannoli, C.; Giraudi, G. A combinatorial approach to obtain affinity media with binding properties towards the aflatoxins. *Anal. Bioanal. Chem.* **2003**, *375*, 994–999. [CrossRef]

202. Heurich, M.; Altintas, Z.; Tothill, I.E. Computational design of peptide ligands for ochratoxin A. *Toxins* **2013**, *5*, 1202–1212. [CrossRef]

203. Giraudi, G.; Anfossi, L.; Baggiani, C.; Giovannoli, C.; Tozzi, C. Solid-phase extraction of ochratoxin A from wine based on a binding hexapeptide prepared by combinatorial synthesis. *J. Chromatogr. A* **2007**, *1175*, 174–180. [CrossRef]

204. Bazin, I.; Andreotti, N.; Hassine, A.I.H.; De Waard, M.; Sabatier, J.M.; Gonzalez, C. Peptide binding to ochratoxin A mycotoxin: A new approach in conception of biosensors. *Biosens. Bioelectron.* **2013**, *40*, 240–246. [CrossRef]

205. Liu, B.; Peng, J.; Wu, Q.; Zhao, Y.; Shang, H.; Wang, S. A novel screening on the specific peptide by molecular simulation and development of the electrochemical immunosensor for aflatoxin B1 in grains. *Food Chem.* **2022**, *372*, 131322. [CrossRef] [PubMed]

206. Appell, M.; Maragos, C.M.; Kendra, D.F. Molecularly imprinted polymers for mycotoxins. In Proceedings of the ACS Symposium Series, 32nd ACS National Meeting, San Francisco, CA, USA, 10–14, September 2006; American Chemical Society: Washington, DC, USA, 2008; Volume 1001, pp. 152–169.

207. Leibl, N.; Haupt, K.; Gonzato, C.; Duma, L. Molecularly Imprinted Polymers for Chemical Sensing: A Tutorial Review. *Chemosensors* **2021**, *9*, 123. [CrossRef]

208. Urraca, J.L.; Marazuela, M.D.; Moreno-Bondi, M.C. Molecularly imprinted polymers applied to the clean-up of zearalenone and α-zearalenol from cereal and swine feed sample extracts. *Anal. Bioanal. Chem.* **2006**, *385*, 1155–1161. [CrossRef]

209. Urraca, J.L.; Marazuela, M.D.; Merino, E.R.; Orellana, G.; Moreno-Bondi, M.C. Molecularly imprinted polymers with a streamlined mimic for zearalenone analysis. *J. Chromatogr. A* **2006**, *1116*, 127–134. [CrossRef]

210. Baggiani, C.; Giraudi, G.; Vanni, A. A molecular imprinted polymer with recognition properties towards the carcinogenic mycotoxin ochratoxin a. *Bioseparation* **2001**, *10*, 389–394. [CrossRef] [PubMed]

211. Yu, J.C.C.; Lai, E.P.C. Molecularly Imprinted Polymers for Ochratoxin A Extraction and Analysis. *Toxins* **2010**, *2*, 1536–1553. [CrossRef] [PubMed]

212. Sergeyeva, T.A.; Piletska, O.V.; Brovko, O.O.; Goncharova, L.A.; Piletsky, S.A.; El'ska, G.V. Aflatoxin-selective molecularly-imprinted polymer membranes based on acrylate-polyurethane semi-interpenetrating polymer networks. *Ukr. Biokhimichnyi Zhurnal* **2007**, *79*, 109–115.

213. Sergeyeva, T.; Yarynka, D.; Piletska, E.; Lynnik, R.; Zaporozhets, O.; Brovko, O.; Piletsky, S.; El'skaya, A. Fluorescent sensor systems based on nanostructured polymeric membranes for selective recognition of Aflatoxin B1. *Talanta* **2017**, *175*, 101–107. [CrossRef]

214. Munawar, H.; Smolinska-Kempisty, K.; Cruz, A.G.; Canfarotta, F.; Piletska, E.; Karim, K.; Piletsky, S.A. Molecularly imprinted polymer nanoparticle-based assay (MINA): Application for fumonisin B1 determination. *Analyst* **2018**, *143*, 3481–3488. [CrossRef]

215. Maier, N.M.; Buttinger, G.; Welhartizki, S.; Gavioli, E.; Lindner, W. Molecularly imprinted polymer-assisted sample clean-up of ochratoxin a from red wine: Merits and limitations. *J. Chromatogr. B Anal. Technol. Biomed. Life Sci.* **2004**, *804*, 103–111. [CrossRef]

216. Baggiani, C.; Giovannoli, C.; Anfossi, L.; Passini, C.; Baravalle, P.; Giraudi, G. A Connection between the binding properties of imprinted and nonimprinted polymers: A change of perspective in molecular imprinting. *J. Am. Chem. Soc.* **2012**, *134*, 1513–1518. [CrossRef] [PubMed]

217. Pascale, M.; De Girolamo, A.; Visconti, A.; Magan, N.; Chianella, I.; Piletska, E.V.; Piletsky, S.A. Use of itaconic acid-based polymers for solid-phase extraction of deoxynivalenol and application to pasta analysis. *Anal. Chim. Acta* **2008**, *609*, 131–138. [CrossRef] [PubMed]

218. Piletska, E.; Karim, K.; Coker, R.; Piletsky, S. Development of the custom polymeric materials specific for aflatoxin B1 and ochratoxin A for application with the ToxiQuant T1 sensor tool. *J. Chromatogr. A* **2010**, *1217*, 2543–2547. [CrossRef] [PubMed]

219. Piletsky, S.A.; Piletska, E.V.; Karim, K.; Freebairn, K.W.; Legge, C.H.; Turner, A.P.F. Polymer cookery: Influence of polymerization conditions on the performance of molecularly imprinted polymers. *Macromolecules* **2002**, *35*, 7499–7504. [CrossRef]

220. Turner, N.W.; Piletska, E.V.; Karim, K.; Whitcombe, M.; Malecha, M.; Magan, N.; Baggiani, C.; Piletsky, S.A. Effect of the solvent on recognition properties of molecularly imprinted polymer specific for ochratoxin A. *Biosens. Bioelectron.* **2004**, *20*, 1060–1067. [CrossRef]

221. Navarro-Villoslada, F.; Vicente, B.S.; Moreno-Bondi, M.C. Application of multivariate analysis to the screening of molecularly imprinted polymers for bisphenol A. *Anal. Chim. Acta* **2004**, *504*, 149–162. [CrossRef]

222. Weiss, R.; Freudenschuss, M.; Krska, R.; Mizaikoff, B. Improving methods of analysis for mycotoxins: Molecularly imprinted polymers for deoxynivalenol and zearalenone. *Food Addit. Contam.* **2003**, *20*, 386–395. [CrossRef]

223. Díaz-Bao, M.; Regal, P.; Barreiro, R.; Fente, C.A.; Cepeda, A. A facile method for the fabrication of magnetic molecularly imprinted stir-bars: A practical example with aflatoxins in baby foods. *J. Chromatogr. A* **2016**, *1471*, 51–59. [CrossRef]

224. Liang, Y.; He, J.; Huang, Z.; Li, H.; Zhang, Y.; Wang, H.; Rui, C.; Li, Y.; You, L.; Li, K.; et al. An amino-functionalized zirconium-based metal-organic framework of type UiO-66-NH2 covered with a molecularly imprinted polymer as a sorbent for the extraction of aflatoxins AFB1, AFB2, AFG1 and AFG2 from grain. *Microchim. Acta* **2020**, *187*, 32. [CrossRef]

225. Giovannoli, C.; Passini, C.; Di Nardo, F.; Anfossi, L.; Baggiani, C. Determination of ochratoxin A in Italian red wines by molecularly imprinted solid phase extraction and HPLC analysis. *J. Agric. Food Chem.* **2014**, *62*, 5220–5225. [CrossRef]

226. De Smet, D.; Dubruel, P.; Van Peteghem, C.; Schacht, E.; De Saeger, S. Molecularly imprinted solid-phase extraction of fumonisin B analogues in bell pepper, rice and corn flakes. *Food Addit. Contam. Part A* **2009**, *26*, 874–884. [CrossRef]

227. Urraca, J.L.; Huertas-Pérez, J.F.; Cazorla, G.A.; Gracia-Mora, J.; García-Campaña, A.M.; Moreno-Bondi, M.C. Development of magnetic molecularly imprinted polymers for selective extraction: Determination of citrinin in rice samples by liquid chromatography with UV diode array detection. *Anal. Bioanal. Chem.* **2016**, *408*, 3033–3042. [CrossRef]

228. De Smet, D.; Monbaliu, S.; Dubruel, P.; Van Peteghem, C.; Schacht, E.; De Saeger, S. Synthesis and application of a T-2 toxin imprinted polymer. *J. Chromatogr. A* **2010**, *1217*, 2879–2886. [CrossRef]

229. Regal, P.; Díaz-Bao, M.; Barreiro, R.; Fente, C.; Cepeda, A. Design of a molecularly imprinted stir-bar for isolation of patulin in apple and LC-MS/MS detection. *Separations* **2017**, *4*, 11. [CrossRef]

230. Lenain, P.; Diana Di Mavungu, J.; Dubruel, P.; Robbens, J.; De Saeger, S. Development of suspension polymerized molecularly imprinted beads with metergoline as template and application in a solid-phase extraction procedure toward ergot alkaloids. *Anal. Chem.* **2012**, *84*, 10411–10418. [CrossRef] [PubMed]

231. Abou-Hany, R.A.G.; Urraca, J.L.; Descalzo, A.B.; Gómez-Arribas, L.N.; Moreno-Bondi, M.C.; Orellana, G. Tailoring molecularly imprinted polymer beads for alternariol recognition and analysis by a screening with mycotoxin surrogates. *J. Chromatogr. A* **2015**, *1425*, 231–239. [CrossRef] [PubMed]

232. Ali, W.H.; Derrien, D.; Alix, F.; Pérollier, C.; Lépine, O.; Bayoudh, S.; Chapuis-Hugon, F.; Pichon, V. Solid-phase extraction using molecularly imprinted polymers for selective extraction of a mycotoxin in cereals. *J. Chromatogr. A* **2010**, *1217*, 6668–6673. [CrossRef] [PubMed]

233. Lucci, P.; Derrien, D.; Alix, F.; Pérollier, C.; Bayoudh, S. Molecularly imprinted polymer solid-phase extraction for detection of zearalenone in cereal sample extracts. *Anal. Chim. Acta* **2010**, *672*, 15–19. [CrossRef] [PubMed]

234. Pichon, V.; Combès, A. Selective tools for the solid-phase extraction of Ochratoxin A from various complex samples: Immunosorbents, oligosorbents, and molecularly imprinted polymers. *Anal. Bioanal. Chem.* **2016**, *408*, 6983–6999. [CrossRef]

235. Cao, J.; Kong, W.; Zhou, S.; Yin, L.; Wan, L.; Yang, M. Molecularly imprinted polymer-based solid phase clean-up for analysis of ochratoxin A in beer, red wine, and grape juice. *J. Sep. Sci.* **2013**, *36*, 1291–1297. [CrossRef] [PubMed]

236. Bodbodak, S.; Hesari, J.; Peighambardoust, S.H.; Mahkam, M. Selective decontamination of aflatoxin M1 in milk by molecularly imprinted polymer coated on the surface of stainless steel plate. *Int. J. Dairy Technol.* **2018**, *71*, 868–878. [CrossRef]

237. Sun, J.; Guo, W.; Ji, J.; Li, Z.; Yuan, X.; Pi, F.; Zhang, Y.; Sun, X. Removal of patulin in apple juice based on novel magnetic molecularly imprinted adsorbent Fe3O4@SiO2@CS-GO@MIP. *LWT-Food Sci. Technol.* **2020**, *118*, 108854. [CrossRef]

238. Jedziniak, P.; Panasiuk, Ł.; Pietruszka, K.; Posyniak, A. Multiple mycotoxins analysis in animal feed with LC-MS/MS: Comparison of extract dilution and immunoaffinity clean-up. *J. Sep. Sci.* **2019**, *42*, 1240–1247. [CrossRef]

239. Jinap, S.; De Rijk, T.C.; Arzandeh, S.; Kleijnen, H.C.H.; Zomer, P.; Van der Weg, G.; Mol, J.G.J. Aflatoxin determination using in-line immunoaffinity chromatography in foods. *Food Control* **2012**, *26*, 42–48. [CrossRef]

240. Uchigashima, M.; Saigusa, M.; Yamashita, H.; Miyake, S.; Fujita, K.; Nakajima, M.; Nishijima, M. Development of a Novel Immunoaffinity Column for Aflatoxin Analysis Using an Organic Solvent-Tolerant Monoclonal Antibody. *J. Agric. Food Chem.* **2009**, *57*, 8728–8734. [CrossRef]

241. Wilcox, J.; Pazdanska, M.; Milligan, C.; Chan, D.; Macdonald, S.J.; Donnelly, C. Analysis of aflatoxins and ochratoxin a in cannabis and cannabis products by LC–fluorescence detection using cleanup with either multiantibody immunoaffinity columns or an automated system with in-line reusable immunoaffinity cartridges. *J. AOAC Int.* **2021**, *103*, 494–503. [CrossRef] [PubMed]

242. De Girolamo, A.; McKeague, M.; Miller, J.D.; DeRosa, M.C.; Visconti, A. Determination of ochratoxin A in wheat after clean-up through a DNA aptamer-based solid phase extraction column. *Food Chem.* **2011**, *127*, 1378–1384. [CrossRef] [PubMed]

243. Pichon, V.; Brothier, F.; Combès, A. Aptamer-based-sorbents for sample treatment - A review. *Anal. Bioanal. Chem.* **2015**, *407*, 681–698. [CrossRef] [PubMed]

244. Liu, L.; Ma, Y.; Zhang, X.; Yang, X.; Hu, X. A dispersive solid phase extraction adsorbent based on aptamer modified chitosan nanofibers for zearalenone separation in corn, wheat, and beer samples. *Anal. Methods* **2020**, *12*, 5852–5860. [CrossRef] [PubMed]

245. Wang, S.; Niu, R.; Yang, Y.; Zhou, X.; Luo, S.; Zhang, C.; Wang, Y. Aptamer-functionalized chitosan magnetic nanoparticles as a novel adsorbent for selective extraction of ochratoxin A. *Int. J. Biol. Macromol.* **2020**, *153*, 583–590. [CrossRef]

246. Zhang, Q.; Yang, Y.; Zhi, Y.; Wang, X.; Wu, Y.; Zheng, Y. Aptamer-modified magnetic metal-organic framework MIL-101 for highly efficient and selective enrichment of ochratoxin A. *J. Sep. Sci.* **2018**, *42*, 716–724. [CrossRef]

247. Jodlbauer, J.; Maier, N.M.; Lindner, W. Towards ochratoxin A selective molecularly imprinted polymers for solid-phase extraction. *J. Chromatogr. A* **2002**, *945*, 45–63. [CrossRef]

248. Jayasinghe, G.D.T.M.; Domínguez-González, R.; Bermejo-Barrera, P.; Moreda-Piñeiro, A. Ultrasound assisted combined molecularly imprinted polymer for the selective micro-solid phase extraction and determination of aflatoxins in fish feed using liquid chromatography-tandem mass spectrometry. *J. Chromatogr. A* **2020**, *1609*, 460431. [CrossRef] [PubMed]

249. Zhao, M.; Shao, H.; Ma, J.; Li, H.; He, Y.; Wang, M.; Jin, F.; Wang, J.; Abd El-Aty, A.M.; Hacımüftüoğlu, A.; et al. Preparation of core-shell magnetic molecularly imprinted polymers for extraction of patulin from juice samples. *J. Chromatogr. A* **2020**, *1615*, 460751. [CrossRef]

250. Turner, N.W.; Bramhmbhatt, H.; Szabo-Vezse, M.; Poma, A.; Coker, R.; Piletsky, S.A. Analytical methods for determination of mycotoxins: An update (2009-2014). *Anal. Chim. Acta* **2015**, *901*, 12–33. [CrossRef]

251. Zhang, K.; Banerjee, K. A Review: Sample Preparation and Chromatographic Technologies for Detection of Aflatoxins in Foods. *Toxins* **2020**, *12*, 539. [CrossRef] [PubMed]

252. Delaunay, N.; Combès, A.; Pichon, V. Immunoaffinity Extraction and Alternative Approaches for the Analysis of Toxins in Environmental, Food or Biological Matrices. *Toxins* **2020**, *12*, 795. [CrossRef]

253. Li, X.; Li, P.; Zhang, Q.; Zhang, Z.; Li, R.; Zhang, W.; Ding, X.; Chen, X.; Tang, X. A Sensitive Immunoaffinity Column-Linked Indirect Competitive ELISA for Ochratoxin A in Cereal and Oil Products Based on a New Monoclonal Antibody. *Food Anal. Methods* **2013**, *6*, 1433–1440. [CrossRef]

254. Hojo, E.; Matsuura, N.; Kamiya, K.; Yonekita, T.; Morishita, N.; Murakami, H.; Kawamura, O. Development of a rapid and versatile method of enzyme-linked immunoassay combined with immunoaffinity column for aflatoxin analysis. *J. Food Prot.* **2019**, *82*, 1472–1478. [CrossRef]

255. Li, D.; Ying, Y.; Wu, J.; Niessner, R.; Knopp, D. Comparison of monomeric and polymeric horseradish peroxidase as labels in competitive ELISA for small molecule detection. *Microchim. Acta* **2013**, *180*, 711–717. [CrossRef]

256. Lin, Y.; Zhou, Q.; Tang, D.; Niessner, R.; Yang, H.; Knopp, D. Silver Nanolabels-Assisted Ion-Exchange Reaction with CdTe Quantum Dots Mediated Exciton Trapping for Signal-On Photoelectrochemical Immunoassay of Mycotoxins. *Anal. Chem.* **2016**, *88*, 7858–7866. [CrossRef] [PubMed]

257. Li, Y.; Zhang, N.; Wang, H.; Zhao, Q. An immunoassay for ochratoxin A using tetramethylrhodamine-labeled ochratoxin A as a probe based on a binding-induced change in fluorescence intensity. *Analyst* **2020**, *145*, 651–655. [CrossRef]

258. Tang, D.; Zhong, Z.; Niessner, R.; Knopp, D. Multifunctional magnetic bead-based electrochemical immunoassay for the detection of aflatoxin B1 in food. *Analyst* **2009**, *134*, 1554–1560. [CrossRef] [PubMed]

259. Tang, D.; Yu, Y.; Niessner, R.; Miró, M.; Knopp, D. Magnetic bead-based fluorescence immunoassay for aflatoxin B1 in food using biofunctionalized rhodamine B-doped silica nanoparticles. *Analyst* **2010**, *135*, 2661–2667. [CrossRef]

260. Tang, D.; Liu, B.; Niessner, R.; Li, P.; Knopp, D. Target-induced displacement reaction accompanying cargo release from magnetic mesoporous silica nanocontainers for fluorescence immunoassay. *Anal. Chem.* **2013**, *85*, 10589–10596. [CrossRef] [PubMed]

261. Wang, X.; Niessner, R.; Knopp, D. Magnetic Bead-Based Colorimetric Immunoassay for Aflatoxin B1 Using Gold Nanoparticles. *Sensors* **2014**, *14*, 21535–21548. [CrossRef] [PubMed]

262. Wang, X.; Niessner, R.; Tang, D.; Knopp, D. Nanoparticle-based immunosensors and immunoassays for aflatoxins. *Anal. Chim. Acta* **2016**, *912*, 10–23. [CrossRef]

263. Pastucha, M.; Farka, Z.; Lacina, K.; Mikušová, Z.; Skládal, P. Magnetic nanoparticles for smart electrochemical immunoassays: A review on recent developments. *Microchim. Acta* **2019**, *186*, 312. [CrossRef]

264. Zhu, H.; Quan, Z.; Hou, H.; Cai, Y.; Liu, W.; Liu, Y. A colorimetric immunoassay based on cobalt hydroxide nanocages as oxidase mimics for detection of ochratoxin A. *Anal. Chim. Acta* **2020**, *1132*, 101–109. [CrossRef]

265. Zhan, S.; Zheng, L.; Zhou, Y.; Wu, K.; Duan, H.; Huang, X.; Xiong, Y. A gold growth-based plasmonic ELISA for the sensitive detection of fumonisin B1 in maize. *Toxins* **2019**, *11*, 323. [CrossRef]

266. Zhan, S.; Hu, J.; Li, Y.; Huang, X.; Xiong, Y. Direct competitive ELISA enhanced by dynamic light scattering for the ultrasensitive detection of aflatoxin B1 in corn samples. *Food Chem.* **2021**, *342*, 128327. [CrossRef] [PubMed]

267. Beloglazova, N.V.; Speranskaya, E.S.; Wu, A.; Wang, Z.; Sanders, M.; Goftmana, V.V.; Zhang, D.; Goryacheva, I.Y.; Saeger, S.D. Novel multiplex fluorescent immunoassays based on quantum dot nanolabels for mycotoxins determination. *Biosens. Bioelectron.* **2014**, *62*, 59–65. [CrossRef]

268. Na, K.I.; Kim, S.J.; Choi, D.S.; Min, W.K.; Kim, S.G.; Seo, J.H. Extracellular production of functional single-chain variable fragment against aflatoxin B 1 using Escherichia coli. *Lett. Appl. Microbiol.* **2019**, *68*, 241–247. [CrossRef]

269. Rangnoi, K.; Rüker, F.; Wozniak-Knopp, G.; Cvak, B.; O'Kennedy, R.; Yamabhai, M. Binding Characteristic of Various Antibody Formats against Aflatoxins. *ACS Omega* **2021**, *6*, 25258–25268. [CrossRef] [PubMed]

270. Van Houwelingen, A.; De Saeger, S.; Rusanova, T.; Waalwijk, C.; Beekwilder, J. Generation of recombinant alpaca VHH antibody fragments for the detection of the mycotoxin ochratoxin A. *World Mycotoxin J.* **2008**, *1*, 407–417. [CrossRef]

271. Tullila, A.; Nevanen, T.K. Utilization of Multi-Immunization and Multiple Selection Strategies for Isolation of Hapten-Specific Antibodies from Recombinant Antibody Phage Display Libraries. *Int. J. Mol. Sci.* **2017**, *18*, 1169. [CrossRef] [PubMed]

272. Choi, G.-H.; Lee, D.-H.; Min, W.-K.; Cho, Y.-J.; Kweon, D.-H.; Son, D.-H.; Park, K.; Seo, J.-H. Cloning, expression, and characterization of single-chain variable fragment antibody against mycotoxin deoxynivalenol in recombinant Escherichia coli. *Protein Expr. Purif.* **2004**, *35*, 84–92. [CrossRef]

273. Lauer, B.; Ottleben, I.; Jacobsen, H.J.; Reinard, T. Production of a single-chain variable fragment antibody against fumonisin B1. *J. Agric. Food Chem.* **2005**, *53*, 899–904. [CrossRef]

274. Zhou, H.R.; Pestka, J.J.; Hart, L.P. Molecular cloning and expression of recombinant phage antibody against fumonisin B1. *J. Food Prot.* **1996**, *59*, 1208–1212. [CrossRef] [PubMed]

275. Cheng, H.; Chen, Y.; Yang, Y.; Chen, X.; Guo, X.; Du, A. Characterization of anti-citrinin specific ScFvs selected from non-immunized mouse splenocytes by eukaryotic ribosome display. *PLoS ONE* **2015**, *10*, e0131482. [CrossRef]

276. Lu, Q.; Li, X.; Zhao, J.; Zhu, J.; Luo, Y.; Duan, H.; Ji, P.; Wang, K.; Liu, B.; Wang, X.; et al. Nanobody horseradish peroxidase and -EGFP fusions as reagents to detect porcine parvovirus in the immunoassays. *J. Nanobiotechnology* **2020**, *18*, 7. [CrossRef] [PubMed]

277. Sheng, Y.; Wang, K.; Lu, Q.; Ji, P.; Liu, B.; Zhu, J.; Liu, Q.; Sun, Y.; Zhang, J.; Zhou, E.M.; et al. Nanobody-horseradish peroxidase fusion protein as an ultrasensitive probe to detect antibodies against Newcastle disease virus in the immunoassay. *J. Nanobiotechnology* **2019**, *17*, 35. [CrossRef]

278. Alsulami, T.; Nath, N.; Flemming, R.; Wang, H.; Zhou, W.; Yu, J.H. Development of a novel homogeneous immunoassay using the engineered luminescent enzyme NanoLuc for the quantification of the mycotoxin fumonisin B1. *Biosens. Bioelectron.* **2021**, *177*, 112939. [CrossRef] [PubMed]

279. Barthelmebs, L.; Jonca, J.; Hayat, A.; Prieto-Simon, B.; Marty, J.L. Enzyme-Linked Aptamer Assays (ELAAs), based on a competition format for a rapid and sensitive detection of Ochratoxin A in wine. *Food Control* **2011**, *22*, 737–743. [CrossRef]

280. Li, Y.; Sun, L.; Zhao, Q. Aptamer-Structure Switch Coupled with Horseradish Peroxidase Labeling on a Microplate for the Sensitive Detection of Small Molecules. *Anal. Chem.* **2019**, *91*, 2615–2619. [CrossRef] [PubMed]

281. Wu, L.; Zhou, M.; Wang, Y.; Liu, J. Nanozyme and aptamer- based immunosorbent assay for aflatoxin B1. *J. Hazard. Mater.* **2020**, *399*, 123154. [CrossRef]

282. Sun, L.; Li, Y.; Wang, H.; Zhao, Q. An aptamer assay for aflatoxin B1 detection using Mg2+ mediated free zone capillary electrophoresis coupled with laser induced fluorescence. *Talanta* **2019**, *204*, 182–188. [CrossRef]

283. He, D.; Wu, Z.; Cui, B.; Xu, E. Aptamer and gold nanorod–based fumonisin B1 assay using both fluorometry and SERS. *Microchim. Acta* **2020**, *187*, 215. [CrossRef]

284. Xing, K.Y.; Peng, J.; Shan, S.; Liu, D.F.; Huang, Y.N.; Lai, W.H. Green Enzyme-Linked Immunosorbent Assay Based on the Single-Stranded Binding Protein-Assisted Aptamer for the Detection of Mycotoxin. *Anal. Chem.* **2020**, *92*, 8422–8426. [CrossRef]

285. Li, P.; Deng, S.; Zech Xu, Z. Toxicant substitutes in immunological assays for mycotoxins detection: A mini review. *Food Chem.* **2021**, *344*, 128589. [CrossRef]

286. Huang, D.T.; Fu, H.J.; Huang, J.J.; Luo, L.; Lei, H.T.; Shen, Y.D.; Chen, Z.J.; Wang, H.; Xu, Z.L. Mimotope-Based Immunoassays for the Rapid Analysis of Mycotoxin: A Review. *J. Agric. Food Chem.* **2021**, *69*, 11743–11752. [CrossRef] [PubMed]

287. Xiong, Y.; Leng, Y.; Li, X.; Huang, X.; Xiong, Y. Emerging strategies to enhance the sensitivity of competitive ELISA for detection of chemical contaminants in food samples. *TrAC-Trends Anal. Chem.* **2020**, *126*, 115861. [CrossRef]

288. Mukhtar, H.; Ma, L.; Pang, Q.; Zhou, Y.; Wang, X.; Xu, T.; Hammock, B.D.; Wang, J. Cyclic peptide: A safe and effective alternative to synthetic aflatoxin B1-competitive antigens. *Anal. Bioanal. Chem.* **2019**, *411*, 3881–3890. [CrossRef] [PubMed]

289. Chen, Y.; Zhang, S.; Hong, Z.; Lin, Y.; Dai, H. A mimotope peptide-based dual-signal readout competitive enzyme-linked immunoassay for non-toxic detection of zearalenone. *J. Mater. Chem. B* **2019**, *7*, 6972–6980. [CrossRef]

290. Peltomaa, R.; Fikacek, S.; Benito-Peña, E.; Barderas, R.; Head, T.; Deo, S.; Daunert, S.; Moreno-Bondi, M.C. Bioluminescent detection of zearalenone using recombinant peptidomimetic Gaussia luciferase fusion protein. *Microchim. Acta* **2020**, *187*, 547. [CrossRef]

291. Lu Hou, S.; Ma, Z.E.; Meng, H.; Xu, Y.; He, Q. hua Ultrasensitive and green electrochemical immunosensor for mycotoxin ochratoxin A based on phage displayed mimotope peptide. *Talanta* **2019**, *194*, 919–924. [CrossRef] [PubMed]

292. Peltomaa, R.; Agudo-Maestro, I.; Más, V.; Barderas, R.; Benito-Peña, E.; Moreno-Bondi, M.C. Development and comparison of mimotope-based immunoassays for the analysis of fumonisin B1. *Anal. Bioanal. Chem.* **2019**, *411*, 6801–6811. [CrossRef] [PubMed]

293. Yan, J.; Shi, Q.; You, K.; Li, Y.; He, Q. Phage displayed mimotope peptide-based immunosensor for green and ultrasensitive detection of mycotoxin deoxynivalenol. *J. Pharm. Biomed. Anal.* **2019**, *168*, 94–101. [CrossRef]

294. Peltomaa, R.; Farka, Z.; Mickert, M.J.; Brandmeier, J.C.; Pastucha, M.; Hlaváček, A.; Martínez-Orts, M.; Canales, Á.; Skládal, P.; Benito-Peña, E.; et al. Competitive upconversion-linked immunoassay using peptide mimetics for the detection of the mycotoxin zearalenone. *Biosens. Bioelectron.* **2020**, *170*, 112683. [CrossRef] [PubMed]

295. Piletsky, S.A.; Piletska, E.V.; Chen, B.; Karim, K.; Weston, D.; Barrett, G.; Lowe, P.; Turner, A.P.F. Chemical grafting of molecularly imprinted homopolymers to the surface of microplates. Application of artificial adrenergic receptor in enzyme-linked assay for β-agonists determination. *Anal. Chem.* **2000**, *72*, 4381–4385. [CrossRef]

296. Wang, S.; Xu, Z.; Fang, G.; Zhang, Y.; Liu, B.; Zhu, H. Development of a Biomimetic Enzyme-Linked Immunosorbent Assay Method for the Determination of Estrone in Environmental Water using Novel Molecularly Imprinted Films of Controlled Thickness as Artificial Antibodies. *J. Agric. Food Chem.* **2009**, *57*, 4528–4534. [CrossRef] [PubMed]

297. Chianella, I.; Guerreiro, A.; Moczko, E.; Caygill, J.S.; Piletska, E.V.; De Vargas Sansalvador, I.M.P.; Whitcombe, M.J.; Piletsky, S.A. Direct replacement of antibodies with molecularly imprinted polymer nanoparticles in ELISA - Development of a novel assay for vancomycin. *Anal. Chem.* **2013**, *85*, 8462–8468. [CrossRef]

298. Cenci, L.; Piotto, C.; Bettotti, P.; Bossi, A.M. Study on molecularly imprinted nanoparticle modified microplates for pseudo-ELISA assays. *Talanta* **2018**, *178*, 772–779. [CrossRef]

299. Munawar, H.; Safaryan, A.H.M.; De Girolamo, A.; Garcia-Cruz, A.; Marote, P.; Karim, K.; Lippolis, V.; Pascale, M.; Piletsky, S.A. Determination of Fumonisin B1 in maize using molecularly imprinted polymer nanoparticles-based assay. *Food Chem.* **2019**, *298*, 125044. [CrossRef]

300. Jia, B.; Liao, X.; Sun, C.; Fang, L.; Zhou, L.; Kong, W. Development of a quantum dot nanobead-based fluorescent strip immunosensor for on-site detection of aflatoxin B1 in lotus seeds. *Food Chem.* **2021**, *356*, 129614. [CrossRef] [PubMed]

301. Wang, J.; Chen, Q.; Jin, Y.; Zhang, X.; He, L.; Zhang, W.; Chen, Y. Surface enhanced Raman scattering-based lateral flow immunosensor for sensitive detection of aflatoxin M1 in urine. *Anal. Chim. Acta* **2020**, *1128*, 184–192. [CrossRef] [PubMed]

302. Geballa-Koukoula, A.; Gerssen, A.; Nielen, M.W.F. Direct analysis of lateral flow immunoassays for deoxynivalenol using electrospray ionization mass spectrometry. *Anal. Bioanal. Chem.* **2020**, *412*, 7547–7558. [CrossRef] [PubMed]

303. Hao, L.; Chen, J.; Chen, X.; Ma, T.; Cai, X.; Duan, H.; Leng, Y.; Huang, X.; Xiong, Y. A novel magneto-gold nanohybrid-enhanced lateral flow immunoassay for ultrasensitive and rapid detection of ochratoxin A in grape juice. *Food Chem.* **2021**, *336*, 127710. [CrossRef]

304. Chen, Y.; Fu, Q.; Xie, J.; Wang, H.; Tang, Y. Development of a high sensitivity quantum dot-based fluorescent quenching lateral flow assay for the detection of zearalenone. *Anal. Bioanal. Chem.* **2019**, *411*, 2169–2175. [CrossRef]

305. Chen, X.; Miao, X.; Ma, T.; Leng, Y.; Hao, L.; Duan, H.; Yuan, J.; Li, Y.; Huang, X.; Xiong, Y. Gold Nanobeads with Enhanced Absorbance for Improved Sensitivity in Competitive Lateral Flow Immunoassays. *Foods* **2021**, *10*, 1488. [CrossRef]

306. Qie, Z.; Yan, W.; Gao, Z.; Meng, W.; Xiao, R.; Wang, S. Ovalbumin antibody-based fluorometric immunochromatographic lateral flow assay using CdSe/ZnS quantum dot beads as label for determination of T-2 toxin. *Microchim. Acta* **2019**, *186*, 816. [CrossRef] [PubMed]

307. Li, Y.; Liu, L.; Kuang, H.; Xu, C. Visible and eco-friendly immunoassays for the detection of cyclopiazonic acid in maize and rice. *J. Food Sci.* **2020**, *85*, 105–113. [CrossRef]

308. Cai, P.; Wang, R.; Ling, S.; Wang, S. A high sensitive platinum-modified colloidal gold immunoassay for tenuazonic acid detection based on monoclonal IgG. *Food Chem.* **2021**, *360*, 130021. [CrossRef] [PubMed]

309. Yin, M.; Hu, X.; Sun, Y.; Xing, Y.; Xing, G.; Wang, Y.; Li, Q.; Wang, Y.; Deng, R.; Zhang, G. Broad-spectrum detection of zeranol and its analogues by a colloidal gold-based lateral flow immunochromatographic assay in milk. *Food Chem.* **2020**, *321*, 126697. [CrossRef] [PubMed]

310. Xu, S.; Zhang, G.; Fang, B.; Xiong, Q.; Duan, H.; Lai, W. Lateral Flow Immunoassay Based on Polydopamine-Coated Gold Nanoparticles for the Sensitive Detection of Zearalenone in Maize. *ACS Appl. Mater. Interfaces* **2019**, *11*, 31283–31290. [CrossRef]

311. Calabria, D.; Calabretta, M.M.; Zangheri, M.; Marchegiani, E.; Trozzi, I.; Guardigli, M.; Michelini, E.; Di Nardo, F.; Anfossi, L.; Baggiani, C.; et al. Recent Advancements in Enzyme-Based Lateral Flow Immunoassays. *Sensors* **2021**, *21*, 3358. [CrossRef] [PubMed]

312. Tripathi, P.; Upadhyay, N.; Nara, S. Recent advancements in lateral flow immunoassays: A journey for toxin detection in food. *Crit. Rev. Food Sci. Nutr.* **2018**, *58*, 1715–1734. [CrossRef]

313. Guo, X.; Yuan, Y.; Liu, J.; Fu, S.; Zhang, J.; Mei, Q.; Zhang, Y. Single-Line Flow Assay Platform Based on Orthogonal Emissive Upconversion Nanoparticles. *Anal. Chem.* **2021**, *93*, 3010–3017. [CrossRef]

314. Tang, D.; Sauceda, J.C.; Lin, Z.; Ott, S.; Basova, E.; Goryacheva, I.; Biselli, S.; Lin, J.; Niessner, R.; Knopp, D. Magnetic nanogold microspheres-based lateral-flow immunodipstick for rapid detection of aflatoxin B2 in food. *Biosens. Bioelectron.* **2009**, *25*, 514–518. [CrossRef]

315. Jin, Y.; Chen, Q.; Luo, S.; He, L.; Fan, R.; Zhang, S.; Yang, C.; Chen, Y. Dual near-infrared fluorescence-based lateral flow immunosensor for the detection of zearalenone and deoxynivalenol in maize. *Food Chem.* **2021**, *336*, 127718. [CrossRef]

316. Charlermroj, R.; Phuengwas, S.; Makornwattana, M.; Sooksimuang, T.; Sahasithiwat, S.; Panchan, W.; Sukbangnop, W.; Elliott, C.T.; Karoonuthaisiri, N. Development of a microarray lateral flow strip test using a luminescent organic compound for multiplex detection of five mycotoxins. *Talanta* **2021**, *233*, 122540. [CrossRef]

317. Li, R.; Meng, C.; Wen, Y.; Fu, W.; He, P. Fluorometric lateral flow immunoassay for simultaneous determination of three mycotoxins (aflatoxin B1, zearalenone and deoxynivalenol) using quantum dot microbeads. *Microchim. Acta* **2019**, *186*, 748. [CrossRef]

318. Di Nardo, F.; Alladio, E.; Baggiani, C.; Cavalera, S.; Giovannoli, C.; Spano, G.; Anfossi, L. Colour-encoded lateral flow immunoassay for the simultaneous detection of aflatoxin B1 and type-B fumonisins in a single Test line. *Talanta* **2019**, *192*, 288–294. [CrossRef] [PubMed]

319. Wang, D.; Zhu, J.; Zhang, Z.; Zhang, Q.; Zhang, W.; Yu, L.; Jiang, J.; Chen, X.; Wang, X.; Li, P. Simultaneous lateral flow immunoassay for multi-class chemical contaminants in maize and peanut with one-stop sample preparation. *Toxins* **2019**, *11*, 56. [CrossRef] [PubMed]

320. Zhang, W.; Tang, S.; Jin, Y.; Yang, C.; He, L.; Wang, J.; Chen, Y. Multiplex SERS-based lateral flow immunosensor for the detection of major mycotoxins in maize utilizing dual Raman labels and triple test lines. *J. Hazard. Mater.* **2020**, *393*, 122348. [CrossRef]

321. Goryacheva, O.A.; Guhrenz, C.; Schneider, K.; Beloglazova, N.V.; Goryacheva, I.Y.; De Saeger, S.; Gaponik, N. Silanized Luminescent Quantum Dots for the Simultaneous Multicolor Lateral Flow Immunoassay of Two Mycotoxins. *ACS Appl. Mater. Interfaces* **2020**, *12*, 24575–24584. [CrossRef]

322. Liu, Z.; Hua, Q.; Wang, J.; Liang, Z.; Li, J.; Wu, J.; Shen, X.; Lei, H.; Li, X. A smartphone-based dual detection mode device integrated with two lateral flow immunoassays for multiplex mycotoxins in cereals. *Biosens. Bioelectron.* **2020**, *158*, 112178. [CrossRef]

323. Tang, X.; Li, P.; Zhang, Q.; Zhang, Z.; Zhang, W.; Jiang, J. Time-Resolved Fluorescence Immunochromatographic Assay Developed Using Two Idiotypic Nanobodies for Rapid, Quantitative, and Simultaneous Detection of Aflatoxin and Zearalenone in Maize and Its Products. *Anal. Chem.* **2017**, *89*, 11520–11528. [CrossRef] [PubMed]

324. Xu, Y.; Yang, H.; Huang, Z.; Li, Y.; He, Q.; Tu, Z.; Ji, Y.; Ren, W. A peptide/maltose-binding protein fusion protein used to replace the traditional antigen for immunological detection of deoxynivalenol in food and feed. *Food Chem.* **2018**, *268*, 242–248. [CrossRef]

325. Yan, J.X.; Hu, W.J.; You, K.H.; Ma, Z.E.; Xu, Y.; Li, Y.P.; He, Q.H. Biosynthetic Mycotoxin Conjugate Mimetics-Mediated Green Strategy for Multiplex Mycotoxin Immunochromatographic Assay. *J. Agric. Food Chem.* **2020**, *68*, 2193–2200. [CrossRef]

326. Wang, L.; Chen, W.; Ma, W.; Liu, L.; Ma, W.; Zhao, Y.; Zhu, Y.; Xu, L.; Kuang, H.; Xu, C. Fluorescent strip sensor for rapid determination of toxins. *Chem. Commun.* **2011**, *47*, 1574–1576. [CrossRef]

327. Wang, L.; Ma, W.; Chen, W.; Liu, L.; Ma, W.; Zhu, Y.; Xu, L.; Kuang, H.; Xu, C. An aptamer-based chromatographic strip assay for sensitive toxin semi-quantitative detection. *Biosens. Bioelectron.* **2011**, *26*, 3059–3062. [CrossRef]

328. Zhao, Z.; Wang, H.; Zhai, W.; Feng, X.; Fan, X.; Chen, A.; Wang, M. A lateral flow strip based on a truncated aptamer-complementary strand for detection of type-B aflatoxins in nuts and dried figs. *Toxins* **2020**, *12*, 136. [CrossRef]

329. Wu, S.; Liu, L.; Duan, N.; Li, Q.; Zhou, Y.; Wang, Z. Aptamer-Based Lateral Flow Test Strip for Rapid Detection of Zearalenone in Corn Samples. *J. Agric. Food Chem.* **2018**, *66*, 1949–1954. [CrossRef]

330. Wu, S.; Liu, L.; Duan, N.; Wang, W.; Yu, Q.; Wang, Z. A test strip for ochratoxin A based on the use of aptamer-modified fluorescence upconversion nanoparticles. *Microchim. Acta* **2018**, *185*, 497. [CrossRef]

331. Wei, C.W.; Cheng, J.Y.; Huang, C.T.; Yen, M.H.; Young, T.H. Using a microfluidic device for 1 µl DNA microarray hybridization in 500 s. *Nucleic Acids Res.* **2005**, *33*, e78. [CrossRef]

332. Zhu, C.; Zhang, G.; Huang, Y.; Yang, S.; Ren, S.; Gao, Z.; Chen, A. Dual-competitive lateral flow aptasensor for detection of aflatoxin B1 in food and feedstuffs. *J. Hazard. Mater.* **2018**, *344*, 249–257. [CrossRef] [PubMed]

333. Thévenot, D.R.; Toth, K.; Durst, R.A.; Wilson, G.S. Electrochemical biosensors: Recommended definitions and classification. *Biosens. Bioelectron.* **2001**, *16*, 121–131. [CrossRef]

334. Székács, I.; Adányi, N.; Szendrő, I.; Székács, A. Direct and Competitive Optical Grating Immunosensors for Determination of Fusarium Mycotoxin Zearalenone. *Toxins* **2021**, *13*, 43. [CrossRef] [PubMed]

335. Evtugyn, G.; Porfireva, A.; Kulikova, T.; Hianik, T. Recent Achievements in Electrochemical and Surface Plasmon Resonance Aptasensors for Mycotoxins Detection. *Chemosensors* **2021**, *9*, 180. [CrossRef]

336. Zhang, N.; Liu, B.; Cui, X.; Li, Y.; Tang, J.; Wang, H.; Zhang, D.; Li, Z. Recent advances in aptasensors for mycotoxin detection: On the surface and in the colloid. *Talanta* **2021**, *223*, 121729. [CrossRef] [PubMed]

337. Rico-Yuste, A.; Abouhany, R.; Urraca, J.L.; Descalzo, A.B.; Orellana, G.; Moreno-Bondi, M.C. Eu(III)-Templated molecularly imprinted polymer used as a luminescent sensor for the determination of tenuazonic acid mycotoxin in food samples. *Sensors Actuators, B Chem.* **2021**, *329*, 129256. [CrossRef]

338. Elfadil, D.; Lamaoui, A.; Della Pelle, F.; Amine, A.; Compagnone, D. Molecularly imprinted polymers combined with electrochemical sensors for food contaminants analysis. *Molecules* **2021**, *26*, 4607. [CrossRef] [PubMed]

339. Santana Oliveira, I.; da Silva Junior, A.G.; de Andrade, C.A.S.; Lima Oliveira, M.D. Biosensors for early detection of fungi spoilage and toxigenic and mycotoxins in food. *Curr. Opin. Food Sci.* **2019**, *29*, 64–79. [CrossRef]

340. Mujahid, A.; Mustafa, G.; Dickert, F. Label-Free Bioanalyte Detection from Nanometer to Micrometer Dimensions—Molecular Imprinting and QCMs †. *Biosensors* **2018**, *8*, 52. [CrossRef]

341. Tothill, I. Biosensors and nanomaterials and their application for mycotoxin determination. *World Mycotoxin J.* **2011**, *4*, 361–374. [CrossRef]

342. Liu, D.; Li, W.; Zhu, C.; Li, Y.; Shen, X.; Li, L.; Yan, X.; You, T. Recent progress on electrochemical biosensing of aflatoxins: A review. *TrAC-Trends Anal. Chem.* **2020**, *133*, 115966. [CrossRef]

343. Zhao, W.W.; Xu, J.J.; Chen, H.Y. Photoelectrochemical Immunoassays. *Anal. Chem.* **2018**, *90*, 615–627. [CrossRef] [PubMed]

344. Evtugyn, G.; Porfireva, A.; Shamagsumova, R.; Hianik, T. Advances in Electrochemical Aptasensors Based on Carbon Nanomaterials. *Chemosensors* **2020**, *8*, 96. [CrossRef]

345. Tang, J.; Huang, Y.; Zhang, C.; Liu, H.; Tang, D. Amplified impedimetric immunosensor based on instant catalyst for sensitive determination of ochratoxin A. *Biosens. Bioelectron.* **2016**, *86*, 386–392. [CrossRef]

346. Zong, C.; Jiang, F.; Wang, X.; Li, P.; Xu, L.; Yang, H. Imaging sensor array coupled with dual-signal amplification strategy for ultrasensitive chemiluminescence immunoassay of multiple mycotoxins. *Biosens. Bioelectron.* **2021**, *177*, 112998. [CrossRef]

347. Lin, Y.; Zhou, Q.; Tang, D.; Niessner, R.; Knopp, D. Signal-On Photoelectrochemical Immunoassay for Aflatoxin B1 Based on Enzymatic Product-Etching MnO2 Nanosheets for Dissociation of Carbon Dots. *Anal. Chem.* **2017**, *89*, 5637–5645. [CrossRef]

348. Tang, X.; Wu, J.; Wu, W.; Zhang, Z.; Zhang, W.; Zhang, Q.; Zhang, W.; Chen, X.; Li, P. Competitive-Type Pressure-Dependent Immunosensor for Highly Sensitive Detection of Diacetoxyscirpenol in Wheat via Monoclonal Antibody. *Anal. Chem.* **2020**, *92*, 3563–3571. [CrossRef] [PubMed]

349. Jiang, K.; Nie, D.; Huang, Q.; Fan, K.; Tang, Z.; Wu, Y.; Han, Z. Thin-layer MoS 2 and thionin composite-based electrochemical sensing platform for rapid and sensitive detection of zearalenone in human biofluids. *Biosens. Bioelectron.* **2019**, *130*, 322–329. [CrossRef]

350. Pei, F.; Feng, S.; Wu, Y.; Lv, X.; Wang, H.; Chen, S.M.; Hao, Q.; Cao, Y.; Lei, W.; Tong, Z. Label-free photoelectrochemical immunosensor for aflatoxin B1 detection based on the Z-scheme heterojunction of g-C3N4/Au/WO3. *Biosens. Bioelectron.* **2021**, *189*, 113373. [CrossRef] [PubMed]

351. Tang, Z.; Liu, X.; Su, B.; Chen, Q.; Cao, H.; Yun, Y.; Xu, Y.; Hammock, B.D. Ultrasensitive and rapid detection of ochratoxin A in agro-products by a nanobody-mediated FRET-based immunosensor. *J. Hazard. Mater.* **2020**, *387*, 121678. [CrossRef]

352. Liu, X.; Wen, Y.; Wang, W.; Zhao, Z.; Han, Y.; Tang, K.; Wang, D. Nanobody-based electrochemical competitive immunosensor for the detection of AFB1 through AFB1-HCR as signal amplifier. *Microchim. Acta* **2020**, *187*, 352. [CrossRef]

353. Badie Bostan, H.; Danesh, N.M.; Karimi, G.; Ramezani, M.; Mousavi Shaegh, S.A.; Youssefi, K.; Charbgoo, F.; Abnous, K.; Taghdisi, S.M. Ultrasensitive detection of ochratoxin A using aptasensors. *Biosens. Bioelectron.* **2017**, *98*, 168–179. [CrossRef] [PubMed]

354. Wang, F.; Han, Y.; Wang, S.; Ye, Z.; Wei, L.; Xiao, L. Single-Particle LRET Aptasensor for the Sensitive Detection of Aflatoxin B1 with Upconversion Nanoparticles. *Anal. Chem.* **2019**, *91*, 11856–11863. [CrossRef] [PubMed]

355. Jahangiri–Dehaghani, F.; Zare, H.R.; Shekari, Z. Measurement of aflatoxin M1 in powder and pasteurized milk samples by using a label–free electrochemical aptasensor based on platinum nanoparticles loaded on Fe–based metal–organic frameworks. *Food Chem.* **2020**, *310*, 125820. [CrossRef]

356. Ahmadi, A.; Danesh, N.M.; Ramezani, M.; Alibolandi, M.; Lavaee, P.; Emrani, A.S.; Abnous, K.; Taghdisi, S.M. A rapid and simple ratiometric fluorescent sensor for patulin detection based on a stabilized DNA duplex probe containing less amount of aptamer-involved base pairs. *Talanta* **2019**, *204*, 641–646. [CrossRef] [PubMed]

357. Jiang, D.; Huang, C.; Shao, L.; Wang, X.; Jiao, Y.; Li, W.; Chen, J.; Xu, X. Magneto-controlled aptasensor for simultaneous detection of ochratoxin A and fumonisin B1 using inductively coupled plasma mass spectrometry with multiple metal nanoparticles as element labels. *Anal. Chim. Acta* **2020**, *1127*, 182–189. [CrossRef]

358. Zhong, H.; Yu, C.; Gao, R.; Chen, J.; Yu, Y.; Geng, Y.; Wen, Y.; He, J. A novel sandwich aptasensor for detecting T-2 toxin based on rGO-TEPA-Au@Pt nanorods with a dual signal amplification strategy. *Biosens. Bioelectron.* **2019**, *144*, 111635. [CrossRef] [PubMed]

359. He, Y.; Tian, F.; Zhou, J.; Zhao, Q.; Fu, R.; Jiao, B. Colorimetric aptasensor for ochratoxin A detection based on enzyme-induced gold nanoparticle aggregation. *J. Hazard. Mater.* **2020**, *388*, 121758. [CrossRef] [PubMed]

360. Wu, H.; Wang, H.; Wu, J.; Han, G.; Liu, Y.; Zou, P. A novel fluorescent aptasensor based on exonuclease-assisted triple recycling amplification for sensitive and label-free detection of aflatoxin B1. *J. Hazard. Mater.* **2021**, *415*, 125584. [CrossRef]

361. Yao, Y.; Wang, H.; Wang, X.; Wang, X.; Li, F. Development of a chemiluminescent aptasensor for ultrasensitive and selective detection of aflatoxin B1 in peanut and milk. *Talanta* **2019**, *201*, 52–57. [CrossRef]

362. Zhao, X.; Wang, Y.; Li, J.; Huo, B.; Huang, H.; Bai, J.; Peng, Y.; Li, S.; Han, D.; Ren, S.; et al. A fluorescence aptasensor for the sensitive detection of T-2 toxin based on FRET by adjusting the surface electric potentials of UCNPs and MIL-101. *Anal. Chim. Acta* **2021**, *1160*, 338450. [CrossRef]

363. Wu, Z.; He, D.; Cui, B.; Jin, Z.; Xu, E.; Yuan, C.; Liu, P.; Fang, Y.; Chai, Q. Trimer-based aptasensor for simultaneous determination of multiple mycotoxins using SERS and fluorimetry. *Microchim. Acta* **2020**, *187*, 495. [CrossRef]

364. Hao, L.; Wang, W.; Shen, X.; Wang, S.; Li, Q.; An, F.; Wu, S. A Fluorescent DNA Hydrogel Aptasensor Based on the Self-Assembly of Rolling Circle Amplification Products for Sensitive Detection of Ochratoxin A. *J. Agric. Food Chem.* **2020**, *68*, 369–375. [CrossRef] [PubMed]

365. Abnous, K.; Danesh, N.M.; Ramezani, M.; Alibolandi, M.; Nameghi, M.A.; Zavvar, T.S.; Taghdisi, S.M. A novel colorimetric aptasensor for ultrasensitive detection of aflatoxin M1 based on the combination of CRISPR-Cas12a, rolling circle amplification and catalytic activity of gold nanoparticles. *Anal. Chim. Acta* **2021**, *1165*, 338549. [CrossRef] [PubMed]

366. Ong, J.Y.; Pike, A.; Tan, L.L. Recent Advances in Conventional Methods and Electrochemical Aptasensors for Mycotoxin Detection. *Foods* **2021**, *10*, 1437. [CrossRef]

367. Evtugyn, G.; Hianik, T. Electrochemical immuno- and aptasensors for mycotoxin determination. *Chemosensors* **2019**, *7*, 10. [CrossRef]

368. Beitollahi, H.; Tajik, S.; Dourandish, Z.; Zhang, K.; Van Le, Q.; Jang, H.W.; Kim, S.Y.; Shokouhimehr, M. Recent Advances in the Aptamer-Based Electrochemical Biosensors for Detecting Aflatoxin B1 and Its Pertinent Metabolite Aflatoxin M1. *Sensors* **2020**, *20*, 3256. [CrossRef] [PubMed]

369. Goud, K.Y.; Reddy, K.K.; Satyanarayana, M.; Kummari, S.; Gobi, K.V. A review on recent developments in optical and electrochemical aptamer-based assays for mycotoxins using advanced nanomaterials. *Microchim. Acta* **2020**, *187*, 29. [CrossRef] [PubMed]

370. Rahimi, F.; Roshanfekr, H.; Peyman, H. Ultra-sensitive electrochemical aptasensor for label-free detection of Aflatoxin B1 in wheat flour sample using factorial design experiments. *Food Chem.* **2021**, *343*, 128436. [CrossRef]

371. Mu, Z.; Ma, L.; Wang, J.; Zhou, J.; Yuan, Y.; Bai, L. A target-induced amperometic aptasensor for sensitive zearalenone detection by CS@AB-MWCNTs nanocomposite as enhancers. *Food Chem.* **2021**, *340*, 128128. [CrossRef] [PubMed]

372. Gao, J.; Yao, X.; Chen, Y.; Gao, Z.; Zhang, J. Near-Infrared Light-Induced Self-Powered Aptasensing Platform for Aflatoxin B1 Based on Upconversion Nanoparticles-Doped Bi2S3Nanorods. *Anal. Chem.* **2021**, *93*, 677–682. [CrossRef]

373. Guo, X.; Wen, F.; Zheng, N.; Li, S.; Fauconnier, M.L.; Wang, J. A qPCR aptasensor for sensitive detection of aflatoxin M1. *Anal. Bioanal. Chem.* **2016**, *408*, 5577–5584. [CrossRef]

374. Sanzani, S.M.; Reverberi, M.; Fanelli, C.; Ippolito, A. Detection of ochratoxin a using molecular beacons and real-time PCR thermal cycler. *Toxins* **2015**, *7*, 812–820. [CrossRef]

375. Tang, J.; Cheng, Y.; Zheng, J.; Li, J.; Sun, Y.; Peng, S.; Zhu, Z. Target-engineered photo-responsive DNA strands: A novel signal-on photoelectrochemical biosensing platform for ochratoxin A. *Anal. Methods* **2019**, *11*, 5638–5644. [CrossRef]

376. Jia, F.; Liu, D.; Dong, N.; Li, Y.; Meng, S.; You, T. Interaction between the functionalized probes: The depressed efficiency of dual-amplification strategy on ratiometric electrochemical aptasensor for aflatoxin B1. *Biosens. Bioelectron.* **2021**, *182*, 113169. [CrossRef]

377. He, L.; Shen, Z.; Wang, J.; Zeng, J.; Wang, W.; Wu, H.; Wang, Q.; Gan, N. Simultaneously responsive microfluidic chip aptasensor for determination of kanamycin, aflatoxin M1, and 17β-estradiol based on magnetic tripartite DNA assembly nanostructure probes. *Microchim. Acta* **2020**, *187*, 176. [CrossRef] [PubMed]

378. Wang, J.; Wang, Y.; Liu, S.; Wang, H.; Zhang, X.; Song, X.; Yu, J.; Huang, J. Primer remodeling amplification-activated multisite-catalytic hairpin assembly enabling the concurrent formation of Y-shaped DNA nanotorches for the fluorescence assay of ochratoxin A. *Analyst* **2019**, *144*, 3389–3397. [CrossRef]

379. Zhang, M.; Wang, Y.; Sun, X.; Bai, J.; Peng, Y.; Ning, B.; Gao, Z.; Liu, B. Ultrasensitive competitive detection of patulin toxin by using strand displacement amplification and DNA G-quadruplex with aggregation-induced emission. *Anal. Chim. Acta* **2020**, *1106*, 161–167. [CrossRef]

380. Yin, J.; Liu, Y.; Wang, S.; Deng, J.; Lin, X.; Gao, J. Engineering a universal and label-free evaluation method for mycotoxins detection based on strand displacement amplification and G-quadruplex signal amplification. *Sensors Actuators, B Chem.* **2018**, *256*, 573–579. [CrossRef]

381. Wang, Y.; Wang, Y.; Liu, S.; Sun, W.; Zhang, M.; Jiang, L.; Li, M.; Yu, J.; Huang, J. Toehold-mediated DNA strand displacement-driven super-fast tripedal DNA walker for ultrasensitive and label-free electrochemical detection of ochratoxin A. *Anal. Chim. Acta* **2021**, *1143*, 21–30. [CrossRef]

382. Yin, N.; Yuan, S.; Zhang, M.; Wang, J.; Li, Y.; Peng, Y.; Bai, J.; Ning, B.; Liang, J.; Gao, Z. An aptamer-based fluorometric zearalenone assay using a lighting-up silver nanocluster probe and catalyzed by a hairpin assembly. *Microchim. Acta* **2019**, *186*, 765. [CrossRef] [PubMed]

383. Zhang, N.; Li, J.; Liu, B.; Zhang, D.; Zhang, C.; Guo, Y.; Chu, X.; Wang, W.; Wang, H.; Yan, X.; et al. Signal enhancing strategies in aptasensors for the detection of small molecular contaminants by nanomaterials and nucleic acid amplification. *Talanta* **2022**, *236*, 122866. [CrossRef]

384. Wang, Y.; Song, W.; Zhao, H.; Ma, X.; Yang, S.; Qiao, X.; Sheng, Q.; Yue, T. DNA walker-assisted aptasensor for highly sensitive determination of Ochratoxin A. *Biosens. Bioelectron.* **2021**, *182*, 113171. [CrossRef] [PubMed]

385. Lowdon, J.W.; Diliën, H.; Singla, P.; Peeters, M.; Cleij, T.J.; van Grinsven, B.; Eersels, K. MIPs for commercial application in low-cost sensors and assays – An overview of the current status quo. *Sensors Actuators B Chem.* **2020**, *325*, 128973. [CrossRef] [PubMed]

386. Mao, L.; Xue, X.; Xu, X.; Wen, W.; Chen, M.M.; Zhang, X.; Wang, S. Heterostructured CuO-g-C3N4 nanocomposites as a highly efficient photocathode for photoelectrochemical aflatoxin B1 sensing. *Sensors Actuators B Chem.* **2021**, *329*, 129146. [CrossRef]

387. Akgönüllü, S.; Armutcu, C.; Denizli, A. Molecularly imprinted polymer film based plasmonic sensors for detection of ochratoxin A in dried fig. *Polym. Bull.* **2021**, *78*, 1–9. [CrossRef]

388. Pacheco, J.G.; Castro, M.; Machado, S.; Barroso, M.F.; Nouws, H.P.A.; Delerue-Matos, C. Molecularly imprinted electrochemical sensor for ochratoxin A detection in food samples. *Sensors Actuators B Chem.* **2015**, *215*, 107–112. [CrossRef]

389. Wang, Q.; Chen, M.; Zhang, H.; Wen, W.; Zhang, X.; Wang, S. Solid-state electrochemiluminescence sensor based on RuSi nanoparticles combined with molecularly imprinted polymer for the determination of ochratoxin A. *Sensors Actuators B Chem.* **2016**, *222*, 264–269. [CrossRef]

390. Li, W.; Diao, K.; Qiu, D.; Zeng, Y.; Tang, K.; Zhu, Y. A highly-sensitive and selective antibody-like sensor based on molecularly imprinted poly (L-arginine) on COOH-MWCNTs for electrochemical recognition and detection of deoxynivalenol. *Food Chem.* **2021**, *350*, 129229. [CrossRef]

391. Radi, A.E.; Eissa, A.; Wahdan, T. Molecularly Imprinted Impedimetric Sensor for Determination of Mycotoxin Zearalenone. *Electroanalysis* **2020**, *32*, 1788–1794. [CrossRef]

392. Mao, L.; Ji, K.; Yao, L.; Xue, X.; Wen, W.; Zhang, X.; Wang, S. Molecularly imprinted photoelectrochemical sensor for fumonisin B 1 based on GO-CdS heterojunction. *Biosens. Bioelectron.* **2019**, *127*, 57–63. [CrossRef] [PubMed]

393. Zhang, W.; Xiong, H.; Chen, M.; Zhang, X.; Wang, S. Surface-enhanced molecularly imprinted electrochemiluminescence sensor based on Ru@SiO2 for ultrasensitive detection of fumonisin B1. *Biosens. Bioelectron.* **2017**, *96*, 55–61. [CrossRef]

394. Hu, X.; Liu, Y.; Xia, Y.; Zhao, F.; Zeng, B. A novel ratiometric electrochemical sensor for the selective detection of citrinin based on molecularly imprinted poly(thionine) on ionic liquid decorated boron and nitrogen co-doped hierarchical porous carbon. *Food Chem.* **2021**, *363*, 130385. [CrossRef]

395. Huang, Q.; Zhao, Z.; Nie, D.; Jiang, K.; Guo, W.; Fan, K.; Zhang, Z.; Meng, J.; Wu, Y.; Han, Z. Molecularly imprinted poly(thionine)-based electrochemical sensing platform for fast and selective ultratrace determination of patulin. *Anal. Chem.* **2019**, *91*, 4116–4123. [CrossRef] [PubMed]

396. Hatamluyi, B.; Rezayi, M.; Beheshti, H.R.; Boroushaki, M.T. Ultra-sensitive molecularly imprinted electrochemical sensor for patulin detection based on a novel assembling strategy using Au@Cu-MOF/N-GQDs. *Sensors Actuators B. Chem.* **2020**, *318*, 128219. [CrossRef]

397. Hu, X.; Xia, Y.; Liu, Y.; Zhao, F.; Zeng, B. Determination of patulin using dual-dummy templates imprinted electrochemical sensor with PtPd decorated N-doped porous carbon for amplification. *Microchim. Acta* **2021**, *188*, 148. [CrossRef]

398. Gupta, G.; Bhaskar, A.S.B.; Tripathi, B.K.; Pandey, P.; Boopathi, M.; Rao, L.V.L.; Singh, B.; Vijayaraghavan, R. Supersensitive detection of T-2 toxin by the in situ synthesized π-conjugated molecularly imprinted nanopatterns. An in situ investigation by surface plasmon resonance combined with electrochemistry. *Biosens. Bioelectron.* **2011**, *26*, 2534–2540. [CrossRef]

399. Selvam, S.P.; Kadam, A.N.; Maiyelvaganan, K.R.; Prakash, M.; Cho, S. Novel SeS2-loaded Co MOF with Au@PANI comprised electroanalytical molecularly imprinted polymer-based disposable sensor for patulin mycotoxin. *Biosens. Bioelectron.* **2021**, *187*, 113302. [CrossRef] [PubMed]

400. Miller, J.N.; Niessner, R.; Knopp, D. Enzyme and Immunoassays. In *Ullmann's Encyclopedia of Industrial Chemistry*; Wiley: Hoboken, NJ, USA, 2021; pp. 1–36.

401. Zhang, L.; Zhang, Z.; Tian, Y.; Cui, M.; Huang, B.; Luo, T.; Zhang, S.; Wang, H. Rapid, simultaneous detection of mycotoxins with smartphone recognition-based immune microspheres. *Anal. Bioanal. Chem.* **2021**, *413*, 3683–3693. [CrossRef]

402. Oswald, S.; Karsunke, X.Y.Z.; Dietrich, R.; Märtlbauer, E.; Niessner, R.; Knopp, D. Automated regenerable microarray-based immunoassay for rapid parallel quantification of mycotoxins in cereals. *Anal. Bioanal. Chem.* **2013**, *405*, 6405–6415. [CrossRef]

403. Mahmoudpour, M.; Ezzati Nazhad Dolatabadi, J.; Torbati, M.; Homayouni-Rad, A. Nanomaterials based surface plasmon resonance signal enhancement for detection of environmental pollutions. *Biosens. Bioelectron.* **2019**, *127*, 72–84. [CrossRef] [PubMed]

404. Wei, T.; Ren, P.; Huang, L.; Ouyang, Z.; Wang, Z.; Kong, X.; Li, T.; Yin, Y.; Wu, Y.; He, Q. Simultaneous detection of aflatoxin B1, ochratoxin A, zearalenone and deoxynivalenol in corn and wheat using surface plasmon resonance. *Food Chem.* **2019**, *300*, 125176. [CrossRef]

405. Liu, S.; Liu, X.; Pan, Q.; Dai, Z.; Pan, M.; Wang, S. A Portable, Label-Free, Reproducible Quartz Crystal Microbalance Immunochip for the Detection of Zearalenone in Food Samples. *Biosensors* **2021**, *11*, 53. [CrossRef]

406. Susilo, M.D.D.; Jayadi, T.; Kusumaatmaja, A.; Nugraheni, A.D. Development of molecular imprinting polymer nanofiber for aflatoxin B1 detection based on quartz crystal microbalance. In Proceedings of the Materials Science Forum, 9th International Conference on Materials Science and Engineering Technology (ICMSET 2020) held virtually in Kyoto, 9–11 October 2020; Trans Tech Publications Ltd.: Bäch, Switzerland, 2021; Volume 1023, pp. 103–111. [CrossRef]

407. Ricciardi, C.; Castagna, R.; Ferrante, I.; Frascella, F.; Luigi Marasso, S.; Ricci, A.; Canavese, G.; Lorè, A.; Prelle, A.; Lodovica Gullino, M.; et al. Development of a microcantilever-based immunosensing method for mycotoxin detection. *Biosens. Bioelectron.* **2013**, *40*, 233–239. [CrossRef]

408. Maragos, C.M. Signal amplification using colloidal gold in a biolayer interferometry-based immunosensor for the mycotoxin deoxynivalenol. *Food Addit. Contam. Part A Chem. Anal. Control. Expo. Risk Assess.* **2012**, *29*, 1108–1117. [CrossRef] [PubMed]

409. Lux, G.; Langer, A.; Pschenitza, M.; Karsunke, X.; Strasser, R.; Niessner, R.; Knopp, D.; Rant, U. Detection of the carcinogenic water pollutant benzo [a]pyrene with an electro-switchable biosurface. *Anal. Chem.* **2015**, *87*, 4538–4545. [CrossRef] [PubMed]

410. Rant, U. Sensing with electro-switchable biosurfaces. *Bioanal. Rev.* **2012**, *4*, 97–114. [CrossRef]

411. Wang, Y.; Jiang, J.; Fotina, H.; Zhang, H.; Chen, J. Advances in Antibody Preparation Techniques for Immunoassays of Total Aflatoxin in Food. *Molecules* **2020**, *25*, 4113. [CrossRef]

412. Weller, M.G. Quality Issues of Research Antibodies. *Anal. Chem. Insights* **2016**, *2016*, 21–27. [CrossRef]

413. O'Kennedy, R.; Fitzgerald, S.; Murphy, C. Don't blame it all on antibodies – The need for exhaustive characterisation, appropriate handling, and addressing the issues that affect specificity. *TrAC-Trends Anal. Chem.* **2017**, *89*, 53–59. [CrossRef]

414. Directive, E. Directive 2010/63/EU of the European Parliament and of the council of 22 September 2010 on the protection of animals used for scientific purposes. *Off. J. Eur. Union* **2010**, *276*, 33–79.

Review

Contamination, Detection and Control of Mycotoxins in Fruits and Vegetables

Mina Nan [1,2], Huali Xue [1,*] and Yang Bi [3,*]

1   College of Science, Gansu Agricultural University, Lanzhou 730070, China; nanmn@gsau.edu.cn
2   Basic Experiment Teaching Center, Gansu Agricultural University, Lanzhou 730070, China
3   College of Food Science and Engineering, Gansu Agricultural University, Lanzhou 730070, China
*   Correspondence: xuehual@gsau.edu.cn (H.X.); biyang@gsau.edu.cn (Y.B.);
     Tel.: +86-931-763-1212 (H.X.); +86-931-763-1113 (Y.B.)

Abstract: Mycotoxins are secondary metabolites produced by pathogenic fungi that colonize fruits and vegetables either during harvesting or during storage. Mycotoxin contamination in fruits and vegetables has been a major problem worldwide, which poses a serious threat to human and animal health through the food chain. This review systematically describes the major mycotoxigenic fungi and the produced mycotoxins in fruits and vegetables, analyzes recent mycotoxin detection technologies including chromatography coupled with detector (i.e., mass, ultraviolet, fluorescence, etc.) technology, electrochemical biosensors technology and immunological techniques, as well as summarizes the degradation and detoxification technologies of mycotoxins in fruits and vegetables, including physical, chemical and biological methods. The future prospect is also proposed to provide an overview and suggestions for future mycotoxin research directions.

Keywords: patulin; Alternaria toxin; ochratoxin A; trichothecenes; adsorption; degradation

Key Contribution: The major mycotoxigenic fungi and mycotoxins in fruits and vegetables were presented. The recent mycotoxin detection methods including chromatographic technology, electrochemical biosensor technology and immunological techniques were discussed. Moreover, the physical, chemical and biological technologies to degrade and detoxify mycotoxin from fruits and vegetables were also summarized in this review.

Citation: Nan, M.; Xue, H.; Bi, Y. Contamination, Detection and Control of Mycotoxins in Fruits and Vegetables. *Toxins* **2022**, *14*, 309. https://doi.org/10.3390/toxins14050309

Received: 14 March 2022
Accepted: 24 April 2022
Published: 27 April 2022

Publisher's Note: MDPI stays neutral with regard to jurisdictional claims in published maps and institutional affiliations.

1. Introduction

Fruits and vegetables are important nutritional sources for human and are one of the most important parts of the human diet. However, the production loss of fruits and vegetables caused by postharvest diseases was 35–55%, and in developing countries the loss even reached 55% [1]. Except for pesticide residues and heavy metals, the contamination of mycotoxins is another key reason that can cause a loss of production and a threat to the health and lives of consumers [2,3]. Mycotoxins are a secondary metabolite produced by filamentous fungi, which can naturally occur during growth, harvest, storage, transportation and processing [4]. About 150 fungi have the ability to produce more than 400 mycotoxins, such as trichothecenes (TCs), ochratoxin A (OTA), patulin (PAT), Alternaria toxins (ATs), etc. [5]. All these toxins have different chemical structures, but in common most of them have the toxic effects of DNA damage and cytotoxicity. At a low concentration, they can cause lesions in the liver, kidneys and gastrointestinal tract, and they even have carcinogenic, teratogenic and mutagenic effects [6,7]. The accumulation of mycotoxins in fruits and vegetables is not only a potential threat to human and animal health but also results in serious economic losses. Therefore, it is very important to minimize the potential risks to humans and economic losses through the monitoring, detection and control of mycotoxins.

In this review, we focus on the main mycotoxins (TCs, OTA, PAT and ATs) produced by major mycotoxigenic fungi (*Alternaria*, *Aspergillus*, *Penicillium* and *Fusarium*) in fruits and vegetables. The mycotoxin detection methods including chromatography coupled with a detector (i.e., mass, ultraviolet, fluorescence, etc.) technology and electrochemical biosensors technology are summarized, as well as the mycotoxin degradation and detoxification strategies in fruits and vegetables.

2. Occurrence, Contamination and Toxicity of Mycotoxins

Different mycotoxins have been found not only in fruits and vegetables but also in their derivative products [8–12]. The mycotoxins commonly found in fruits and vegetables can be divided into four major categories: PAT, TCs, OTA and ATs. The toxicity, occurrence and contamination status of these mycotoxins are related to the fungi and hosts, as well as the environmental conditions [4]. The main route of infection of mycotoxigenic fungi to fruit and vegetable is that they can successfully colonize the host through wounds or natural orifices on the surface of fruits and vegetables before harvest. During the growth period, the host has strong resistance to exogenous, biological or abiotic stress. However, due to the large energy consumption at the postharvest stage, the resistance effect of fruits and vegetables is significantly weakened. Once the temperature and humidity conditions are appropriate, the pathogen can grow and colonize rapidly. Then, the secondary metabolite pathogenic mycotoxins are produced in the medium and late stages of growth [1]. The main factors related to the occurrence of fungi and toxin production in fruits and vegetables are shown in Figure 1.

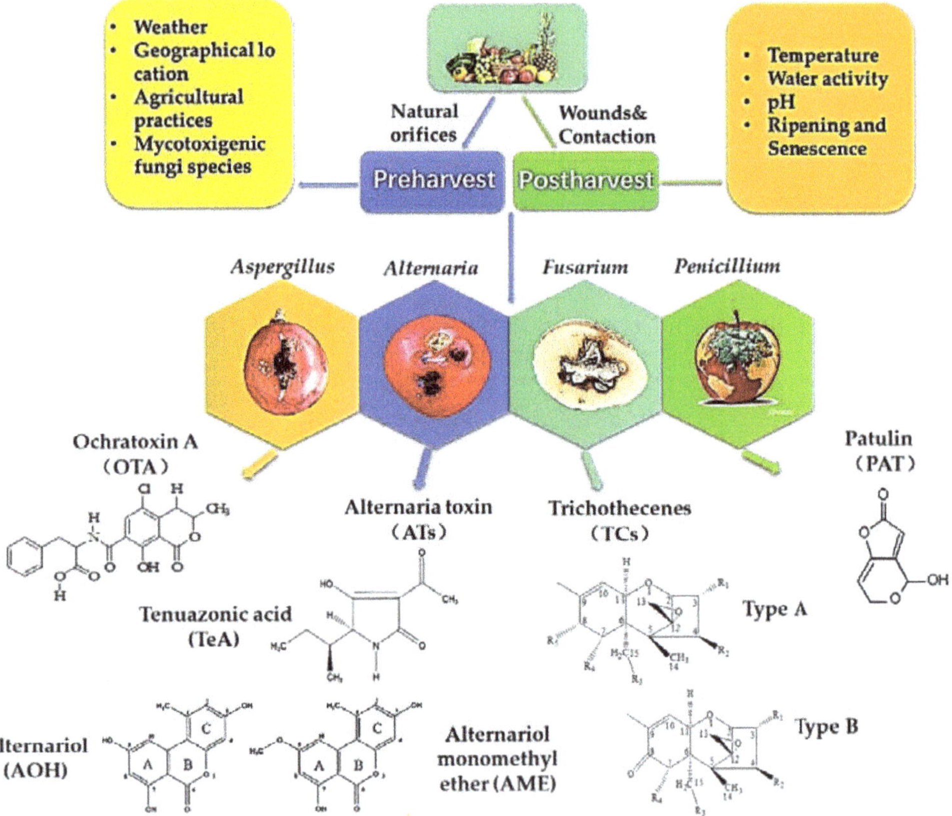

Figure 1. Mycotoxin structure and factors affecting fungi and toxin production in fruits and vegetables.

2.1. Ochratoxin A (OTA)

OTA was first isolated from *Aspergillus ochraceus* in South Africa in 1965 [13]. Since then, it was found that many species of *Aspergillus* and *Penicillium* fungi can produce OTA in different hosts. These fungi that are capable of producing OTA mainly include nine species of fungi belonging to two sections of *Aspergillus* and two species of *Penicillium*. OTA-producing species belonging to the *Circumdati* section were *A. ochraceus*, *A. westerdijkiae* and *A. steynii*. Moreover, six OTA-producing species in the *Aspergillus* section *Nigri* are *A. carbonarius*, *A. niger*, *A. lacticoffeatus*, *A. sclerotioniger* and *A. tubingensis* [14]. Because *Penicillium* species are more diverse in numbers and habitat range than *Aspergillus*, it is difficult to distinguish subgenus of *Penicillium* from each other. Although many species of *Penicillium* (e.g., *P. cyclopium*, *P. viridicatum*, *P. chrysogenum*) have been reported as OTA producers, they were classified as *P. viridicatum* in genus [15]. At present, the two widely known and accepted OTA producers of *Penicillium* are *P. verrucosum* and *P. nordicum* [15,16]. However, species of *Penicillium* are often isolated from meat and its products, while species of *Aspergillus* are more likely to infect plants and then produce OTA during their pre-harvest and postharvest processes [14]. In addition to grains and coffee, OTA is often found in grapes and red peppers, especially in grapes and their derivative products [17]. The most important OTA-producing fungi in grapes and wine are *A. niger* and *A. carbonarius*. The major pathogenic fungi that can produce OTA are summarized in Figure 2.

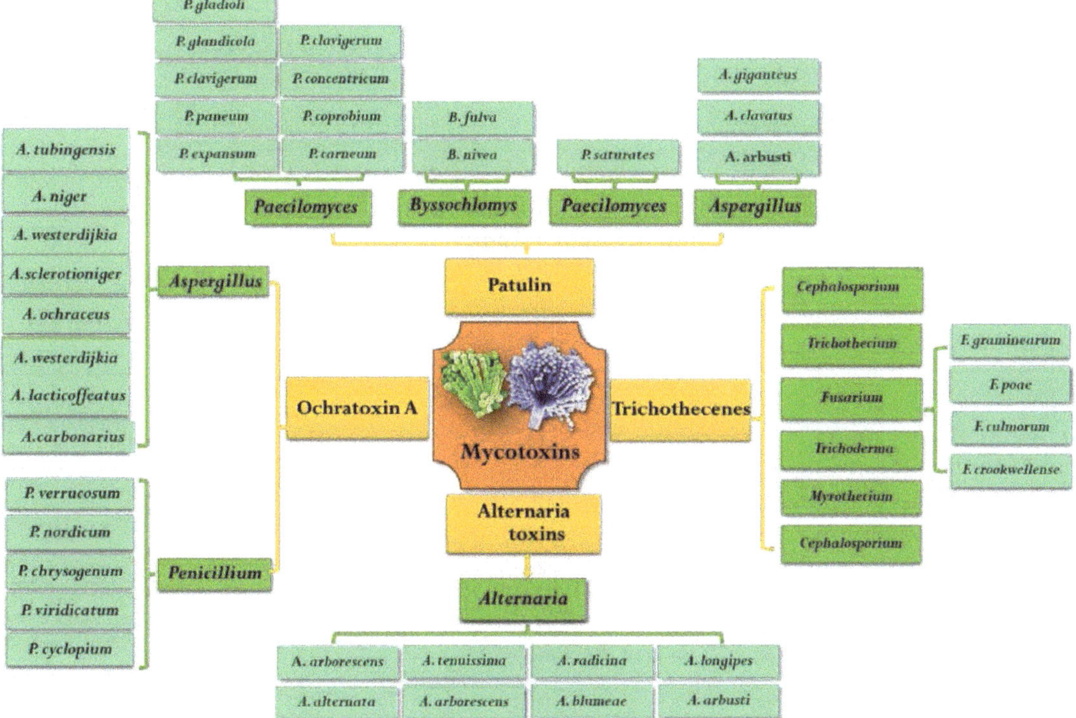

Figure 2. Common species of mycotoxigenic fungi in fruits and vegetables.

Grapes are more likely to be infected by OTA-producing fungi than a variety of fruits and vegetables. Since the first report on the occurrence of OTA in wines and grape juices by Zimmerli and Dick in 1996, many reports have studied OTA contamination in grapes and their products worldwide [18]. The differences in the type of grape products, processing technology, grape varieties and origin of raw material always cause significant differences in the content of OTA [19]. Silva et al. studied 100 wine samples purchased from different Portuguese retail outlets, and 4% of red wine and 1% of white wine were contaminated with OTA, which were produced in different region [20]. Kochman et al. analyzed OTA in red wine from Poland, Spain and France, but unexpectedly, the OTA content in all of the samples significantly exceeded the maximum limit established of 2 μg/L (EU, 2006); the average concentration of OTA even reached 6.629 μg/L in Polish wines, 6.848 μg/L in Spanish wines and 6.711 μg/L in French wines [21]. In China, the maximum concentration of OTA in red wine was 5.65 μg/L [22]. Another study showed a frequency of contamination of 85% (35 out of 41) in American wines with a highest level of approximately 8.6 ± 4.8 μg/L [23]. However, the same research found that samples of OTA content exceeding 1 μg/L were 54.8% in red wine, 40.0% in white wines, 58.1% in dry wine and 30% in sweet wines. OTA levels in red wines are always higher than that in rosés and white wines, because for red wine making the maceration is an important and necessarily step. Since the skins of grapes were the most contaminated tissue of grapes, the maceration facilitated OTA going from the skin to the must, resulting in an increase in toxin content [19].

Grape juices exposed to OTA have been found in different regions of the world. According to previous reports, Mehri et al. investigated the prevalence and concentration of OTA and estimated that the global pooled prevalence of OTA in grape juice was 36% [24]. In Germany, only 7/38 white grape juices were low in OTA content (less than 10 ng/L) and in other samples were 10–1300 ng/L, while 65/73 red grape juice contained OTA between 10–5300 ng/L [25]. In Poland, 26.7% (16/66) of grape beverages tested positive for OTA at the highest value of 0.7 μg/kg [26]. Wei et al. analyzed 41 grape juices and found that all of the samples were exposed to OTA, the incidence of positives was 100% and the OTA concentration of six grape juice samples was between 0.1 and 0.2 μg/kg [27]. In Iran, it was found that the OTA level of 39 out of 70 samples (55.7%) was higher than 2 μg/L, and the upper bound mean was 2.6 μg/L [28]. A total of 1401 fruit juice samples that were produced in Argentina from 2005 to 2013 were evaluated, confirming the identity of only 22 OTA-positive juices, and the concentration of OTA was from 0.03 to 3.6 μg/L [29]. In conclusion, most of the above reports with high OTA concentration were before 2000, and OTA-positive juices seemed to decrease significantly after 2000. The changes of OTA exposition rate and maximum concentration seemed to be related to the progress of OTA detection technology and the strict regulatory limit on OTA in the market.

Although there was a small amount of research about OTA focusing on fruits, vegetables and their processed products other than grapes, a few recent reports of OTA in fruits and vegetables are summarized in Table 1. Sanzani et al. isolated OTA-producing fungi species of *A. westerdijkiae*, *A. ochraceus* and *A. occultus* from tomato; however, the OTA content in the tomato infected by these fungi was not determined in this study [30]. One study from China analyzed the capacity of two *Penicillium* strains to produce OTA in strawberries. OTA was detected in *Penicillium* fermentation broth (2.22 μg/L), but OTA was not detected in spoilage strawberries [31]. However, OTA was found in dried fruits such as figs, apricots, dates, mulberries and raisins [32–41], though the OTA concentration in dried apricots and dates was much lower compared with dried figs and raisin (Table 1). The difference of OTA content was related to many factors, including the species of fungi, the sugar content, pH, temperature and water activity of the host [42]. Fruits with high sugar levels were conducive to mold growth and induce OTA production, and heat treatment or forced air was easier for eliminating toxigenic fungi than natural drying, so as to reduce the OTA contamination [43].

Table 1. Contamination of ochratoxin A in fruits and vegetables.

	Foodstuffs	Detection Method	Country	Positives (%)	OTA (µg/Kg)	Ref.
Fruits and products	Peppers	HPLC-FD	Poland	37.5%	1.7–2.4	[26]
	Grapes	ELISA	Italy	30.4%	0.02–9.2	[30]
	Strawberries	HPLC-FD	China	-	-	[32]
	Apple juice	HPLC-FD	Saudi Arabia	29.41%	100–200	[33]
Dried fruits	Raisins	HPLC-FD	Poland	47%	1.1–34	[26]
	Figs	HPLC-FD	Turkey	48%	0.1–15.3	[34]
	Figs	ELISA	Iran	10.4	2.3–14.2	[35]
	Apricots	ELISA	Iran	6.7%	2.8	[35]
	Raisins	ELISA	Iran	44.7%	2.9–18.2	[35]
	Dates	ELISA	Iran	10%	1.4–3.6	[35]
	Figs	ELISA	Spain	54.3%	3.15–277	[36]
	Apricots	HPLC-UV–VIS	India	28.57%	194 ± 0.001	[37]
	Raisins	HPLC-FD	Pakistan	23.5%	5.60 ± 1.34	[38]
	Apricots	HPLC-FD	Pakistan	23.1%	3.10 ± 0.70	[38]
	Plums	HPLC-FD	Pakistan	25%	3.90 ± 0.95	[38]
	Figs	HPLC-FD	Pakistan	21.4%	2.10 ± 0.79	[38]
	Mulberries	HPLC-FD	Iran	45.5%	0.4–3.4	[39]
	Dates	HPLC-FD	Iran	22.3%	0.5–2.1	[39]
	Figs	HPLC-FD	Iran	45.5%	0.4–12.2	[39]
	Apricots	HPLC-FD	Iran	50%	0.75–5.5	[39]
Dried vegetables	Packed red peppers	HPLC-FD	Turkey	87.1%	0.6–1.0	[40]
	Unpacked red peppers	HPLC-FD	Turkey	100%	1.1–31.7	[40]
	Packed red peppers	HPLC-FD	Korea	48%	0.23–56.30	[41]
	Unpacked red peppers	HPLC-FD	Korea	4%	0.15–0.20	[41]

Note: "-" means no relevant report, and "0" means no detection; HPLC-FD: high-performance liquid chromatography with fluorescence detection; ELISA: enzyme-linked immunosorbent assay; HPLC-UV–VIS: high-performance liquid chromatography with variable wavelength absorbance detector.

OTA-contaminated foods, especially plant foods, were related to the occurrence of many diseases in animals and humans, and it was a potential health threat with nephrotoxicity, hepatotoxicity, teratogenicity, immunotoxicity and neurotoxicity [43]. More and more research results indicated that even a very low level of OTA intake will accumulate in the human body [44,45]. The acute toxicity of OTA showed wide variability, which is related to animal sex, species and sample size. The acute dietary 96 h LC_{50} of OTA for sea bass was 9.23 mg/kg BW [46], for rats and mice it was about 46–58 mg/kg BW and that for pigs, cats, rabbits and dogs was 0.2–1 mg/kg BW [47], while LC_{50} for primates and humans was not obtained. However, it has been found that the pathogenesis of OTA is mostly chronic and subchronic toxicity, and the major mechanisms of OTA toxicity were the inhibition of protein synthesis, promotion of membrane peroxidation, disruption of calcium homeostasis, inhibition of mitochondrial respiration and DNA damage [48]. OTA contamination in food has been reported to induce hepatocellular tumors (well differentiated trabecular adenomas), renal cell tumors (renal cystic adenomas and solid renal tumors) and proliferative liver nodules [49]. Therefore, the International Agency for Research on Cancer (IARC) has classified it as an IIB group [50]. Erceg et al. found that OTA produces mild cytotoxic effects in HESCs by inhibiting cell attachment, survival and proliferation [43]. OTA administration induced renal injury of mice and increased histological injury, renal fibrosis molecules and activated the NOD-like receptor protein 3 inflammasome and induced pyroptosis [51,52]. As a neurotoxin, many neurodegenerative diseases, such as Parkinson's disease and Alzheimer's disease, are related to the apoptosis of nerve cells and the obstruction of nerve tissue regeneration caused by OTA [53].

2.2. Patulin (PAT)

PAT is an α, β-unsaturated γ-lactone, which was first isolated and detected from *Penicillium griseofulvum* in 1943 [54]. Early studies found that patulin has broad-spectrum antimicrobial activity and can also even inhibit the growth of more than 70 kinds of bacteria, but after the 1960s it was reclassified as a mycotoxin [55,56]. PAT was well known as a secondary metabolite produced by *Penicillium, Aspergillus, Byssochlomys* and *Pacelomyces* (Figure 2) [57]. It was reported that more than 60 species fungi belonging to 30 genera can produce PAT, mainly including 13 *Penicillium* genus and 4 *Aspergillus* genus [1]. The specific PAT pathogens were *P. expansum, P. patulin, P. claviforme, P. novaezeelandiae, P. lapiclosum, P. granulatum, P. cyclopium, P. chrysogenum, P. roguefooti, A. clavatu, A. Nivea, A. terreu, A. giganteus, Byssochlamys nivea, Brachypsectra fulva* and *Paecilomyces saturatus* [1,58]. Among them, *P. expansum* and *P. patulin* have an enormous capacity for PAT production in fruits and vegetables and were considered as wound pathogens.

The contamination of PAT is mainly derived from apples, pears, apricots, peaches, blueberries, plums, strawberries and their products, such as juice and jam [59]. PAT has become an important index to judge the quality and safety of apple juice. At present, more than 100 countries and regions in the world have formulated the maximum content limit of patulin in fruit juice and other processed products. China, the United States (U.S.) and the European Union (EU) have set a maximum limit of PAT in fruit juice at 50 µg/L, and the World Health Organization (WHO) and Food and Agriculture Organization (FAO) recommend that the daily intake of PAT per body weight should not exceed 0.4 µg/kg [58,60]. PAT in fruits is mainly caused by diseases such as blue mold rot induced by *P. expansum*, which is the major postharvest disease and problem for apples and apple juice [61]. *P. expansum* often infects fruits through wounds and produces toxins during the transportation and storage periods. As a lipid soluble small molecule, PAT has strong diffusion ability, which can spread into the surrounding healthy pulp tissue. When the rotten fruit is used as raw material to produce fruit juice and other products, it is difficult to eliminate the PAT contamination through conventional food processing due to its stable chemical properties. Thus, PAT will not disappear with product processing but will go further into the fruit products [1]. Hammami et al. detected 45 apples and apple products and found that the PAT levels of 5 out of 12 fresh apples ranged from 3.85 to 17.35 µg/kg, and 100% of the apple juice was contaminated by PAT with an average content of 35.37 µg/kg and 5% higher than 50 µg/kg (EU limit). PAT was found in all baby apple juices, ranging from 7.7–61.3 µg/kg [62]. In Tunisia, 11 out of 30 apple juices presented a contamination of PAT, while 12/30 were contaminated in mixed juices, and the average contents of PAT were 80 and 55 µg/kg, respectively [63]. Other research has also showed that the percentage of PAT-positive samples was 80% in concentrated juice, 64% in apple juice, 33% in apple jam, 50% in mixed juice and 20% in compote, with average content levels of 158.1 µg/L, 45.7 µg/L, 302 µg/L, 28.5 µg/L and 32.3 µg/L, respectively [64].

Some studies have suggested that PAT is difficult to generate in fresh fruits and vegetables. However, due to the wide range of hosts of *P. expansum*, once *P. expansum* infects the appropriate host, conidia will continue to infect the hosts, resulting in brown spots, more extensive rotten and accompanied by the biosynthesis of PAT [58]. The contamination of PAT was mainly found in apples, pears and their products, which has become an important quality and safety evaluation index of apple juice [59,62]. In recent years, more and more reports have detected PAT occurrence in fresh fruits and vegetables. Although, samples from Belgium and Argentina showed that the proportion of PAT-positive samples was low, and the content of PAT in most samples was lower than 50 µg/L [29,65]. In Pakistan, the positive levels of PAT were relatively high, with the content exceeding the European maximum limit [66]. It was reported that PAT content in the rotten parts of fruits infected by fungi was as the high as 1000 µg/kg, and PAT in the whole fruit batch was 21–746 µg/kg, while PAT in fruit products made from rotten fruits reached 0.79–140 µg/kg [3,67]. However, another study showed that the PAT-positive percentage in fresh tomatoes was 10.8%, but none of the 173 tomato-derived products from different countries tested positive [65]. The obtained levels of PAT in other fruits and vegetables other than apples are list in Table 2.

Table 2. Contamination of patulin in fruits and vegetables.

	Foodstuffs	Detection Method	Country	Positives (%)	PAT (µg/Kg)	Ref.
Fresh fruits	Tomatoes	HPLC-UV	Belgium	10.8	-	[65]
	Sweet bell peppers	HPLC-UV	Belgium	11.4	-	[65]
	Onions	HPLC-UV	Belgium	-	-	[65]
	Apricots	HPLC-UV-VIS	Argentina	4.5	0.7	[29]
	Grapes	HPLC-UV-VIS	Argentina	10	28.3	[29]
	Pears	HPLC-UV-VIS	Argentina	10.7	54	[29]
	Peaches	HPLC-UV-VIS	Argentina	9.7	5	[29]
	Pineapples	HPLC-UV-VIS	Argentina	-	-	[29]
	Oranges	HPLC-UV-VIS	Argentina	50	0.1	[29]
	Seedless grapes	HPLC-UV	Pakistan	70	286.1	[66]
	Red globe grapes	HPLC-UV	Pakistan	75	921.1	[66]
	Flame grapes	HPLC-UV	Pakistan	66.7	190.1	[66]
	Pineapples	HPLC-UV	Pakistan	81.8	254.1	[66]
	Pears	HPLC-UV	Pakistan	66.7	232.1	[66]
	Tomatoes	HPLC-UV	Pakistan	80	410.2	[66]
Dried fruits	Figs	HPLC-UV	China	65	87.6	[68]
	Longans	HPLC-UV	China	90.5	68.4	[68]
	Apricots, dates, plums, peaches and bananas	HPLC-UV	China	8.3	7.4	[68]
	Apricots	HPLC-UV	Iraq	100	0.008–2.84	[69]
	Grapes	HPLC-UV	Iraq	100	0.0198–30.5	[69]
Fruit products	Pear juice	HPLC-UV	Tunisia	47.61%	62.5	[64]
	Pear jams	HPLC-UV	Tunisia	43.75%	123.7	[64]
	Fruit juice	HPLC-UV	Pakistan	58.3	110.3	[66]
	Smoothie of tomatoes, mint and carrots	HPLC-UV	Pakistan	44.4	50.7	[66]
	Smoothie of pineapple and watermelon	HPLC-UV	Pakistan	42.9	60.6	[66]
	Smoothie of oranges, carrots and mint	HPLC-UV	Pakistan	50	110.4	[66]
	Smoothie of banana, mangoes and strawberry	HPLC-UV	Pakistan	50	20.3	[66]
	Hawthorn products	HPLC-UV	China	10	5.1	[68]
	Fruit juice	HPLC-UV	China	15	5.4	[68]
	Fruit jams	HPLC-UV	China	10	5.0	[68]

Note: "-" means no relevant report, and "0" means no detection; HPLC-UV: high-performance liquid chromatography with ultraviolet detector.

A large number of studies about the negative effects of PAT on the human body have been carried out in the past 50 years. The results have indicated that PAT has acute, chronic and cellular-level toxic effects. The symptoms of acute PAT poisoning are agitation, spasms, edema, convulsions, dyspnea, pulmonary congestion, ulceration, hyperemia and distension of the gastrointestinal [70]. In rodents, the LD_{50} of PAT ranged from 0.1 to 295 mg/kg BW/day, but in poultry and monkeys it was 0.1 and 0.5 mg/kg BW/day [60]. In terms of chronic toxic properties, patulin was reported as a mutagen, carcinogen and teratogen, and these effects often accompanied each other [58]. The toxicology of PAT is concentrated on the oxidative damage and the oxidative stress-induced apoptosis via the intrinsic mitochondrial pathway [60,71]. PAT treatment decreased the activities of antioxidant enzymes and increased the level of ROS in mouse kidney [72]. The inhibition of mitosis by PAT is always accompanied by binuclear cell formation, DNA synthesis damage and chromosomal disorder, which are the main causes of PAT genotoxicity [73]. Chromosome aberrations and gene mutations induced by PAT were found in different cells (C3H mouse mammary carcinoma cell line, V79 cells, FM3A cells and mouse lymphoma L5178Y cells) [70]. PAT may disrupt the composition of the gut microbiota and damage the intestinal barrier, and it has been proven to be related to different intestinal diseases, such as endotoxemia, inflammation, intestinal lesions and endotoxemia [60,74].

2.3. Alternaria Toxins (ATs)

Alternaria is a notorious pathogen of fruits and vegetables, which can infect tomatoes, apples, citrus and olives and grows at low temperatures, even at −3 °C during storage and transportation [75]. Since the first Alternaria toxins were isolated from *Alternaria kikuchiana* by Tanaka in 1933, more than 70 ATs have been reported, but only 30 ATs with toxic effects on animals have been reported [76,77]. ATs can be divided into four major categories according to their chemical structure and properties: ① dibenzopyrones and related compounds mainly including alternariol (AOH), alternariol monomethyl ether (AME) and alternanene (ALT); ② tetramic acids mainly including tenuazonic acid (TeA) and its isomer iso-TeA; ③ perylene derivative, mainly including alterotoxin I (ATX-I), alterotoxin II (ATX-II) and alterotoxin III (ATX-III); ④ other structures, mainly including tentoxin (TEN) [78]. TeA, AOH and AME are the most common non-host-specific toxins in fruits and vegetables. ATs have frequently been associated with *Alternaria* isolates, the major mycotoxin-producing species was considered to be *A. alternata* and other *Alternaria* species: *A. arborescens*, *A. blumeae*, *A. tenuissima*, *A. tenuissima*, *A. arborescens*, *A. longipes*, *A. radicina*, *A. dauci*, *A. infectoria*, etc. (Figure 2) [79].

Tomato is an economically important crop that is not only consumed as a fresh crop, but it is also the major ingredient of many processed foods. *Alternaria* causes tomato black spots and always accumulates ATs, resulting in serious economic losses. At present, the percentage of Ats-positive samples in tomato and its products is relatively high, especially that of TeA [80]. As early as 1981, Stinson et al. investigated tomatoes, apples, oranges and lemons infected by *Alternaria* in the U.S. market and found that the main ATs in tomatoes was TeA, with the highest content of 139 mg/kg and positive rate of 57.9%; moreover, small amounts of AOH, AME and ALT were detected [81]. TeA was found in all of the tomato concentrates, 78–100% of tomato sauce samples, 80% of pastes and 50–100% juices, and the maximum concentration of TeA was 100–462 μg/kg. In addition, the detected percentages of AOH, AME and TEN in tomato product samples were 28–86%, 20–78% and 21–64%, respectively [82]. Hickert et al. analyzed nine Alternaria toxins in Germany and found the frequencies of contamination of AOH, AME and TeA in tomato products was 70.6%, 79.4% and 91.2%, and the average amounts were 13.0, 2.5 and 200.0 μg/kg, respectively [83]. In China, it was found that the TeA, AOH, TEN and AME levels in 31 ketchup samples were 10.2–1787 (100%), 2.5–300 (45.2%),1.53–15.8 μg/kg (83.9%) and 0.32–8 μg/kg (90.3%), and only nine tomato juice samples were found without AOH, but TeA, TEN and AME-positive juices were 100.0%, 33.3% and 77.8%, and the concentrations were 7.4–278, 1.85–5.7 and 0.2–5.8 μg/kg [84]. The European Food Safety Authority (EFSA) proposed the threshold for toxicological concern (TTC) method to assess the potential risk of ATs in food to human health according to ATs levels in different foods from 2010–2015 [85]. The upper bound of AOH was 17.1 mg/kg in tomato puree, 17.4 mg/kg in tomato sauce and 17.4 mg/kg sun-dried tomatoes. The high levels of TeA reported were 351 mg/kg in dried tomato soup, 233 mg/kg in sun-dried tomatoes, 212 mg/kg in tomato puree and 54 mg/kg in fresh tomatoes, tomato sauce, tomato ketchup and tomato juice [86].

Fruits, vegetables and their products are rich in nutrients, and they are easily infected by *Alternaria* under suitable conditions. Tournas et al. studied the ability of *A. alternata* to produce AME and AOH at different temperatures, which showed that grapes and apples could produce AME and AOH when stored at a low temperature, and the toxin gradually accumulated with the extension of storage time [87]. The main ATs in apples are AOH and AME, followed by ATX-I and small amounts of ALT and TEA [80,88]. The levels of the five ATs were screened in 11 fresh apples and 7 apple juices, but none were detected [88]. AOH, AME, TeA and ALT were found in different parts of apple, and the proportion of TeA-positive samples was the highest, followed by ALT, AOH and AME. The positive percentages of TeA in the rotten part, junction of disease, and even in the healthy part were

66.35%, 42.31% and 5.77, and those of ALT were 52.88%, 47.17% and 2.88%, respectively. The contamination of AOH and AME was less serious [9]. Some research found the none of AOH and AME tested positive in grape juice, carrot juice and vegetable juice, but AOH and AME in wine were in higher levels [89,90]. The contamination of ATs in fruits, vegetables and their products in recent years is shown in Table 3.

Table 3. Contamination of Alternaria toxins in fruits and vegetables.

	Foodstuffs	Detection Method	Country	AOH		AME		TeA		Ref.
				Positives (%)	AOH (μg/Kg)	Positives (%)	AME (μg/Kg)	Positives (%)	TeA (μg/Kg)	
Fresh fruits	Apples	LC–MS/MS	China	27.88	6.71–8517	16.35	4.97–2623	66.35	36–145276	[9]
	Apples	LC–MS/MS	Netherlands	-	0	-	0	-	0	[88]
	Tomatoes	LC–MS/MS	Netherlands	-	0	-	0	-	0	[88]
	Grapes	UPLC–MS/MS	China	26.8%	0.09–7.15	3.6%	0.11–0.15	28.6%	0.25–46.97	[91]
Dried fruits	Raisins	UPLC–MS/MS	China	5.3	3.5~15.6	19.3	0.3~13.5	35.1	6.9~594.4	[91]
	Dates	UPLC–MS/MS	China	-	-	-	-	34.0	9.6~4411.4	[91]
	Apricots	UPLC–MS/MS	China	-	-	5.4	0.5~2.1	37.5	10.4~1231.8	[91]
	Figs	LC–MS/MS	Netherlands	0	-	0	-	100	25–2345	[88]
	Wolfberries	UPLC–MS/MS	China	3.7	5.9~27.4	7.4	0.2~15.0	64.8	23.8~5665.3	[91]
Fruits products	Apple juice	LC–MS/MS	China	2.44	3.70	4.88	<LOQ	9.76	<LOQ	[9]
	Apple jams	LC–MS/MS	China	23.53	<LOQ-4.4	-	-	-	-	[9]
	Apple vinegar	LC–MS/MS	China	2.94	<LOQ	-	-	2.94	14.5	[9]
	Juice	EIA	Germany	56.5	0.65–16	43.5	0.14–4.9	52.2	21–250	[92]
	Red wine	LC-UV and LC–MS/MS	Canada	83.3	0.03–19.4	83.3	0.01–0.23	-	-	[93]
	Tomato sauce	HPLC–MS/MS	Belgium	86	<LOQ-42	78	<LOQ-3.8	84.3	7.7–330.6	[94]
	Tomato concentrate	HPLC–MS/MS	Belgium	85	<LOQ-31	67	<LOQ-6.1	100	<LOQ-174	[94]
	Tomato juice	HPLC–MS/MS	Belgium	71	<LOQ-7.0	54	<LOQ-3.3	100	3.7–333.1	[94]
	Trockenbeerenauslese	LC–MS/MS	Germany	66.7	1.2–4.9	66.7	0.1–0.3	-	-	[95]
	Vegetable products	LC–MS/MS	Germany	50	2.6–25	60	0.1–5			[95]

Note: "-" means no relevant report, and "0" means not detected; EIA: implementation in enzyme immunoassay.

Most ATs have low acute toxicity. However, some previous research has suggested that the most acutely toxic molecule in the group of ATs is TeA [88,95]. TeA is the only ATs listed in the Registry of Toxic Effects of Chemical Substances by the National Institute for Occupational Safety and Health [96]. In female and male mice, the LD_{50} of TeA was 81 and 186 mg/kg, respectively, and in rats it ranged from 168 to 180 mg/kg [97]. TeA has been recently characterized as responsible for dizziness, salivation and vomiting, followed by tachycardia, massive bleeding of the esophagus and gastrointestinal tract, circulatory failure and motor dysfunction [98]. Based on in vitro toxicological studies using mammalian and bacterial cells, it was found that AME and AOH have genotoxicity, carcinogenicity, cytotoxicity and mutagenicity [99]. There is a synergistic effect between AOH and AME, and the toxic effects of AOH or AME on HeLa cells are weaker than that of their mixture. Fehr et al. believed that the toxic mechanisms of AOH and AME are to inhibit the activity of DNA topoisomerase II-alpha isoform and cause cell DNA helix deformation [100]. A high concentration of AOH involving DNA polymerase β mutation and overexpression of DNA polymerase ß causes genetic instability and tumors [80]. In addition, AOH and AME are also related to protein synthesis inhibition, sphingolipid metabolism disorder and photosynthetic phosphorylation [80,101]. Alternaria toxins in cereal grains have been suggested to have a role in the high levels of human esophageal cancer in South Africa and in the Shanxi province of China. Thus, the TTC value of AME and AOH was set to 2.5 ng/kg of body weight per day, and for TEN and TeA it was 1500 ng/kg per day [102].

2.4. Trichothecenes (TCs)

Trichothecenes are a class of sesquiterpenoid mycotoxins with a double bond at $C_{9,10}$, and an epoxide ring at $C_{12,13}$ [82]. According to their diverse structures, more than 200 different TCs have been subdivided into four basic categories: ① Type A has hydroxyl (-OH) or ester (—COOR) at C-8, such as HT-2 toxin, T-2 toxin, diacetoxyscirpenol (DAS), 15-monoacetoxyscirpenol (MAS) and neosolaniol (NEO). ② Type B has carbonyl (—C=O) at C-8, such as deoxynivalenol (DON), 3-acetyl- deoxynivalenol (3-ADON), 15-acetyl-deoxynivalenol (15-ADON), nivalenol (NIV) and fusarenon X (Fus-X). ③ Type C has a second epoxy group on C-7, C-8, C-9 and C-10, such as baccharin and crotocin. ④ Type D contains a single large ring structure at C-4 and C-15, such as satratoxin and roridin [102]. The most common and toxic groups are type A and B, especially DON, also known as vomitoxin [103]. TCs are produced by different fungal species of Fusarium, Trichoderma, Myrothecium, Trichothecium, Cephalosporium, Verticillium, Verticimonosporium and Stachybotrys [104]. Generally, low temperature, high humidity and acidic pH were conducive to the growth of TCs-producing fungi, and the TCs-producing capability was also enhanced [105].

TCs are not only found in cereals (wheat, barley and corn), they also endanger cash crops (potato, muskmelon and apple), as well as livestock products (meat, milk and eggs). Among fruits and vegetables, TCs are widely tested positive for in the dry rot of potato tubers, fusarium rot of muskmelon, core rot of apples and other fruits and vegetables [106]. Xue et al. detected the presence of 3-ADON, Fus-X, T-2 and DAS in the dry rot of potato tubers inoculated with *F. sulphureum*, *F. sambucinum* and *F. solani* and suggested that TCs exist not only in lesions but also in adjacent asymptomatic tissues, and the concentrations of these four toxins were much higher than the maximum level established by the European Union [107]. The maximum permissible limits of DON were 750 µg/kg in flour and 500 µg/kg in breads. Moreover, EU legislations set the maximum residue limits of ZEN, which was 100–200 µg/kg in unprocessed cereals, 75 µg/kg in processed cereals, 20 µg/kg in processed cereal foods and 50 µg/kg in cereal snacks, accordingly [82]. Zhang et al. detected 20 types of mycotoxins in grapes and wines and found 6% of grapes and 7% of wine were detected to be ZEN-positive with concentrations ranging from 0.29–0.36 µg/Kg to <LOQ-1.85 µg/Kg, respectively [108]. The levels of T-2 toxin in the core rot lesions of apple were in the range 7.1–128.4 µg/kg, and NEO was 14.69 µg/kg [109]. The same research group found that NEO accumulated in muskmelon fruits inoculated with *F. sulphureum*, and postharvest ozone treatment or acetylsalicylic acid treatment effectively reduced the concentration of NEO [110,111]. At present, the research of TCs in fruits and vegetables is still in the initial stage, and there are few relevant reports, but that does not mean the toxins will not accumulate in fruits and vegetables. The contamination of the TCs to fruits and vegetables is worthy of extensive and deep research in the future.

For type A and type B, a ring opening can reduce the toxicity of TCs and turn them into low-toxicity or even non-toxic products [112]. The main toxicity of TCs is strong inhibition of protein and nucleic acid synthesis in the body, which results in immunosuppression in human and animal health [6]. In addition, the toxic effect of TCs on humans and animals was related to animal species, age, dosage and the type of TCs. The oral acute LD_{50} of T-2 for mammals and poultry was in the range of 4–10 mg/kg [113]. After oral intake of T-2 in animal diet, it can be absorbed in the digestive tract, which shows toxic effects on the digestive system, causing damage to digestive tract mucosa and finally reducing the absorption rate of nutrients, showing symptoms such as vomiting, diarrhea and extensive visceral bleeding [114,115]. For humans, as a lipophilic substance T-2 easily penetrates into the skin and causes skin irritation at low doses, cell membrane damage at high doses and apoptosis of lymph glands and hematopoietic cells [116,117]. T-2 with strong cytotoxicity can be quickly absorbed by the digestive system and enters the liver, other organs and muscle tissues 3–4 h later. Human feeding leucocyte deficiency (ATA) has also been found to be related to the consumption of T-2 toxin, because in vivo T-2 toxin can affect human hematopoietic stem cells [118].

The toxicity of DON was lower than that of T-2, but its contamination was more extensive. Consuming DON-containing food will cause headaches, nausea, abdominal pain, anemia and decreased immunity, and long-time consumption will increase the risk of carcinogenesis, teratogenesis, toxic renal damage, reproductive disorder and immunosuppression [119]. The cytotoxicity of DON is mainly manifested by a C12,13-epoxy group, which can act on bone marrow hematopoietic cells, lymphocytes and blymphocytes to produce cellular immunotoxicity and cause programmed cell death. DON can also destroy the structure of ribosome to damage its function and inhibit the active center of peptidyl transferase located on the 60S subunit of ribosome to affect the synthesis of protein in cells [120]. According to the research of long-term consumption of DON-contaminated food, it was shown that DON can affect human hematopoietic stem cells and result in human esophageal cancer [121]. Pigs are highly sensitive to DON, which causes mycosis and porcine leukocyte deficiency in Asian and European countries [122]. When the DON content in pig feed exceeded 1 mg/kg, the feed intake of the pigs decreased significantly and was occasionally accompanied by vomiting. The toxic mechanism of low-dose DON on pigs was mainly linked to the inhibition of immune-related genes, which leads to a series of symptoms [123]. However, cattle, sheep and adult chickens and ducks would eat food with DON, because microorganisms in rumen can quickly convert DON into epoxy-DON, which reduces the toxicity [124].

3. Determination Technique of Mycotoxins in Fruits and Vegetables

The strict maximum content limits of mycotoxin all over the world have greatly promoted the development of mycotoxin detection technology. In addition, the levels of mycotoxins in fruits and vegetables are very low, which puts forward higher technical requirements for the accuracy and sensitivity of the detection. The food matrix has complex components and trace amounts of mycotoxins, so it is difficult to analyze the content of mycotoxins directly. Therefore, except enzyme-linked immunosorbent assay, most detection methods require efficient sample pretreatment techniques for the separation and enrichment of mycotoxins [12]. Generally, for mycotoxins with different chemical structures and matrices, it is necessary to select correct sample pretreatment or clean-up methods. The common extraction methods include liquid–liquid extraction, solid phase extraction (SPE), supercritical fluid extraction (SFE), gel permeation chromatography (GPC) and immunoaffinity cleanup (IAC) [2]. At present, mycotoxin detection methods mainly include chromatographical techniques, immunological techniques and biosensors, which can derive many different technologies according to the pretreatment methods, detection methods and detection principles [125]. The quantitative procedures for mycotoxins have developed rapidly and have different advantages and disadvantages. However, mycotoxins always have different chemical structures and properties, so it is difficult to analyze multi-mycotoxins at the same time, which is a common problem that must be solved [126]. The characteristics of different technologies for mycotoxin determination are shown in Figure 3.

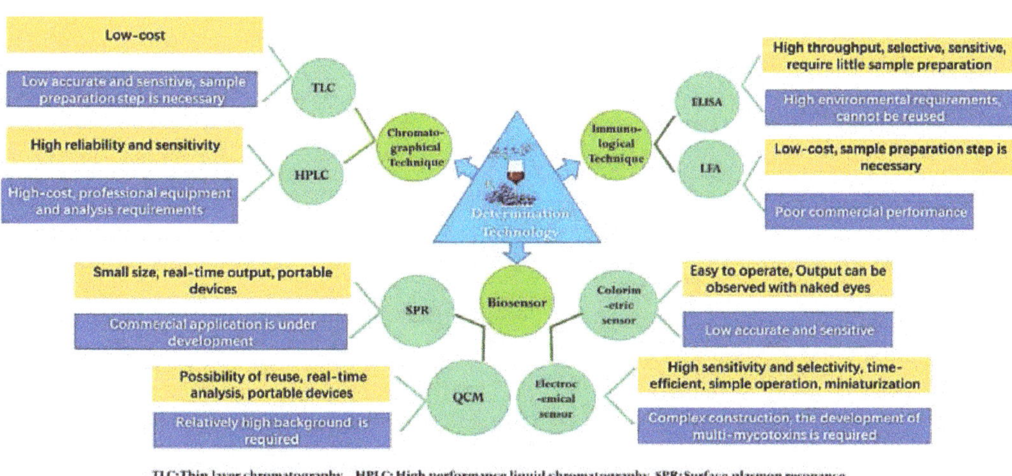

Figure 3. Scheme for mycotoxin determination technologies.

3.1. Chromatographical Technique

Techniques based on chromatography include thin-layer chromatography (TLC), high-performance liquid chromatography (HPLC) and gas chromatography (GC), which are always coupled with ultraviolet (UV), diode array (DAD), fluorescence (FD), diode array (DAD) and mass spectrometry (MS) detector and use immunoaffinity clean-up as pretreatment [127]. HPLC-FD has been utilized for the detection of ZEA, OTA and DON. HPLC-PDA was reported for analysis of ATs, and HPLC-PDA-FD was applied for DON, OTA and ZEA determination [37,64,93]. During the mycotoxin analysis, because chromatography can provide accurate and reliable results, many organizations around the world, such as the International Organization for Standardization (ISO), the Association of Official Analytical Chemists (AOAC) and the European Committee for Standardization (CEN), mostly choose chromatography as the standard method for mycotoxin detection [128–130]. As early as 1993, the ISO issued the detection method standard of practical HPLC-UV for the detection of PAT in apple juice [128]. The CEN set standard analysis methods of PAT in apple products and OTA in grape products and dried fruits. These standards adopt liquid–liquid distribution purification, solid-phase extraction column purification or immunoaffinity column combined with HPLC chromatography. The AOAC also formulated three PAT test standards, including the TLC method and HPLC and IAC-HPLC for OTA determination [131]. Although chromatography is widely accepted by international organizations, they generally have the limits of being time-consuming, needing a professional tester and expensive laboratory equipment (Figure 3).

Liquid chromatography–tandem mass spectrometry (LC–MS/MS) has the advantages of the chromatographic separation and molecular identification of mass spectrometry, which has become the main method for multi-mycotoxins analysis in recent years [80]. Based on isotope dilution, five ATs in tomato-based samples were quantified by LC–MS/MS, with a limit of quantification (LOD) lower than 1.0 µg/kg and a recovery of 52.7–111%. The possible matrix effects were compensated by the corresponding isotopically labeled internal standards [132]. In Belgium, UPLC–MS/MS was applied to quantitate six different ATs in 129 commercial fruit and vegetable juices and tomato products with the preparation and extraction protocol of QuEChERS (quick, easy, cheap, effective, rugged and safe). The quantitation limits (0.7–5.7 µg/kg), repeatability (RSDr < 15.7%), reproducibility (RSDR < 29 17.9%) and recovery (87.0–110.6%) were acceptable [98]. AOH and AFB1 were extracted from fruit juice and fruit juice/milk beverages and then controlled by dispersive liquid–liquid microextraction (DLLME) and determined by HPLC–MS/MS-IT [133]. A rapid solid-phase extraction (SPE) method to the LC–MS/MS system was developed to determine nine mycotoxins in five types of commonly used betel nut products, with an LOQ of 0.25–10 µg/L, an LOD of 5–20 µg/L and recovery in the range of 70.1–113.9% [134]. The interference of the matrix in the multiple detection of mycotoxins has been greatly reduced, and the detection sensitivity and accuracy of LC–MS/MS has been improved, but the detection cost was still high. Developing a fast, low-cost method suitable for on-site detection of mycotoxins is the research hotspot of the LC–MS/MS method in the future.

3.2. Biosensor Technique

3.2.1. Antibody and Aptamer

The biosensor technique has become a low-cost and fast alternative to traditional chromatographic methods in the field of mycotoxin analysis. The biosensor technique refers to the technique of using biological substances (such as antibodies, antigens, aptamers, peptides, cells, etc.) as recognition elements to transform biochemical reactions into quantitative physical and chemical signals through appropriate energy exchangers (electrochemical, piezoelectric, spectral, thermal and surface plasmon resonance) [135]. Biosensors can be divided into four groups based on the physicochemical properties of mycotoxins and the type of transduction: optical sensors (SPR), piezoelectric sensors (QCM), colorimetric and electrochemical sensors [136]. The antibody is identified through an in vivo process, induced by the biological-system-producing immune response to the target molecule in animals, and is obtained by further screening [131]. Therefore, the antibody has high specificity and affinity for the target, and the antibody-based immunochemical assays often have high selectivity, sensitivity and a low detection limit. Up to now, many mycotoxin antibodies have been screened and gradually commercialized, such as AFB1, OTA, DON, ZEN, PAT, etc. [137]. As a chemically synthesized single-stranded nucleic acid, an aptamer capable of binding and highly specific recognition to the target was obtained in vitro by systematic evolution of the ligand by the exponential enrichment process (SELEX) [126]. So far, six mycotoxin aptamers have been screened, which involved OTA, PAT, ZEN, FB1 and T-2 [138], and some sequences of these aptamers are shown in Table 4. Up to now, both antibodies and aptamers have been widely used as mycotoxin recognition elements and constructed the different types of sensors mentioned above.

Table 4. Sequence of mycotoxin aptamers.

Mycotoxin	Sequence	Dissociation Constant (nmol/L)	Reference
OTA	GATCGGGTGTGGGTGGCGTAAAGGGAGCAT-CGGACA	200	[139]
PAT	GGCCCGCCAACCCGCATCATCTACACTGAT ATTTTACCT T	21.83 ± 5.022	[138]
ZEN	CGTGCTACCGTGAAATACCAGCTTATTCAATTC-TACCAGCTTTG AGGCTCGATCCAGCT-TATTCAATTATACCAGCTTATTCAATTAT-ACCAGCACAATCGTAATCAGTTAG	15.2 ± 3.4	[138]
FB1	ATACCAGCTTATTCAATTAATCGCATTACCTT-ATACCAGCTTATTCAATTACGTCTGCACATAC-CAGCTTATTCAATTAGATAGTAAGTGCAATCT	100 ± 30	[139]
T-2	GTATATCAAGCATCGCGTGTTTACACATGCGA-GAGGTGAA	20.8 ± 3.1	[140]

3.2.2. SPR, QCM and Colorimetric Sensor

A surface plasmon resonance sensor is always applied to characterize and quantify the biomolecular interactions. A surface plasmon is excited in the refractive index at an interface between media of different refractive indices, and the changes are measured. That means that the changes in the mass concentration of material on the surface are crucial to the sensitivity of SPR sensors [126,141]. Zhu et al. reported that an SPR biosensor involved an anti-OTA aptamer by immobilized a sensor chip, and that was used to quantitatively analyze OTA with a lower detection limit of 0.005 μg/L, and a recovery of 86.5–116.5% in red wine. The principle is that the biotin–aptamer can be captured through streptavidin–biotin interaction [141]. A multi-mycotoxins detection assay based on an SPR instrument and two biosensor chips (one for DON, ZEN and T-2 toxin and the other for OTA, FUMB1 and AFB1) was developed, which was modified by ovalbumin (OVA) conjugates of mycotoxins. The SPR response was recorded when antibodies mixed with samples were added to a chip. The LODs of DON, ZEN, T-2, OTA, FUMB1 and AFB1 were 26, 6, 0.6, 3, 2 and 0.6 μg/kg, respectively [126]. A competitive inhibition iSPR biosensor was prepared through the competition between the immobilized mycotoxins on a nanostructured iSPR chip and the free mycotoxins in the solution. The LODs were 17 μg/L for DON and 7 μg/L for OTA [142]. Although the SPR biosensor has been widely studied in academic research, in commercial application the professional requirements of technology and data analysis should be optimized, and cheaper labeling reagents should be prepared and applied to cut costs.

The quartz crystal microbalance is a biosensor platform that has the component of a thin quartz crystal disk in between two gold electrodes, one of which is typically coated with an antigen conjugate to sense mycotoxins [143]. A label-free QCM-based immunosensor was developed by direct immobilization of OTA to amine-bearing sensor surfaces using 1-ethyl-3-(3-dimethylaminopropyl) carbodiimide (EDC)/N-hydroxysuccinimide (NHS) chemistry, which can measure OTA in red wine with a range of 17.2–200 µg/L [144]. Karczmarczyk et al. applied a secondary antibody conjugated to gold nanoparticles to amplify the signal of a QCM-based sensor, which has a lower LOD of 0.16 µg/L for OTA determination in red wine [145]. However, commercialization of a QCM biosensor is limited by the thickness of the quartz plates, but thinner materials with higher mass sensitivities are quite expensive.

Colorimetric sensors based on aptamer have simplicity, rapidity, lower cost and are more suitable for on-site detection. A T-2 colorimetric aptasensor was developed based on the aggregation of gold nanoparticles (AuNPs) in the presence of T-2. When an aptamer bound with T-2 would release AuNPs and make the color change from red to purple blue, the LOD of this method was 57.8 pg/mL [146]. Yang et al. also used the stability against the salt-induced aggregation effect of AuNPs in the presence of OTA's aptamer to build a colorimetric sensor and proved that the aptamers form G-quadruplexes after binding OTA [147]. He et al. proposed an innovative colorimetric method based on enzyme-induced gold nanoparticle aggregation, which can monitor OTA concentrations in red wine and grape juice. OTA aptamer, biotinylated cDNA and SA–ALP were modified on DNA-ALP-immobilized MBs. In the presence of OTA, the binding of OTA with aptamers switched the aptamer structure from an aptamer–cDNA duplex to an aptamer–OTA complex, and then ALP-cDNA was liberated and converted AAP to AA. MnO_2 was reduced by AA and produced Mn^{2+}, which leads to aggregation of AuNPs, and a color change from brownish red to blue was detected [148]. Due to the dependence of the chemical properties of mycotoxin, colorimetric sensing is vulnerable to the interference of the environment, which affects the accuracy of the detection results. By improving the stability of gold nanoparticles, the accuracy of the detection method would be increased based on the color change of its surface plasma corresponding to the spectral shift.

3.2.3. Electrochemical Sensor

Electrochemical sensing can be divided into label and label-free approaches, most of which apply cyclic voltammetry (CV), differential pulse voltammetry (DPV), square wave voltammetry (SWV) and electrochemical impedance spectroscopy (EIS) to monitor the change of current, circuit potential or resistance of the detection system [86]. There are many kinds of substances that can be used to label aptamer sensors, such as radioactive or fluorescent dyes, metal complexes, quantum dots, nanoparticles and enzymes. This label attached to the target molecule, aptamer or antibody always has its own common feature of easy conjugation and detection, but the labeling and immobilization steps are time-consuming and expensive [149]. Some researchers have developed an aptamer sensor for detecting OTA under flow conditions. An automated flow-based aptasensor was utilized for OTA detection with a flow injection system with a detection limit of 0.05 µg/L. Biotin-labeled OTA and free OTA compete to bind the aptamer on the surface of the printed electrode in the central flow cell, which reduces the analysis time [150]. Abnous et al. modified methylene blue on the double-stranded DNA formed by the OTA aptamer and its cDNA. The combination between aptamer and OTA prohibited the binding of cDNA and the aptamer and the release of MB accompanied by the reduction of the signal. The LOD of OTA in grape juice was 58 pM [151]. Based on the good electrical conductivity and large specific surface area of thionine (Thi)-labeled Fe_3O_4 nanoparticles (Fe_3O_4NPs)/rGO and rigid structure of tetrahedral DNA nanostructures (TDNs), an electrochemical aptasensor for PAT was constructed to analyze PAT in apple juice. Once PAT bound with the aptamer, the aptamer was released from the TDNs-Apt, and Fe_3O_4NPs/rGO with Thi was introduced to the electrode surface, resulting in significant Thi signal changes. The proposed aptasensor showed an excellent LOD of 30.4 fg/mL and a recovery in apple juice of 96.9–105.8% [152]. According to most studies of labeled electrochemical sensors, it could be found that voltammetry was the common approach to detect mycotoxins by measuring the variation of the system current.

The label-free electrochemical sensor can directly measure the electroactive substances or the current generated by the oxidation and reduction of probe molecules, which corresponds to the concentration of analyte in the sample. The detection signal of the label-free aptamer sensor is directly related to the interaction between the target and the aptamer, and it can avoid false-positive results and difficulty in the labeling process [149]. A label-free sensor was developed based on anti-PAT-BSA IgG, which was immobilized on the graphene oxide/gold nanocomposite-coated glass carbon electrode (GCE). The combination between anti-PAT-BSA IgG and PAT reduced the spatial hindrance effect of the sensor and resulted in a decrease in the electron transfer resistance. The current changes exhibited a linear relationship with the patulin concentration with an LOD of 5 g/L [153]. Xu et al. modified GCE with black phosphorus nanosheets (BP NSs) and an electrostatic-adsorption-modified PAT aptamer on the surface. The same research also found a wider linear range and LODs could be obtained by modifying AuNPs and thiolated PAT aptamer on the BP NS-GCE [154]. AuNPs and thiolated OTA aptamer were immobilized on the surface of the gold electrode through layer-by-layer self-assembly. This label-free impedimetric aptasensor was applied to quantify the OTA concentration in grapes and their products. The OTA aptamer changed its random coil structure into the antiparallel G-quadruplex conformation in presence of OTA, because the electrostatic and steric repulsion the diffusion of $[\mathrm{Fe(CN)_6}]^{3-}$ into the surface of the electrode was hindered, and the relative normalized electron-transfer resistance values were proportional to the concentration of OTA [155]. In a further study, the same team proved that AuNPs can effectively improve the specific gold surface area and fix more aptamers. When it comes to regaining of the aptasensor, it was found that the modification density of thiolated aptamer was maintained relatively constantly while that of the amino aptamer decreased [156]. Because the sensitivity improvement of the label-free electrochemical mostly depends on the aptamer modification levels, it is extremely significant to screen and prepare novel electrode modified materials that can increase the specific surface area, especially novel nanomaterials. Different from the labeled sensor, EIS is more suitable for monitoring the resistance changes caused by the aptamer structural transformation modified on the electrode surface. In order to promote the development of electrochemical sensors for mycotoxins, the sample pretreatment process and immobilization of molecular recognition elements should be further simplified, and simultaneous detection technologies for multi-mycotoxins analysis need to be further developed.

3.3. Immunological Technique

As an important rapid quantification detection technique, immunochemical assays include major examples such enzyme-linked immunosorbent assay (ELISA), dipsticks, flow-through membranes and LFDs. Antibodies and antigens were very effective immunochemical tools that were wildly applied in the mycotoxin detection filed [86]. For example, the principle of ELISA is the interaction of the antigen–antibody complex with the chromogenic substrates and measuring the developed color by spectrophotometry. There are many enzymes applied to replace the immunoassay in combination with a substrate, among them horse radish peroxidase (HRP) and alkaline phosphatase (ALP) are widely used [157]. ELISA has been developed for ZEN, AF's, OTA, T-2 and FUM monitoring with a short incubation period time of 15 min, but most are between 1 and 2 h [152]. Wang et al. designed a hapten of TeA mycotoxin by computer-assisted modeling to produce the highly specific camel polyclonal antibody and established an indirect competitive chemiluminescence enzyme immunoassay (icCLEIA) that has high sensitivity, high specificity and acceptable recoveries for TeA detection in fruit juices [158]. Pei et al. presented a pELISA measurement of OTA based on the competing antigen of OTA-labeled urease, which can hydrolyze urea into ammonia. The ammonia molecules increased the pH, so silver ions were reduced by the formyl group from glucose to generate a silver shell and resulted in the changing of the solution color from blue to brownish red. The LOD of OTA was 40 pg/mL, 15.6- and 14.3-folds lower than those of HRP based ELISA [159]. A sensitive and novel multiplex nano-array based on ELISA has been developed allowing for the simultaneous semi-quantitative analysis of ZEA, T-2 and FUM in around 70 min due to the cross-reactivity profiles of the antibodies used. The conjugates of three mycotoxins were nano-spotted into the single wells of a microplate, and a calibration curve for the concentration was analyzed by a competitive assay format [160]. For ELISA, because this technology mostly depends on the biological activity of antibodies and antigens, they are very sensitive to temperature and pH variations and undergo irreversible denaturation. The application of regenerable and stable cognitive elements as a substitute for antibodies is an important means to improve the stability and feasibility of this technology.

Lateral flow immunoassays (LFA) take the cellulose or glass fiber as the solid phase and move the sample smoothly and homogenously on the chromatographic material with the help of capillarity. Antibody-labeling tag conjugates are usually used as the labeled reagents that could interact with the pathogens in the moving liquid sample. Visual detection results can be obtained in a short time through direct color markers or enzymatic color reaction [138,161]. LFA can monitor the detection results by the naked eye, which has become an excellent on-site detection technology. The labeling tags mainly include quantum dots, AuNPs and luminescent nanoparticles [143]. OTA in grape juice and wine was determined by a silver nanoparticle-based fluorescence-quenching lateral flow immunoassay with a competitive format (cLFIA). As background fluorescence signals, Ru(phen)32+-doped silica nanoparticles (RuNPs) were sprayed on the test and control line zones, and its fluorescence quenchers were AgNPs. The proposed method exhibited an acceptable LOD of 0.06 g/L and average recoveries of 88.0–110.0% in red grape wine and 92.0–110.0% in grape juice [162]. A multiplex immunochromatographic test strip based on 25 nm AuNPs was designed for simultaneous detection of FB1, DON and ZEN in 15 min. The different mycotoxin conjugates (FB1–BSA, DON–BSA and ZEN–BSA) and goat anti-mouse IgG were separately spotted onto the NC membranes as three T lines, which dried together with antibody–AuNP conjugates sprayed on the conjugate pad. In the presence of the target mycotoxins, a part of the antibody–AuNP conjugates will react with the respective antigens in the sample pad to competitively inhibit the interaction of the T line and the resulting change of color [163]. Epoxy-functionalized silica-coated QDs showing green, orange and red color were conjugated with anti-ZEN, anti-DON and anti-T-2 mAb for multiplex lateral flow immunoassay (LFIA) preparation. The LFIA can detect ZEN, DON and T-2 toxin in 15 min, and the false-negative rate is lower than 5% [164]. Although the sensitivity and accuracy of LFA have been improved, reducing nonspecific adsorption and avoiding the interference of background fluorescent substances still limit the development of this technology. Research on the modification and immobilization technology of antigens and antibodies on the surface of a microfluidic chip can greatly improve the sensitivity and repeatability of detection.

4. Degradation Technique of Mycotoxins in Fruits and Vegetables

Patulin, ochratoxin, Alternaria toxins and trichothecenes are the major mycotoxins in fruits and vegetables, whose contamination is inevitable. These mycotoxins are resistant to heating, and the common heating operation is difficult to remove or degrade. Therefore, developing an efficient removal strategy is a challenge for researchers. Mycotoxin removal involves adsorption and degradation in terms of mechanisms. Adsorption treatment is to remove mycotoxin by applying an absorbent with a large specific surface area. Degradation treatment is to remove mycotoxin by some strategies to destroy the chemical structure of the mycotoxins, especially the toxic group in the chemical structure. Physical, chemical and biological methods are the main strategies to degrade mycotoxins in fruits and vegetables.

4.1. Physical Method

4.1.1. Physical Adsorption

Absorption treatment is based on adsorbents with a tremendous specific surface area, which is a widely used physical strategy to remove mycotoxins. Among the many commercial adsorbents used for removal of mycotoxins, activated carbon, macroporous resin and diatomite are the commonly used physical adsorbents. Activated carbon treatment significantly reduces patulin content in apple juice, and more importantly, no significant changes to the nutrition and quality of the apple juice, such as soluble solids content, reducing sugar and total acid, were observed [165]. The composite carbon adsorbent (CCA) can be prepared by absorbing ultrafine activated carbon particles onto granular quartz and packing them in a fixed bed adsorption column that has excellent adsorptive properties such as a huge carbonaceous surface area, excellent bed porosity and higher bulk density, which could efficiently enhance the adsorption rate for patulin in both aqueous solutions and apple juice; however, the appearance and flavor of the apple juice were affected [166]. In fact, some porous chemicals can achieve the same effect as activated carbon. For instance, the contaminated apple juice was treated with 0.5% polyvinylpolypyrrolidone (PVPP) or 0.5% beta-cyclodextrin (β-CD) for 24 h, which not only reduced the patulin level in the apple juice but also decreased the total phenols content, avoiding browning of the apple juice. Moreover, activated carbon is also applied to absorb ochratoxin (OTA) and deoxynivalenol (DON); nevertheless, when activated carbon absorb the two mycotoxins, some nutrient substances were simultaneously absorbed, which markedly influences the quality and nutritional value [167].

Macroporous resin is widely applied to absorb patulin in apple juice, and macroporous resin can be divided into different types according to their source and pore size. Liu et al. compared eight types of macroporous resin of LSA-900B, XDA-600, LS-803, LS-806, LSF-500, HPD-850, DM-2 and DM-3 for adsorption of patulin from apple juice and suggested that LSA-900B indicated the highest absorbent effect (92.55% at 50 °C); more importantly, the quality parameters such as absorbance, color value, light transmittance and turbidity were remarkably improved [168]. XDA-600 macroporous resin also displayed a higher adsorption efficiency for patulin in pear juice at 25 °C, in which the adsorption behavior is suited to the adsorption thermodynamic Freundlich isotherm and line with the quasi-secondary rate equation [169].

Attapulgite is regarded as a green material in the 21st century due to the advantages of extensive resources, low cost, nontoxic nature, environmental friendliness and unique three-dimensional spatial and large specific surface area, which gives attapulgite an unusual adsorption property. Liu et al. screened the optimal adsorption condition, which was 0.0800 g of attapulgite at 40 °C for 22 h based on the Box–Behnken experimental design [170]. Zhang et al. suggested that the attapulgite adsorption process follows the Freundlich isotherms model and the pseudo first-order kinetics model [171]. In fact, the ion exchange and chelation processes were associated with attapulgite adsorption for patulin, which leads to a slight decrease in the contents of total acid, reducing sugar, viscosity, soluble solids and total phenols. The effort attempted to overcome the influence on juice quality by modification and developed a magnetic attapulgite (Fe_3O_4@ATP) that was prepared by precipitation and the spreading of Fe_3O_4 nanoparticles on attapulgite (ATP). Then, the magnetic adsorbent was used to remove mycotoxins from contaminated peanut oil, and the results suggested that the magnetic adsorbent of Fe_3O_4@ATP displayed an excellent removal efficiency of 86.82% for mycotoxin contamination [172].

A microporous ceramic membrane has been employed for the removal toxic or harmful contamination. However, the removal efficiency is not always satisfying. In order to improve the adsorption efficiency, Nan et al. developed a nano-MgO-modified diatomite ceramic membrane (MCM) with a high positive charge and successfully applied it to remove OTA in wine (Figure 4). The main adsorption principle is that the nano-MgO MCM displays a positive charge from pH 2 to 12.8, after dispersing nano-MgO particles on the surface of the ceramic membrane, and OTA is negatively charged. Therefore, it is taking advantage of the electrostatic adsorption between the positively charged nano-MgO MCM and the negatively charged OTA to effectively remove OTA in wine. At the same time, the isotherm adsorption was verified to suit to Langmuir model, with a maximum adsorption capacity of 806 ng/g at 25 °C [173]. More importantly, no significant changes were observed on the quality and flavor of the wines.

Figure 4. Schematic of the removal device with a nano-MgO/diatomite ceramic membrane.

Molecular imprinting technology (MIT) has gained more and more attention from researchers due to its broad range of potential application, especially in controlling contamination. Molecularly imprinted polymers (MIP) are a kind of cross-linked polymer, which can bind the target compound with high specificity. MIP could be synthesized with specific recognition sites complementary in shape, size and functional group to template molecules [174]. Amino groups could be introduced onto the silica surface with 3-aminopropyltriethoxysilane, and then azo initiation onto the silica surface was achieved by the reaction of the surface amino groups with 4,4'-azobis(4-cyanopentanoic acid). The synthesized MIP was then performed in the presence of 6-hydroxynicotinic acid as template substitute, functional and cross-linking monomers, and the result showed that the synthesized MIP had specific adsorption qualities for patulin, with a 93.97% adsorption rate and a 0.654 µg/mg adsorption capacity [175]. Similarly, Sun et al. achieved a more effective adsorbent for patulin, Fe_3O_4@SiO_2@CS-GO@MIP, and during the synthesis process, Fe_3O_4 was added into the reaction to make the MIP adsorbent magnetic. Finally, chitosan (CS) and SiO_2 were introduced to improve the biocompatibility, stability and dispersibility of the coated MIP adsorbent, and the results indicated that the proposed adsorbent had an efficient adsorption for patulin in apple juice, with an adsorption capacity of 7.11 mg/g and an adsorption rate of 90% after 24 h of adsorption [176].

4.1.2. Physical Removal and Detoxification

Most mycotoxins are heat-stable and form toxic degradation products. Although several detoxification methods have been developed, only a few have been accepted for practical use. The common treatments for removal or detoxification of mycotoxins include heating, microwaving and irradiation, etc. These treatments can inactivate or degrade mycotoxins by destroying the structure of the mycotoxins.

(1) Heating

In general, mycotoxins are relatively stable during heating, and the usual boiling and autoclaving treatments do not easily destroy the structure [177]. The degree of mycotoxin degradation by heating processes largely depends on factors such as temperature, moisture content and time period. OTA, the most widespread contaminant worldwide, can lead to human carcinogen with its potent nephrotoxicity, and a 90% OTA reduction was achieved when heated at 200–250 °C for a longer treatment period [178]. Alternaria toxins mainly include AOH, AME and TeA, and only a few AOH and AME were degraded when heating at 100 °C for 90 min, while the higher degradation efficiency of AOH and AME could be achieved when heating at 121 °C for 60 min. Similarly, the TeA concentration reduced by 50% after 90 min of heating [179]. Trichothecenes are usually stable at 120 °C; however, trichothecenes will partly decompose when the temperature is above 200 °C [180]. However, overall, heating treatment not only needs more energy but also leads to a lot of nutrient destruction. Therefore, heating treatment is difficult to widely apply to remove and degrade mycotoxins in fruits and vegetables.

(2)   Microwave

Microwaves, electromagnetic waves with a certain frequency of 103–105 MHz, can come into contact with contaminants to cause a molecular dipole swing, producing high frequency fluctuation to cause a heat effect. A medium heat microwave treatment of patulin (concentration ranges from 100 to1000 µg/L) for 90 s results in an almost complete degradation of 100%. Microwave-treated patulin in apricot juice for 3 min to leads to a reduction of 39%; however, the degradation efficiency reached 95% after 15 min of microwave treatment, and more importantly, no significant changes to nutrition and flavor were found [181]. Microwave treatment is more suitable for the acidic conditions of apple juice because it can better preserve the critical nutrients of the juice. Therefore, microwaving can be widely employed to sterilize and condense apple juice.

(3)   Irradiation

Irradiation contains ionic types (such as, X-rays, γ-rays and ultraviolet light) and nonionic types (such as radio waves, microwaves, infrared radiation and visible light). For ionic irradiation, the structure of mycotoxin will be destroyed even treated at a lower temperature, which is severe harmful for humans. For nonionic irradiation, only when the sample is exposed for a long time to cause an increase in temperature can the mycotoxin structure be destroyed; however, nonionic irradiation will not pose harm or threat to humans. Most mycotoxins can be decomposed into nontoxic or less toxic middle products after exposure to nonionic irradiation. Ionic irradiation displays better effects on liquid food than solid food, and the reason may be that it has weak penetration with solid food. The application of γ-rays significantly reduced mycotoxin contamination in juice due to its high efficiency, quickness and there being no secondary pollution. γ-ray exposure significantly reduced the patulin content in citrus juice, and with the increase of irradiation dose, the degradation effect was improved [182]. When the irradiation dose was 8 kGy, the degradation rate of patulin in apple juice could reach 100%. γ-ray treatment also destroys the OTA structure, when the exposure dose was 6 kGy, and the degradation rate of OTA was 96.2% [183]. Similarly, ^{60}Co-γ irradiation has excellent degradation efficiency on TeA, and a lower dose irradiation has more influence on TeA than a higher dose irradiation; more importantly, no significant influence was found on the juice quality such as the amino acids, reducing sugars and titratable acids. However, the chromatic value and clarity of juice were improved [184]. The Food and Drug Administration (FDA) regulations state that γ-ray treatment is safe for humans when the dose is less than 10 kGy. Therefore, γ irradiation treatment is an effective method to remove and degrade mycotoxin from vegetables and fruits.

UV irradiation has been widely used as an effective approach to remove mycotoxins because of its lower secondary pollution and lesser influence on degradation production. UV irradiation showed remarkable degradation efficiency on patulin from apple juice. When the initial concentration of patulin was 500 µg/L, it was almost completely degraded after 90 min of UV-C irradiation of 255 nm at 25 °C, pH 7.0; the same degradation effect could be achieved after 90 min of irradiation when the temperature is increased to 65 °C. However, when the pH was 4.0, the patulin degradation rate reached 100% after 40 min of irradiation [185]. Moreover, different wavelengths of UV-C have distinct degradation efficiencies, which is following that 222 nm > 282 nm > 254 nm. After the treatment of UV-C for 222 nm, the L*, a* and b* values were reduced, as well as the content of vitamin C, and the structure of OTA was broken; similarly, environmental temperature and pH significantly influence the UV degradation of OTA. High temperature accelerates OTA degradation, for instance, and UV-C (255 nm) exposure for 6.5 min at 45 °C achieves the same effect as 12 min at 15 °C [186]. In addition, UV irradiation also has the significant degradation effect on trichothecenes, and the factors such as irradiation time, distance and medium pH significantly influence trichothecene degradation efficiency. The degradation of DON and T-2 has a positive relationship with irradiation time and a negative relationship with irradiation distance, and the degradation efficiency reduces as the medium pH increases [187].

At present, irradiation technology does not have widespread application because of the acceptance by consumers. However, with the development of irradiation technology and the management system of food security is extensively applied in the food industry, as well as the further study of the irradiation degradation of mycotoxins, and the application of irradiation technology to remove mycotoxin will be more widespread.

4.2. Chemical Degradation

Treatment with chemicals can efficiently remove mycotoxins from fruits and vegetables and their derivative products by using strong oxidant, acid, base and other chemical substances, which decompose mycotoxins by chemical reactions to destroy the structure of the mycotoxins. Ozone, one kind of strong oxidant, has the ability to oxidize the double bond of mycotoxin structure and is widely applied to degrade patulin, OTA and trichothecene. For instance, the patulin was completely degraded after 7–12 mg/L of ozone treatment for 30 min the ozone almost completely degraded the patulin content from 50–500 µg/L in apple juice and the degradation effect is increasing with the extension of the processing time. Ozone treatment had no significant effects on the juice quality of soluble solids, pH and total acid; however, it obviously decreased the color, malic acid, ascorbic acid and total phenols of the apple juice [188]. OTA was effectively degraded after 50 mg/L of ozone combination electron beam irradiation treatment for 180 s [189]. Ozone treatment also reduced neosolaniol (NEO) content (one kind of trichothecene) from muskmelon fruits. When 1.10 mg/L was ozone-treated for 120 min, the content of NEO from fusarium rot of muskmelon fruit was significantly decreased. The possible involved mechanism is that ozone firstly attacks the C9,10 double bond in the NEO structure by electrophilic addition, leading to the change in oxidation state for the C9,10 double bond, then the remaining NEO are left intact at this stage of the reaction, and the two expected keto aldehydes are generated by isomerization at C11 after formation of the aldehyde at the site of C10 [111]. During the ozone processing of the degradation of mycotoxin from juice, few effects on pH, vitamin C, soluble solids content and color value were found; therefore, ozone is widely used in the food processing industry to remove patulin from juice and improve the safety of juice, which is recognized as an effective, safe and cheap method to remove mycotoxins from fruits and vegetables and their juice products.

Glow discharge plasma (GDP) is a novel electrochemical technology in which plasma is sustained by DC glow discharges between a pointed electrode and surface of liquid electrolyte [190]. The possible mechanism is various active substances, such as hydrogen peroxide (H_2O_2) and hydroxyl radicals ($\cdot OH$), that can be generated during electrode discharge and diffuse into the media solution to oxidize the organic molecules. GDP completely destroys organic molecules and produces CO_2 and H_2O. Therefore, GDP is widely used to degrade organic contaminants. Pu et al. developed a GDP reactor and firstly used it to degrade patulin in juice and T-2 toxins in water. When the DC voltage was 550 V and the current range was from 145 to155 mA, GDP effectively degraded patulin by 96.63% in 5 min, and no significant effect on the quality and nutrients of the apple juice were observed. The same degradation efficiency was shown in T-2 toxins after 40 min of GDP treatment, and cytotoxicity tests have shown no toxic effect [191].

In addition, ascorbic acid (vitamin C), sulfur dioxide, thiamine (vitamin B1), vitamin B6 and calcium pantothenate were efficiently employed for degrading mycotoxins. For instance, 482 mg/L of ascorbic acid reduced the patulin content to 30% after 34 d of incubation [192]. Acetic acid, propionic acid and other fatty acids with low carbon chains treated the juice contaminated with OTA, and the OTA content was markedly decreased. The treatment of OTA with sodium hypochlorite and formaldehyde significantly destroyed the OTA structure to obtain a product with less toxicity.

Treatment with chemicals, to some extent, effectively degrades and removes mycotoxins from fruits and vegetables and their derivative products; however, rather than complete degradation, there is formation of the degradation products with toxic residues, which is likely to change the nutritional ingredients of the juice. Therefore, chemical methods are somewhat restricted by juice manufacturers.

4.3. Biological Degradation

Biodegradation is considered as one of the promising and reliable approaches for control of mycotoxin contamination. The involved mechanism is to destroy the chemical structure of the mycotoxin by microbiology, or the enzymes secreted by microbiology reacting with the mycotoxin. At present, biological degradation has become a hotspot issue worldwide because of its minimal side effects and it being environmentally friendly. Biological degradation mainly includes bacteria, yeast and enzymes secreted by fungi.

Acinetobacter spp., *Lactobacillus* spp., *Bifidobacterium* spp. and *Streptococcus lactis* are the main bacterial strains for removing mycotoxin. Treatment of OTA with *Acinetobacter* spp. was also found to reduce OTA by 80% for 24 h, and OTα was detected as the degradation byproduct of OTA; however, inactivated *Acinetobacter* spp. had no detoxification ability, the involved mechanism was responsible for enzymatic biodegradation and the degradation efficiency decreased with the descent of the OTA concentration [193]. *Lactobacillus*, as one kind of important gastrointestinal probiotic, plays an indispensable role in the detoxification of mycotoxin from fruits and vegetables [194]. Some strains of lactic acid bacteria such as *Lactobacillus* spp., *Bifidobacterium* spp. and *Streptococcus lactis* were proved to detoxify patulin. Among them, *Bifidobacterium* VM12 displayed the best degradation effect; moreover, a HepG2 toxicity test demonstrated that *Bifidobacterium* VM12 greatly reduced the toxicity of patulin [195]. *L. acidophilus* has also shown a higher removal efficiency on PAT and OTA, and the factors such as the initial concentration of mycotoxin, pH, cell density and strain growth significantly influence the detoxification efficiency of *L. acidophilus* [196]. Some progress on trichothecenes detoxification was made by bacterial strains; the strain of commercialized *A. Eubacterium* (BBSH 797) could deactivate trichothecenes by reducing of the epoxide ring [197]. Treatment of DON with a strain of BBSH 797 produced a de-epoxy DON of DOM-1, whose IC_{50} value was 54-fold lower than that of DON, while the IC50 of de-epoxy NIV is 55-fold lower than that of NIV [198].

In addition, the bacteria strains of LS100 and SS3 that were isolated from chicken intestinal microbes have the ability to degrade 12 trichothecenes, and de-acylation and de-epoxidation are the main degradation pathways [199]. Furthermore, the hydroxylation and carbonylation of molecular structures markedly influence the toxicity of trichothecenes; for instance, the difference between NIV and DON is that NIV has C4-OH, which makes the toxicity of NIV 10 times higher than that of DON [200]. *Baccharis* spp. transformed T-2 toxin, in presence of NADPH, into HT-2 toxin with C3′-OH through a hydroxylation reaction, which mainly occurs in liver microsomes [201].

The yeast strains of *Pichia ohmeri*, *Saccharomyces cerevisiae*, *Rhodotorula* spp. and *Paecilomyces* spp. have shown the capacity to degrade mycotoxins. *S. cerevisiae* can degrade patulin from apple juice under oxygen-free conditions. *R. paludigenum* treatment completely degraded patulin in vitro at 30 °C and pH at 6.0 for 48 h, and the involved action mechanism was mainly attributed to enzymolysis [202]. Li et al. reported that patulin was degraded into desoxypatulinic acid by *R. mucilaginosa* JM19; in addition, the capacity of *R. mucilaginosa* JM19 to degrade patulin was strongly dependent on temperature, cell density and initial patulin concentration [203]. Tang et al. suggested the main action mechanism is that *R. mucilaginosa* secreted a kind of orotate phosphoribosyltransferase (EC: 2.4.2.10) in apple juice during the degradation process, which plays an important role in the degradation of patulin [204]. Xing et al. cloned a short-chain dehydrogenase/reductase (SDR) gene of CgSDR from the yeast strain of *Candida guilliermondii*, which was expressed in *Escherichia coli*, and the recombination expression of CgSDR conferred a strong tolerance and degradation capacity on patulin. The purified CgSDR transformed patulin into

E-ascladiol with the help of the NADPH of a coenzyme. When 150 µg/mL of CgSDR was added into the apple juice contaminated with patulin, the degradation efficiency reached 80%; more importantly, the quality of the apple juice did not decrease during the treatment process [205].

Zhang et al. screened the optimum conditions of *Yarrowia lipolytica* degradation of OTA, which were 28 °C, pH 4.0, 108 spores/mL of yeast concentration and the degradation rate reaching 95.7% in vitro. In vivo, the same concentration of *Y. lipolytica* directly reduced 90% of the OTA content in the lesion part of infected table grapes [206]. The possible mechanism of OTA elimination by *Yarrowia lipolytica* Y-2 was analyzed by Zhang et al., who indicated that OTA degradation by *Y. lipolytica* was attributed to the action of intracellular enzymes, in which two kinds of carboxypeptidase proteins were expressed and reflected the hydrolysis activity of carboxypeptidase [207]. *Hanseniaspora uvarum* could also significantly and efficiently degrade OTA [208].

Porcine pancreatic lipase, a kind of hydrolases, can efficiently catalyze a variety of biological reactions. Tang et al. suggested that calcium carbonate immobilized porcine pancreatic lipase can be applied as an alternative biological strategy for patulin detoxification in apple juice; the optimal degradation condition is 40 °C, 18 h and calcium carbonate immobilized porcine pancreatic lipase of 0.03 g/mL [209]. Moreover, Liu et al. inferred the degradable product of patulin after the fermentation of porcine pancreatic lipase and patulin and indicated that the molecular weight is 159.0594, named $C_7H_{11}O_4$ [210].

5. Conclusions

The diseases of fruits and vegetables caused by fungal infections are widely distributed, which has attracted great attention all over the world. Mycotoxins produced by fungi in fruits and vegetables mainly include ochratoxin A, patulin, Alternaria toxins and trichothecenes, and most of these have the potential to threaten human and animal health. As one of the most important dietary sources for human beings, it is essential to ensure the safety of fruits and vegetables. At present, in only a few varieties of fruits and vegetables have mycotoxins been systematically studied, especially for trichothecenes, which need further research. Accurate and rapid detection technology, risk assessment and degradation technology have become hot issues for the future. In addition to traditional methods that require large-scale instruments, biosensors and immunological techniques have also made a positive improvement and achieved good results. However, there are still some problems to be improved in the detection of mycotoxins. In fruits and vegetables, different mycotoxins often coexist, and it is necessary to develop a technique for the simultaneous detection of multiple mycotoxins. From the perspective of toxin control technology, the available physical and chemical strategies to remove mycotoxins from fruits and vegetables have some issues, such as safety issues, losses in nutrition and quality, environmental contamination and high cost, which limit their application at a large scale. Biodegradation, due to its features of high efficiency, specificity and non-pollution, is considered a powerful and potential alternative strategy and has aroused more and more concern. In addition, most research on detoxification is focused on the screening and isolation of microbial strains, and few studies contribute to the involved detoxification mechanism. Moreover, in future, we need to isolate and identify high-purity enzymes for the degradation of mycotoxins, analyze the protein structure and action mechanism and demonstrate the encoding gene of the degrading enzyme, to be expressed highly with the help of genetic engineering.

Author Contributions: Writing—original draft preparation, M.N.; writing—review and editing, M.N. and H.X.; visualization, H.X.; supervision, Y.B.; project administration, H.X.; funding acquisition, H.X. All authors have read and agreed to the published version of the manuscript.

Funding: This research was funded by the Natural Science Foundation of China (32060566) and the Natural Science Foundation of Gansu Province, China, grant number 20JR5RA022.

Institutional Review Board Statement: Not applicable.

Informed Consent Statement: Not applicable.

Conflicts of Interest: The authors declare no conflict of interest.

References

1. Sanzani, S.M.; Reverberi, M.; Geisen, R. Mycotoxins in harvested fruits and vegetables: Insights in producing fungi, biological role, conducive conditions, and tools to manage postharvest contamination. *Postharvest Biol. Technol.* **2016**, *122*, 95–105. [CrossRef]
2. Yang, J.; Li, J.; Jiang, Y.; Duan, X.; Qu, H.; Yang, B.; Chen, F.; Dharini, S. Natural occurrence, analysis, and prevention of my-cotoxins in fruits and their processed products. *Crit. Rev. Food Sci.* **2014**, *54*, 64–83. [CrossRef] [PubMed]
3. Fernandez-Cruz, M.L.; Mansilla, M.L.; Tadeo, J.L. Mycotoxins in fruits and their processed products: Analysis, occurrence and health implications. *J. Adv. Res.* **2010**, *1*, 113–122. [CrossRef]
4. Barkai-golan, R.; Paster, N. *Mycotoxins in Fruits and Vegetables*; Academic Press: New York, NY, USA, 2008.
5. Ünüsan, N. Systematic review of mycotoxins in food and feeds in Turkey. *Food Control.* **2018**, *97*, 1–14. [CrossRef]
6. Omotayo, O.P.; Omotayo, A.O.; Mwanza, M.; Babalola, O.O. Prevalence of Mycotoxins and Their Consequences on Human Health. *Toxicol. Res.* **2019**, *35*, 1–7. [CrossRef]
7. Da Rocha, M.E.B.; da Chagas Oliveira Freire, F.; Feitosa Maia, F.E.; Florindo Guedes, M.I.; Rondina, D. Mycotoxins and their effects on human and animal health. *Food Control* **2014**, *36*, 159–165. [CrossRef]
8. Zhao, Z.; Yang, X.; Zhao, X.; Chen, L.; Bai, B.; Zhou, C.; Wang, J. Method Development and Validation for the Analysis of Emerging and Traditional Fusarium Mycotoxins in Pepper, Potato, Tomato, and Cucumber by UPLC-MS/MS. *Food Anal. Methods* **2018**, *11*, 1780–1788. [CrossRef]
9. Li, Y.; Zhang, X.; Nie, J.; Bacha, S.A.S.; Yan, Z.; Gao, G. Occurrence and co-occurrence of mycotoxins in apple and apple products from China. *Food Control* **2020**, *118*, 107354. [CrossRef]
10. Pan, T.-T.; Sun, D.-W.; Pu, H.; Wei, Q.-Y. Simple Approach for the Rapid Detection of Alternariol in Pear Fruit by Surface-Enhanced Raman Scattering with Pyridine-Modified Silver Nanoparticles. *J. Agric. Food Chem.* **2018**, *66*, 2180–2187. [CrossRef]
11. López-Puertollano, D.; Agulló, C.; Mercader, J.V.; Abad-Somovilla, A.; Abad-Fuentes, A. Immunoanalytical methods for ochratoxin A monitoring in wine and must based on innovative immunoreagents. *Food Chem.* **2020**, *345*, 128828. [CrossRef]
12. Myresiotis, C.K.; Testempasis, S.; Vryzas, Z.; Karaoglanidis, G.S.; Papadopoulou-Mourkidou, E. Determination of mycotoxins in pomegranate fruits and juices using a QuEChERS-based method. *Food Chem.* **2015**, *182*, 81–88. [CrossRef] [PubMed]
13. Van Der Merwe, K.J.; Steyn, P.S.; Fourie, L.; Scott, D.B.; Theron, J.J. Ochratoxin A, a Toxic Metabolite produced by *Aspergillus ochraceus* Wilh. *Nature* **1965**, *205*, 1112–1113. [CrossRef] [PubMed]
14. Malir, F.; Ostry, V.; Pfohl-Leszkowicz, A.; Malir, J.; Toman, J. Ochratoxin A: 50 Years of Research. *Toxins* **2016**, *8*, 191. [CrossRef] [PubMed]
15. Cabañes, F.J.; Bragulat, M.R.; Castellá, G. Ochratoxin A Producing Species in the Genus *Penicillium*. *Toxins* **2010**, *2*, 1111–1120. [CrossRef] [PubMed]
16. Álvarez, M.; Rodríguez, A.; Peromingo, B.; Núñez, F.; Rodríguez, M. Enterococcus faecium: A promising protective culture to control growth of ochratoxigenic moulds and mycotoxin production in dry-fermented sausages. *Mycotoxin Res.* **2019**, *36*, 137–145. [CrossRef]
17. Amézqueta, S.; Schorr-Galindo, S.; Murillo-Arbizu, M.; González-Peñas, E.; de Cerain, A.L.; Guiraud, J. OTA-producing fungi in foodstuffs: A review. *Food Control* **2012**, *26*, 259–268. [CrossRef]
18. Zimmerli, B.; Dick, R. Ochratoxin A in table wine and grape-juice: Occurrence and risk assessment*. *Food Addit. Contam.* **1996**, *13*, 655–668. [CrossRef]
19. Gil-Serna, J.; Vázquez, C.; González-Jaén, M.T.; Patiño, B. Wine Contamination with Ochratoxins: A Review. *Beverages* **2018**, *4*, 6. [CrossRef]
20. Silva, L.J.; Rodrigues, A.P.; Pereira, A.M.P.T.; Lino, C.; Pena, A. Ochratoxin A in the Portuguese Wine Market, Occurrence and Risk Assessment. *Food Addit. Contam. Part B* **2019**, *12*, 145–149. [CrossRef]
21. Kochman, J.; Jakubczyk, K.; Janda, K. Mycotoxins in red wine: Occurrence and risk assessment. *Food Control* **2021**, *129*, 108229. [CrossRef]
22. Zhang, X.; Chen, L.; Li, J.; Zhu, B.; Ma, L. Occurrence of Ochratoxin A in Chinese wines: Influence of local meteorological parameters. *Eur. Food Res. Technol.* **2012**, *236*, 277–283. [CrossRef]
23. De Jesus, C.L.; Bartley, A.; Welch, A.Z.; Berry, J.P. High Incidence and Levels of Ochratoxin A in Wines Sourced from the United States. *Toxins* **2018**, *10*, 1. [CrossRef] [PubMed]
24. Hajok, I.; Kowalska, A.; Piekut, A.; Ćwieląg-Drabek, M. A risk assessment of dietary exposure to ochratoxin A for the Polish population. *Food Chem.* **2019**, *284*, 264–269. [CrossRef] [PubMed]
25. Iqbal, S.Z. Mycotoxins in food, recent development in food analysis and future challenges; a review. *Curr. Opin. Food Sci.* **2021**, *42*, 237–247. [CrossRef]
26. Saadullah, A.A.M.; A Bdullah, S.K. Fungal contamination of dried vine fruits and ochratoxin a detection in grape juice from Duhok, Iraq. *Rev. Innovaciencia* **2018**, *6*, 1–8. [CrossRef]

27. Wei, D.; Wu, X.; Xu, J.; Dong, F.; Liu, X.; Zheng, Y.; Ji, M. Determination of Ochratoxin A contamination in grapes, processed grape products and animal-derived products using ultra-performance liquid chromatography-tandem mass spectroscopy system. *Sci. Rep.* **2018**, *8*, 2051. [CrossRef]

28. Yusefi, J.; Valaee, M.; Nazari, F.; Maleki, J.; Mottaghianpour, E.; Khosrokhavar, R.; Hosseini, M.J. Occurrence of ochratoxin a in grape juice of Iran. *Iran. J. Pharm. Res. IJPR* **2018**, *17*, 140–146.

29. Oteiza, J.M.; Khaneghah, A.M.; Campagnollo, F.B.; Granato, D.; Mahmoudi, M.R.; Sant'Ana, A.S.; Gianuzzi, L. Influence of production on the presence of patulin and ochratoxin A in fruit juices and wines of Argentina. *LWT* **2017**, *80*, 200–207. [CrossRef]

30. Lucchetta, G.; Bazzo, I.; Cortivo, G.D.; Stringher, L.; Bellotto, D.; Borgo, M.; Angelini, E. Occurrence of Black Aspergilli and Ochratoxin A on Grapes in Italy. *Toxins* **2010**, *2*, 840–855. [CrossRef]

31. Sanzani, S.M.; Djenane, F.; Incerti, O.; Admane, N.; Mincuzzi, A.; Ippolito, A. Mycotoxigenic fungi contaminating greenhouse-grown tomato fruit and their alternative control. *Eur. J. Plant Pathol.* **2021**, *160*, 287–300. [CrossRef]

32. Wei, C.; Dai, X.; Guo, L.; Lei, X. Identification and toxin-producing capability of causing-spoilage fungi in strawberry. *J. Food Saf. Food Qual.* **2017**, *8*, 1721–1726. [CrossRef]

33. Al-Hazmi, N. Determination of Patulin and Ochratoxin A using HPLC in apple juice samples in Saudi Arabia. *Saudi J. Biol. Sci.* **2010**, *17*, 353–359. [CrossRef] [PubMed]

34. Heperkan, D.; Güler, F.K.; Oktay, H. Mycoflora and natural occurrence of aflatoxin, cyclopiazonic acid, fumonisin and ochratoxin A in dried figs. *Food Addit. Contam. Part A* **2012**, *29*, 277–286. [CrossRef] [PubMed]

35. Rahimi, E.; Shakerian, A. Ochratoxin A in dried figs, raisings, apricots, dates on Iranian retail market. *Health* **2013**, *5*, 2077–2080. [CrossRef]

36. Pavón, M.A.; González, I.; Martín, R.; Garcia, T. Competitive direct ELISA based on a monoclonal antibody for detection of Ochratoxin A in dried fig samples. *Food Agric. Immunol.* **2012**, *23*, 83–91. [CrossRef]

37. Gupta, D.; Bala, P.; Sharma, Y.P. Evaluation of Fungal Flora and Mycotoxin Contamination in Whole Dried Apricots (*Prunus armeniaca* L.) from J&K, India. *Proc. Natl. Acad. Sci. India Sect. B Boil. Sci.* **2015**, *87*, 81–87. [CrossRef]

38. Iqbal, S.Z.; Mehmood, Z.; Asi, M.R.; Shahid, M.; Sehar, M.; Malik, N. Co-occurrence of aflatoxins and ochratoxin A in nuts, dry fruits, and nuty products. *J. Food Saf.* **2018**, *38*, e12462. [CrossRef]

39. Heshmati, A.; Zohrevand, T.; Khaneghah, A.M.; Nejad, A.S.M.; Sant'Ana, A.S. Co-occurrence of aflatoxins and ochratoxin A in dried fruits in Iran: Dietary exposure risk assessment. *Food Chem. Toxicol.* **2017**, *106*, 202–208. [CrossRef]

40. Tosun, A.; Ozden, S. Ochratoxin A in red pepper flakes commercialised in Turkey. *Food Addit. Contam. Part B* **2015**, *9*, 46–50. [CrossRef]

41. Ok, H.E.; Chung, S.H.; Lee, N.; Chun, H.S. Simple High-Performance Liquid Chromatography Method for the Simultaneous Analysis of Aflatoxins, Ochratoxin A, and Zearalenone in Dried and Ground Red Pepper. *J. Food Prot.* **2015**, *78*, 1226–1231. [CrossRef]

42. Spadaro, D.; Patharajan, S.; Lore, A.; Gullino, M.L.; Garibaldi, A. Effect of pH, water activity and temperature on the growth and accumulation of ochratoxin A produced by three strains of *Aspergillus carbonarius* isolated from Italian vineyards. *Phytopathol. Mediterr.* **2010**, *49*, 65–73. [CrossRef]

43. Welke, J.E. Fungal and mycotoxin problems in grape juice and wine industries. *Curr. Opin. Food Sci.* **2019**, *29*, 7–13. [CrossRef]

44. Erceg, S.; Mateo, E.M.; Zipancic, I.; Jiménez, F.J.R.; Aragó, M.A.P.; Jiménez, M.; Soria, J.M.; Garcia-Esparza, M.Á. Assessment of toxic efects of ochratoxin A in human embryonic stem Cells. *Toxins* **2019**, *11*, 217. [CrossRef] [PubMed]

45. Sirot, V.; Fremy, J.-M.; Leblanc, J.-C. Dietary exposure to mycotoxins and health risk assessment in the second French total diet study. *Food Chem. Toxicol.* **2012**, *52*, 1–11. [CrossRef] [PubMed]

46. El-Sayed, Y.S.; Khalil, R.; Saad, T.T. Acute toxicity of ochratoxin-A in marine water-reared sea bass (*Dicentrarchus labrax* L.). *Chemosphere* **2009**, *75*, 878–882. [CrossRef]

47. Gresham, A.; Done, S.; Livesey, C.; MacDonald, S.; Chan, D.; Sayers, R.; Clark, C.; Kemp, P. Survey of pigs' kidneys with lesions consistent with PMWS and PDNS and ochratoxicosis. Part 1: Concentrations and prevalence of ochratoxin A. *Veter. Rec.* **2006**, *159*, 737–742. [CrossRef]

48. Ringot, D.; Chango, A.; Schneider, Y.-J.; Larondelle, Y. Toxicokinetics and toxicodynamics of ochratoxin A, an update. *Chem. Interact.* **2006**, *159*, 18–46. [CrossRef]

49. Huff, W.E.; Kubena, L.F.; Harvey, R.B.; Phillips, T.D. Efficacy of Hydrated Sodium Calcium Aluminosilicate to Reduce the Individual and Combined Toxicity of Aflatoxin and Ochratoxin A. *Poult. Sci.* **1992**, *71*, 64–69. [CrossRef]

50. IARC. *Monographs on the Evaluation of Carcinogenic Risks to Humans: Some Naturally Occuring Substances: Food Items and Costituents, Heterocyclic Aromatic Amines and Mycotoxins*; IARC: Lyon, France, 1993; Volume 56, pp. 489–524.

51. Li, H.; Mao, X.; Liu, K.; Sun, J.; Li, B.; Malyar, R.M.; Liu, D.; Pan, C.; Gan, F.; Liu, Y.; et al. Ochratoxin A induces nephrotoxicity in vitro and in vivo via pyroptosis. *Arch. Toxicol.* **2021**, *95*, 1489–1502. [CrossRef]

52. Imaoka, T.; Yang, J.; Wang, L.; McDonald, M.G.; Afsharinejad, Z.; Bammler, T.K.; Van Ness, K.; Yeung, C.K.; Rettie, A.E.; Himmelfarb, J.; et al. Microphysiological system modeling of ochratoxin A-associated nephrotoxicity. *Toxicology* **2020**, *444*, 152582. [CrossRef]

53. Niaz, K.; Shah, S.Z.A.; Khan, F.; Bule, M. Ochratoxin A–induced genotoxic and epigenetic mechanisms lead to Alzheimer disease: Its modulation with strategies. *Environ. Sci. Pollut. Res.* **2020**, *27*, 44673–44700. [CrossRef] [PubMed]

54. Raistrick, H. Patulin in the Common Cold Collaborative Research on a Derivative of Penicillium Patulum Bainier: I. Introduction. *Lancet* **1943**, *242*, 625. [CrossRef]
55. Andersen, B.; Smedsgaard, A.J.; Frisvad, J.C. *Penicillium expansum*: Consistent Production of Patulin, Chaetoglobosins, and Other Secondary Metabolites in Culture and Their Natural Occurrence in Fruit Products. *J. Agric. Food Chem.* **2004**, *52*, 2421–2428. [CrossRef] [PubMed]
56. Saleh, I.; Goktepe, I. The characteristics, occurrence, and toxicological effects of patulin. *Food Chem. Toxicol.* **2019**, *129*, 301–311. [CrossRef]
57. Paster, N.; Barkai-Golan, R. Mouldy fruits and vegetables as a source of mycotoxins: Part 2. *World Mycotoxin J.* **2008**, *4*, 385–396. [CrossRef]
58. Sajid, M.; Mehmood, S.; Yuan, Y.; Yue, T. Mycotoxin patulin in food matrices: Occurrence and its biological degradation strategies. *Drug Metab. Rev.* **2019**, *51*, 105–120. [CrossRef]
59. Wright, S.A. Patulin in food. *Curr. Opin. Food Sci.* **2015**, *5*, 105–109. [CrossRef]
60. Wei, C.; Yu, L.; Qiao, N.; Zhao, J.; Zhang, H.; Zhai, Q.; Tian, F.; Chen, W. Progress in the distribution, toxicity, control, and detoxification of patulin: A review. *Toxicon* **2020**, *184*, 83–93. [CrossRef]
61. Cao, J.; Zhang, H.; Yang, Q.; Ren, R. Efficacy of Pichia caribbica in controlling blue mold rot and patulin degradation in apples. *Int. J. Food Microbiol.* **2013**, *162*, 167–173. [CrossRef]
62. Hammami, W.; Al-Thani, R.; Fiori, S.; Al-Meer, S.; Atia, F.A.; Rabah, D.; Migheli, Q.; Jaoua, S. Patulin and patulin producing *Penicillium* spp. occurrence in apples and apple-based products including baby food. *J. Infect. Dev. Ctries.* **2017**, *11*, 343–349. [CrossRef]
63. Zaied, C.; Abid, S.; Hlel, W.; Bacha, H. Occurrence of patulin in apple-based-foods largely consumed in Tunisia. *Food Control* **2012**, *31*, 263–267. [CrossRef]
64. Zouaoui, N.; Sbaii, N.; Bacha, H.; Abid-Essefi, S. Occurrence of patulin in various fruit juice marketed in Tunisia. *Food Control* **2015**, *51*, 356–360. [CrossRef]
65. Van de Perre, E.; Jacxsens, L.; Van Der Hauwaert, W.; Haesaert, I.; De Meulenaer, B. Screening for the Presence of Patulin in Molded Fresh Produce and Evaluation of Its Stability in the Production of Tomato Products. *J. Agric. Food Chem.* **2013**, *62*, 304–309. [CrossRef] [PubMed]
66. Iqbal, S.Z.; Malik, S.; Asi, M.R.; Selamat, J.; Malik, N. Natural occurrence of patulin in different fruits, juices and smoothies and evaluation of dietary intake in Punjab, Pakistan. *Food Control* **2018**, *84*, 370–374. [CrossRef]
67. Sarubbi, F.; Formisano, G.; Auriemma, G.; Arrichiello, A.; Palomba, R. Patulin in homogenized fruit's and tomato products. *Food Control* **2016**, *59*, 420–423. [CrossRef]
68. Ji, X.; Li, R.; Yang, H.; Qi, P.; Xiao, Y.; Qian, M. Occurrence of patulin in various fruit products and dietary exposure assessment for consumers in China. *Food Control* **2017**, *78*, 100–107. [CrossRef]
69. Aboud, H.M.; Shatha, A.; Shafiq, S.; Jabbar, I.S. Investigation of mycotoxin patulin in some types of dried fruits in Baghdad governorate. *Int. J. Adv. Res.* **2015**, *3*, 1128–1135.
70. Puel, O.; Galtier, P.; Oswald, I.P. Biosynthesis and Toxicological Effects of Patulin. *Toxins* **2010**, *2*, 613–631. [CrossRef] [PubMed]
71. Zhong, Y.; Jin, C.; Gan, J.; Wang, X.; Shi, Z.; Xia, X.; Peng, X. Apigenin attenuates patulin-induced apoptosis in HEK293 cells by modulating ROS-mediated mitochondrial dysfunction and caspase signal pathway. *Toxicon* **2017**, *137*, 106–113. [CrossRef]
72. Boussabbeh, M.; Ben Salem, I.; Belguesmi, F.; Bacha, H.; Abid-Essefi, S. Tissue oxidative stress induced by patulin and protective effect of crocin. *NeuroToxicology* **2016**, *53*, 343–349. [CrossRef]
73. Alves, I.; Oliveira, N.; Laires, A.; Rodrigues, A.; Rueff, J. Induction of micronuclei and chromosomal aberrations by the mycotoxin patulin in mammalian cells: Role of ascorbic acid as a modulator of patulin clastogenicity. *Mutagenesis* **2000**, *15*, 229–234. [CrossRef] [PubMed]
74. Wang, S.; Yang, J.; Zhang, B.; Zhang, L.; Wu, K.; Yang, A.; Li, C.; Wang, Y.; Zhang, J.; Qi, D. Potential Link between Gut Microbiota and Deoxynivalenol-Induced Feed Refusal in Weaned Piglets. *J. Agric. Food Chem.* **2019**, *67*, 4976–4986. [CrossRef] [PubMed]
75. Ostry, V. *Alternaria* mycotoxins: An overview of chemical characterization, producers, toxicity, analysis and occurrence in foodstuffs. *World Mycotoxin J.* **2008**, *1*, 175–188. [CrossRef]
76. Nakashima, T.; Ueno, T.; Fukami, H. Structure elucidation of AK-toxins, host-specific phytotoxic metabolites produced by alternaria kikuchiana tanaka. *Tetrahedron Lett.* **1982**, *23*, 4469–4472. [CrossRef]
77. Dall'Asta, C.; Cirlini, M.; Falavigna, C. Mycotoxins from alternaria: Toxicological implications. In *Advances in Molecular Toxicology*; University of Parma: Parma, Italy, 2014; Volume 8, pp. 107–121. [CrossRef]
78. King, A.D.; Schade, J.E. Alternaria Toxins and Their Importance in Food. *J. Food Prot.* **1984**, *47*, 886–901. [CrossRef]
79. Mamgain, A.; Roychowdhury, R.; Tah, J. Review alternaria pathogenicity and its strategic controls. *Res. J. Biol.* **2013**, *1*, 1–9.
80. Fleck, S.C.; Burkhardt, B.; Pfeiffer, E.; Metzler, M. Alternaria toxins: Altertoxin II is a much stronger mutagen and DNA strand breaking mycotoxin than alternariol and its methyl ether in cultured mammalian cells. *Toxicol. Lett.* **2012**, *214*, 27–32. [CrossRef]
81. Stinson, E.E.; Osman, S.F.; Heisler, E.G.; Siciliano, J.; Bills, D.D. Mycotoxin production in whole tomatoes, apples, oranges, and lemons. *J. Agric. Food Chem.* **1981**, *29*, 790–792. [CrossRef]
82. Agriopoulou, S.; Stamatelopoulou, E.; Varzakas, T. Advances in Occurrence, Importance, and Mycotoxin Control Strategies: Prevention and Detoxification in Foods. *Foods* **2020**, *9*, 137. [CrossRef]

83. Hickert, S.; Bergmann, M.; Ersen, S.; Cramer, B.; Humpf, H.-U. Survey of Alternaria toxin contamination in food from the German market, using a rapid HPLC-MS/MS approach. *Mycotoxin Res.* **2015**, *32*, 7–18. [CrossRef]

84. Zhao, K.; Shao, B.; Yang, D.; Li, F. Natural Occurrence of Four Alternaria Mycotoxins in Tomato- and Citrus-Based Foods in China. *J. Agric. Food Chem.* **2014**, *63*, 343–348. [CrossRef] [PubMed]

85. EFSA on Contaminants in the Food Chain (CONTAM). Scientific opinion on the risks for animal and public health related to the presence of Alternaria toxins in feed and food. *EFSA J.* **2011**, *9*, 2407. [CrossRef]

86. Gonçalves, C.; Tölgyesi, Á.; Bouten, K.; Robouch, P.; Emons, H.; Stroka, J. Determination of Alternaria Toxins in Tomato, Wheat, and Sunflower Seeds by SPE and LC-MS/MS—A Method Validation Through a Collaborative Trial. *J. AOAC Int.* **2021**, *105*, 80–94. [CrossRef]

87. Guiso, N.; von Konig, C.H.W.; Becker, C.; Hallander, H. Fimbrial Typing of *Bordetella pertussis* Isolates: Agglutination with Polyclonal and Monoclonal Antisera. *J. Clin. Microbiol.* **2001**, *39*, 1684–1685. [CrossRef] [PubMed]

88. López, P.; Venema, D.; De Rijk, T.; De Kok, A.; Scholten, J.M.; Mol, H.G.; De Nijs, M. Occurrence of Alternaria toxins in food products in The Netherlands. *Food Control* **2016**, *60*, 196–204. [CrossRef]

89. Asam, S.; Konitzer, K.; Rychlik, M. Precise determination of the Alternaria mycotoxins alternariol and alternariol monomethyl ether in cereal, fruit and vegetable products using stable isotope dilution assays. *Mycotoxin Res.* **2010**, *27*, 23–28. [CrossRef]

90. Guo, W.; Fan, K.; Nie, D.; Meng, J.; Huang, Q.; Yang, J.; Shen, Y.; Tangni, E.K.; Zhao, Z.; Wu, Y.; et al. Development of a QuEChERS-Based UHPLC-MS/MS Method for Simultaneous Determination of Six Alternaria Toxins in Grapes. *Toxins* **2019**, *11*, 87. [CrossRef]

91. Wei, D.; Wang, Y.; Jiang, D.; Feng, X.; Li, J.; Wang, M. Survey of Alternaria Toxins and Other Mycotoxins in Dried Fruits in China. *Toxins* **2017**, *9*, 200. [CrossRef]

92. Ackermann, Y.; Curtui, V.; Dietrich, R.; Gross, M.; Latif, H.; Märtlbauer, E.; Usleber, E. Widespread Occurrence of Low Levels of Alternariol in Apple and Tomato Products, as Determined by Comparative Immunochemical Assessment using Monoclonal and Polyclonal Antibodies. *J. Agric. Food Chem.* **2011**, *59*, 6360–6368. [CrossRef]

93. Scott, P.M.; Lawrence, G.A.; Lau, B.P.Y. Analysis of wines, grape juices and cranberry juices for Alternaria toxins. *Mycotoxin Res.* **2006**, *22*, 142–147. [CrossRef]

94. Walravens, J.; Mikula, H.; Rychlik, M.; Asam, S.; Devos, T.; Ediage, E.N.; Di Mavungu, J.D.; Jacxsens, L.; Van Landschoot, A.; Vanhaecke, L.; et al. Validated UPLC-MS/MS Methods To Quantitate Free and Conjugated *Alternaria* Toxins in Commercially Available Tomato Products and Fruit and Vegetable Juices in Belgium. *J. Agric. Food Chem.* **2016**, *64*, 5101–5109. [CrossRef] [PubMed]

95. Babič, J.; Tavčar-Kalcher, G.; Celar, F.; Kos, K.; Knific, T.; Jakovac-Strajn, B. Occurrence of *Alternaria* and Other Toxins in Cereal Grains Intended for Animal Feeding Collected in Slovenia: A Three-Year Study. *Toxins* **2021**, *13*, 304. [CrossRef] [PubMed]

96. Wu, C.S.; Ma, L.; Jiang, T.; Jiang, L.Y.; Dai, F.F.; Zhang, Y.H. A review on tenuazonic acid, a toxic produced by alternaria. *Food Sci.* **2014**, *35*, 295–301. [CrossRef]

97. Pero, R.W.; Posner, H.; Blois, M.; Harvan, D.; Spalding, J.W. Toxicity of Metabolites Produced by the Alternaria. *Environ. Health Perspect.* **1973**, *4*, 87. [CrossRef] [PubMed]

98. Sauer, D.B.; Seitz, L.M.; Burroughs, R.; Mohr, H.E.; West, J.L.; Milleret, R.J.; Anthony, H.D. Toxicity of Alternaria metabolites found in weathered sorghum grain at harvest. *J. Agric. Food Chem.* **1978**, *26*, 1380–1383. [CrossRef]

99. Escrivá, L.; Oueslati, S.; Font, G.; Manyes, L. *Alternaria* Mycotoxins in Food and Feed: An Overview. *J. Food Qual.* **2017**, *2017*, 1569748. [CrossRef]

100. Fehr, M.; Pahlke, G.; Fritz, J.; Christensen, M.O.; Boege, F.; Altemöller, M.; Podlech, J.; Marko, D. Alternariol acts as a topoisomerase poison, preferentially affecting the IIα isoform. *Mol. Nutr. Food Res.* **2009**, *53*, 441–451. [CrossRef]

101. Víctor-Ortega, M.D.; Lara, F.J.; García-Campaña, A.M.; del Olmo-Iruela, M. Evaluation of dispersive liquid–liquid microextraction for the determination of patulin in apple juices using micellar electrokinetic capillary chromatography. *Food Control* **2013**, *31*, 353–358. [CrossRef]

102. Awuchi, C.G.; Ondari, E.N.; Ogbonna, C.U.; Upadhyay, A.K.; Baran, K.; Okpala, C.O.; Korzeniowska, M.; Guiné, R.P. Mycotoxins Affecting Animals, Foods, Humans, and Plants: Types, Occurrence, Toxicities, Action Mechanisms, Prevention, and Detoxification Strategies—A Revisit. *Foods* **2021**, *10*, 1279. [CrossRef]

103. Wang, Z.; Wu, Q.; Kuča, K.; Dohnal, V.; Tian, Z. Deoxynivalenol: Signaling pathways and human exposure risk assessment—An update. *Arch. Toxicol.* **2014**, *88*, 1915–1928. [CrossRef]

104. Udovicki, B.; Audenaert, K.; De Saeger, S.; Rajkovic, A. Overview on the Mycotoxins Incidence in Serbia in the Period 2004–2016. *Toxins* **2018**, *10*, 279. [CrossRef] [PubMed]

105. Mahato, D.K.; Pandhi, S.; Kamle, M.; Gupta, A.; Sharma, B.; Panda, B.K.; Srivastava, S.; Kumar, M.; Selvakumar, R.; Pandey, A.K.; et al. Trichothecenes in food and feed: Occurrence, impact on human health and their detection and management strategies. *Toxicon* **2022**, *208*, 62–77. [CrossRef] [PubMed]

106. Xue, H.; Bi, I.; Zong, Y.Y.; Pu, L.; Wang, Y.; Li, Y. Progress of mycotoxins contamination and detection in fruits and vegetables and their products. *Food Sci.* **2016**, *37*, 285–290. [CrossRef]

107. Xue, H.-L.; Bi, Y.; Tang, Y.-M.; Zhao, Y.; Wang, Y. Effect of cultivars, Fusarium strains and storage temperature on trichothecenes production in inoculated potato tubers. *Food Chem.* **2013**, *151*, 236–242. [CrossRef] [PubMed]

108. Zhang, B.; Chen, X.; Han, S.-Y.; Li, M.; Ma, T.-Z.; Sheng, W.-J.; Zhu, X. Simultaneous Analysis of 20 Mycotoxins in Grapes and Wines from Hexi Corridor Region (China): Based on a QuEChERS–UHPLC–MS/MS Method. *Molecules* **2018**, *23*, 1926. [CrossRef] [PubMed]

109. Tang, Y.; Xue, H.; Bi, Y.; Li, Y.; Wang, Y.; Zhao, Y.; Shen, K. A method of analysis for T-2 toxin and neosolaniol by UPLC-MS/MS in apple fruit inoculated with *Trichothecium roseum*. *Food Addit. Contam. Part A* **2014**, *32*, 480–487. [CrossRef] [PubMed]

110. Xue, H.; Bi, Y.; Sun, Y.; Hussain, R.; Wang, H.; Zhang, S.; Zhang, R.; Long, H.; Nan, M.; Cheng, X.; et al. Acetylsalicylic acid treatment reduce Fusarium rot development and neosolaniol accumulation in muskmelon fruit. *Food Chem.* **2019**, *289*, 278–284. [CrossRef]

111. Xue, H.L.; Bi, Y.; Hussain, R.; Pu, L.M.; Nan, M.N.; Cheng, X.Y.; Wang, Y.; Li, Y.C. Detection of NEO in muskmelon fruits inoculated with *Fusarium sulphureum* and its control by postharvest ozone treatment. *Food Chem.* **2018**, *254*, 193–200. [CrossRef]

112. Zhu, M.; Cen, Y.; Ye, W.; Li, S.; Zhang, W. Recent Advances on Macrocyclic Trichothecenes, Their Bioactivities and Biosynthetic Pathway. *Toxins* **2020**, *12*, 417. [CrossRef]

113. Cope, R.B. Chapter 75—Trichothecenes. In *Veterinary Toxicology*; Elsevier: Amsterdam, The Netherlands, 2018; Volume 75, pp. 1043–1053. [CrossRef]

114. Sokolović, M.; Garaj-Vrhovac, V.; Šimpraga, B. T-2 Toxin: Incidence and Toxicity in Poultry. *Arch. Ind. Hyg. Toxicol.* **2008**, *59*, 43–52. [CrossRef]

115. Wu, W.; Zhou, H.-R.; Bursian, S.J.; Link, J.E.; Pestka, J.J. Emetic responses to T-2 toxin, HT-2 toxin and emetine correspond to plasma elevations of peptide YY3–36 and 5-hydroxytryptamine. *Arch. Toxicol.* **2015**, *90*, 997–1007. [CrossRef] [PubMed]

116. Kuca, K.; Dohnal, V.; Jezkova, A.; Jun, D. Metabolic Pathways of T-2 Toxin. *Curr. Drug Metab.* **2008**, *9*, 77–82. [CrossRef] [PubMed]

117. Daenicke, S.; Keese, C.; Goyarts, T.; Döll, S. Effects of deoxynivalenol (DON) and related compounds on bovine peripheral blood mononuclear cells (PBMC) in vitro and in vivo. *Mycotoxin Res.* **2010**, *27*, 49–55. [CrossRef] [PubMed]

118. Hymery, N.; Sibiril, Y.; Parent-Massin, D. In vitro effects of trichothecenes on human dendritic cells. *Toxicol. Vitr.* **2006**, *20*, 899–909. [CrossRef]

119. Pestka, J.J. Deoxynivalenol: Mechanisms of action, human exposure, and toxicological relevance. *Arch. Toxicol.* **2010**, *84*, 663–679. [CrossRef]

120. Awad, W.A.; Ghareeb, K.; Dadak, A.; Gille, L.; Staniek, K.; Hess, M.; Böhm, J. Genotoxic effects of deoxynivalenol in broiler chickens fed low-protein feeds. *Poult. Sci.* **2012**, *91*, 550–555. [CrossRef]

121. Pinton, P.; Oswald, I.P. Effect of Deoxynivalenol and Other Type B Trichothecenes on the Intestine: A Review. *Toxins* **2014**, *6*, 1615–1643. [CrossRef]

122. Desjardins, A.; Proctor, R. Molecular biology of Fusarium mycotoxins. *Int. J. Food Microbiol.* **2007**, *119*, 47–50. [CrossRef]

123. Becker, C.; Reiter, M.; Pfaffl, M.W.; Meyer, H.H.D.; Bauer, J.; Meyer, K.H.D. Expression of immune relevant genes in pigs under the influence of low doses of deoxynivalenol (DON). *Mycotoxin Res.* **2011**, *27*, 287–293. [CrossRef]

124. Zhao, Y.R.; Ma, L.Y.; Wang, F.H. Research progress in trichothecenes and removal method. *Sci. Technol. Food Ind.* **2016**, *37*, 383–387.

125. Singh, J.; Mehta, A. Rapid and sensitive detection of mycotoxins by advanced and emerging analytical methods: A review. *Food Sci. Nutr.* **2020**, *8*, 2183–2204. [CrossRef] [PubMed]

126. Rhouati, A.; Yang, C.; Hayat, A.; Marty, J.-L. Aptamers: A Promising Tool for Ochratoxin A Detection in Food Analysis. *Toxins* **2013**, *5*, 1988–2008. [CrossRef] [PubMed]

127. Man, Y.; Liang, G.; Li, A.; Pan, L. Recent Advances in Mycotoxin Determination for Food Monitoring via Microchip. *Toxins* **2017**, *9*, 324. [CrossRef] [PubMed]

128. *ISO 8128-1-1993*; Apple Juice, Apple Juice Concentrates and Drinks Containing Apple Juice-Determination of Patulincontent. Part1: Method Using High-Performance Liquid Chromatography. ISO International Standard (ISO): Geneva, Switzerland, 1993.

129. *EN 15829: 2010*; Foodstuffs-Determination of Ochratoxin A in Currants, Raisins, Sultanas, Mixed Dried Fruit and Dried Figs-HPLC Method with Immunoaffinity Column Cleanup and Fluorescence Detection. The Spanish Association for Standardization and Certification: Madrid, Spain, 2010.

130. *AOAC 2001.01*; Determination of Ochratoxin A in Wine and Beer Immunoaffinitycolumn Cleanup/Liquid Chromatographic Analysis. Association of Official Agricultural Chemists: Washington, DC, USA, 2001.

131. Wang, M.; Jiang, N.; Dai, Y.; Feng, X. Maximum residue levels and testing standards of mycotoxins in fruits in china and abroad. *J. Food Saf. Qual.* **2016**, *7*, 459–467. (In Chinese)

132. Tölgyesi, Á.; Farkas, T.; Bálint, M.; McDonald, T.J.; Sharma, V.K. A Dilute and Shoot Strategy for Determining *Alternaria* Toxins in Tomato-Based Samples and in Different Flours Using LC-IDMS Separation. *Molecules* **2021**, *26*, 1017. [CrossRef]

133. Pallarés, N.; Sebastià, A.; Martínez-Lucas, V.; González-Angulo, M.; Barba, F.J.; Berrada, H.; Ferrer, E. High Pressure Processing Impact on Alternariol and Aflatoxins of Grape Juice and Fruit Juice-Milk Based Beverages. *Molecules* **2021**, *26*, 3769. [CrossRef]

134. Lin, H.-Y.; Agrawal, D.C.; Yang, W.-G.; Chien, W.-J. A simple HPLC-MS/MS method for the analysis of multi-mycotoxins in betel nut. *Int. J. Appl. Sci. Eng.* **2021**, *18*, 1–7. [CrossRef]

135. Bazin, I.; Tria, S.A.; Hayat, A.; Marty, J.-L. New biorecognition molecules in biosensors for the detection of toxins. *Biosens. Bioelectron.* **2017**, *87*, 285–298. [CrossRef]

136. Vidal, J.C.; Bonel, L.; Ezquerra, A.; Hernández, S.; Bertolín, J.R.; Cubel, C.; Castillo, J.R. Electrochemical affinity biosensors for detection of mycotoxins: A review. *Biosens. Bioelectron.* **2013**, *49*, 146–158. [CrossRef]

137. Zhang, X.; Sun, M.; Kang, Y.; Xie, H.; Wang, X.; Song, H.; Li, X.; Fang, W. Identification of a high-affinity monoclonal antibody against ochratoxin A and its application in enzyme-linked immunosorbent assay. *Toxicon* **2015**, *106*, 89–96. [CrossRef]

138. Wang, Y.; Liu, Y.; Fu, R.; Liu, H.; Zhou, J.; Zhao, Q.; Wang, C.; Jiao, B.; He, Y. A review of portable biosensor for the field detection of mycotoxins. *Food Sci.* **2022**, *in press*.

139. Yang, X.-H.; Kong, W.-J.; Yang, M.-H.; Zhao, M.; Ouyang, Z. Application of Aptamer Identification Technology in Rapid Analysis of Mycotoxins. *Chin. J. Anal. Chem.* **2013**, *41*, 297–306. [CrossRef]

140. Chen, X.; Huang, Y.; Duan, N.; Wu, S.; Xia, Y.; Ma, X.; Zhu, C.; Jiang, Y.; Wang, Z. Screening and Identification of DNA Aptamers against T-2 Toxin Assisted by Graphene Oxide. *J. Agric. Food Chem.* **2014**, *62*, 10368–10374. [CrossRef]

141. Zhu, Z.; Feng, M.; Zuo, L.; Zhu, Z.; Wang, F.; Chen, L.; Li, J.; Shan, G.; Luo, S.-Z. An aptamer based surface plasmon resonance biosensor for the detection of ochratoxin A in wine and peanut oil. *Biosens. Bioelectron.* **2015**, *65*, 320–326. [CrossRef] [PubMed]

142. Joshi, S.; Annida, R.M.; Zuilhof, H.; Van Beek, T.A.; Nielen, M.W.F. Analysis of Mycotoxins in Beer Using a Portable Nanostructured Imaging Surface Plasmon Resonance Biosensor. *J. Agric. Food Chem.* **2016**, *64*, 8263–8271. [CrossRef] [PubMed]

143. Nolan, P.; Auer, S.; Spehar, A.; Elliott, C.T.; Campbell, K. Current trends in rapid tests for mycotoxins. *Food Addit. Contam. Part A* **2019**, *36*, 800–814. [CrossRef] [PubMed]

144. Pirinçci, Ş.Ş.; Ertekin, Ö.; Laguna, D.E.; Özen, F.Ş.; Öztürk, Z.Z.; Öztürk, S. Label-Free QCM Immunosensor for the Detection of Ochratoxin A. *Sensors* **2018**, *18*, 1161. [CrossRef]

145. Karczmarczyk, A.; Haupt, K.; Feller, K.H. Development of a QCM-D biosensor for Ochratoxin A detection in red wine. *Talanta* **2017**, *166*, 193–197. [CrossRef]

146. Zhang, W.; Wang, Y.; Nan, M.; Li, Y.; Yun, J.; Wang, Y.; Bi, Y. Novel colorimetric aptasensor based on unmodified gold nanoparticle and ssDNA for rapid and sensitive detection of T-2 toxin. *Food Chem.* **2021**, *348*, 129128. [CrossRef]

147. Yang, C.; Wang, Y.; Marty, J.-L.; Yang, X. Aptamer-based colorimetric biosensing of Ochratoxin A using unmodified gold nanoparticles indicator. *Biosens. Bioelectron.* **2011**, *26*, 2724–2727. [CrossRef]

148. He, Y.; Tian, F.; Zhou, J.; Zhao, Q.; Fu, R.; Jiao, B. Colorimetric aptasensor for ochratoxin A detection based on enzyme-induced gold nanoparticle aggregation. *J. Hazard. Mater.* **2019**, *388*, 121758. [CrossRef] [PubMed]

149. Rhouati, A.; Catanante, G.; Nunes, G.; Hayat, A.; Marty, J.-L. Label-Free Aptasensors for the Detection of Mycotoxins. *Sensors* **2016**, *16*, 2178. [CrossRef] [PubMed]

150. Rhouati, A.; Hayat, A.; Hernandez, D.B.; Meraihi, Z.; Munoz, R.; Marty, J.-L. Development of an automated flow-based electrochemical aptasensor for on-line detection of Ochratoxin A. *Sensors Actuators B Chem.* **2013**, *176*, 1160–1166. [CrossRef]

151. Abnous, K.; Danesh, N.M.; Alibolandi, M.; Ramezani, M.; Taghdisi, S.M. Amperometric aptasensor for ochratoxin A based on the use of a gold electrode modified with aptamer, complementary DNA, SWCNTs and the redox marker Methylene Blue. *Mikrochim. Acta* **2017**, *184*, 1151–1159. [CrossRef]

152. He, B.; Lu, X. An electrochemical aptasensor based on tetrahedral DNA nanostructures as a signal probe carrier platform for sensitive detection of patulin. *Anal. Chim. Acta* **2020**, *1138*, 123–131. [CrossRef]

153. Song, X.; Wang, D.; Kim, M. Development of an immuno-electrochemical glass carbon electrode sensor based on graphene oxide/gold nanocomposite and antibody for the detection of patulin. *Food Chem.* **2020**, *342*, 128257. [CrossRef]

154. Xu, J.; Qiao, X.; Wang, Y.; Sheng, Q.; Yue, T.; Zheng, J.; Zhou, M. Electrostatic assembly of gold nanoparticles on black phosphorus nanosheets for electrochemical aptasensing of patulin. *Mikrochim. Acta* **2019**, *186*, 238. [CrossRef]

155. Nan, M.; Bi, Y.; Xue, H.; Xue, S.; Long, H.; Pu, L.; Fu, G. Rapid Determination of Ochratoxin A in Grape and Its Commodities Based on a Label-Free Impedimetric Aptasensor Constructed by Layer-by-Layer Self-Assembly. *Toxins* **2019**, *11*, 71. [CrossRef]

156. Nan, M.-N.; Bi, Y.; Xue, H.-L.; Long, H.-T.; Xue, S.-L.; Pu, L.-M.; Prusky, D. Modification performance and electrochemical characteristics of different groups of modified aptamers applied for label-free electrochemical impedimetric sensors. *Food Chem.* **2020**, *337*, 127761. [CrossRef]

157. Meulenberg, E.P. Immunochemical Methods for Ochratoxin A Detection: A Review. *Toxins* **2012**, *4*, 244–266. [CrossRef]

158. Wang, F.; Wan, D.-B.; Shen, Y.-D.; Tian, Y.-X.; Xiao, Z.-L.; Xu, Z.-L.; Yang, J.-Y.; Sun, Y.-M.; Hammock, B.D.; Wang, H. Development of a chemiluminescence immunoassay for detection of tenuazonic acid mycotoxin in fruit juices with a specific camel polyclonal antibody. *Anal. Methods* **2021**, *13*, 1795–1802. [CrossRef] [PubMed]

159. Pei, K.; Xiong, Y.; Xu, B.; Wu, K.; Li, X.; Jiang, H.; Xiong, Y. Colorimetric ELISA for ochratoxin A detection based on the urease-induced metallization of gold nanoflowers. *Sensors Actuators B Chem.* **2018**, *262*, 102–109. [CrossRef]

160. Mcnamee, S.; Bravin, F.; Rosar, G.; Elliott, C.T.; Campbell, K. Development of a nanoarray capable of the rapid and simultaneous detection of zearalenone, T2-toxin and fumonisin. *Talanta* **2017**, *164*, 368–376. [CrossRef] [PubMed]

161. Kim, G.; Lim, J.; Mo, C. A Review on Lateral Flow Test Strip for Food Safety. *J. Biosyst. Eng.* **2015**, *40*, 277–283. [CrossRef]

162. Jiang, H.; Li, X.; Xiong, Y.; Pei, K.; Nie, L.; Xiong, Y. Silver Nanoparticle-Based Fluorescence-Quenching Lateral Flow Immunoassay for Sensitive Detection of Ochratoxin A in Grape Juice and Wine. *Toxins* **2017**, *9*, 83. [CrossRef]

163. Hou, S.; Ma, J.; Cheng, Y.; Wang, H.; Sun, J.; Yan, Y. One-step rapid detection of fumonisin B1, dexyonivalenol and zearalenone in grains. *Food Control* **2020**, *117*, 107107. [CrossRef]

164. Foubert, A.; Beloglazova, N.V.; Gordienko, A.; Tessier, M.D.; Drijvers, E.; Hens, Z.; De Saeger, S. Development of a Rainbow Lateral Flow Immunoassay for the Simultaneous Detection of Four Mycotoxins. *J. Agric. Food Chem.* **2016**, *65*, 7121–7130. [CrossRef]

165. Leggott, N.L.; Shephard, G.S.; Stockenström, S.; Staal, E.; Van Schalkwyk, D.J. The reduction of patulin in apple juice by three different types of activated carbon. *Food Addit. Contam.* **2001**, *18*, 825–829. [CrossRef]
166. Huebner, H.J.; Mayura, K.; Pallaroni, L.; Ake, C.L.; Lemke, S.L.; Herrera, P.; Phillips, T.D. Development and Characterization of a Carbon-Based Composite Material for Reducing Patulin Levels in Apple Juice. *J. Food Prot.* **2000**, *63*, 106–110. [CrossRef]
167. Carrasco-Sánchez, V.; Marican, A.; Vergara-Jaque, A.; Folch-Cano, C.; Comer, J.; Laurie, V.F. Polymeric substances for the removal of ochratoxin A from red wine followed by computational modeling of the complexes formed. *Food Chem.* **2018**, *265*, 159–164. [CrossRef]
168. Liu, H.F.; Han, S.Y.; Sheng, W.Y.; Zhu, X.; Jiang, Y.M. Adsorption patulin from apple juice with 8 Kinds of resin. *J. Gansu Agric. Univ.* **2010**, *3*, 126–130. (In Chinese)
169. Wang, Y.H.; Ma, C.Z.; Guo, Z.Y.; Zheng, Y.Y.; Xue, H.L. Study on the kineticmodel of adsorption patulin in pear juice by macroporous resin. *J. Food Saf. Qual.* **2021**, *11*, 472–478. (In Chinese)
170. Liu, Q.L.; Zhang, S.; Zhang, R.; Zhai, T.T.; Fan, P.F.; Xue, H.L. Study on the adsorption of patulin in pear juice by attapulgite. *J. Food Saf. Qual.* **2019**, *10*, 2575–2580. (In Chinese)
171. Zhang, S.; Zhang, R.; Xue, H.L.; Ma, Y.Y.; Bi, Y.; Zong, Y.Y. Adsorption efficiency of attapulgite towards patulin in pear juice. *Food Sci.* **2019**, *40*, 57–63. (In Chinese) [CrossRef]
172. Ji, J.; Xie, W. Removal of aflatoxin B1 from contaminated peanut oils using magnetic attapulgite. *Food Chem.* **2020**, *339*, 128072. [CrossRef]
173. Nan, M.-N.; Bi, Y.; Qiang, Y.; Xue, H.-L.; Yang, L.; Feng, L.-D.; Pu, L.-M.; Long, H.-T.; Prusky, D. Electrostatic adsorption and removal mechanism of ochratoxin A in wine via a positively charged nano-MgO microporous ceramic membrane. *Food Chem.* **2021**, *371*, 131157. [CrossRef] [PubMed]
174. Fu, H.; Xu, W.; Wang, H.; Liao, S.; Chen, G. Preparation of magnetic molecularly imprinted polymer for selective identification of patulin in juice. *J. Chromatogr. B* **2020**, *1145*, 122101. [CrossRef]
175. Zhao, D.; Jia, J.; Yu, X.; Sun, X. Preparation and characterization of a molecularly imprinted polymer by grafting on silica supports: A selective sorbent for patulin toxin. *Anal. Bioanal. Chem.* **2011**, *401*, 2259–2273. [CrossRef]
176. Sun, J.; Guo, W.; Ji, J.; Li, Z.; Yuan, X.; Pi, F.; Zhang, Y.; Sun, X. Removal of patulin in apple juice based on novel magnetic molecularly imprinted adsorbent Fe$_3$O$_4$@SiO$_2$@CS-GO@MIP. *LWT* **2019**, *118*, 108854. [CrossRef]
177. Aiko, V.; Mehta, A. Occurrence, detection and detoxification of mycotoxins. *J. Biosci.* **2015**, *40*, 943–954. [CrossRef]
178. Dahal, S.; Lee, H.J.; Gu, K.; Ryu, D. Heat Stability of Ochratoxin A in an Aqueous Buffered Model System. *J. Food Prot.* **2016**, *79*, 1748–1752. [CrossRef] [PubMed]
179. Combina, M.; Dalcero, A.; Varsavsky, E.; Torres, A.M.; Etcheverry, M.; Rodriguez, M.; Gonzalez, Q.H. Effect of heat treatments on stability of altemariol, alternariol monomethyl ether and tenuazonic acid in sunflower flour. *Mycotoxin Res.* **1999**, *15*, 33–38. [CrossRef] [PubMed]
180. Blaser, A.R.; Starkopf, L.; Deane, A.M.; Poeze, M.; Starkopf, J. Comparison of Different Definitions of Feeding Intolerance: A Retrospective Observational Study. *Clin. Nutr.* **2015**, *34*, 956–961. [CrossRef] [PubMed]
181. Pramanik, T.; Padan, S.K.; Gupta, R.; Bedi, P.; Singh, G. Comparative efficacy of microwave, visible light and ultrasound irradiation for green synthesis of dihydropyrimidinones in fruit juice medium. Recent Advances in Fundamental and Applied Sciences. In *AIP Conference Proceedings*; AIP Publishing LLC.: New York, NY, USA, 2017; Volume 1860, p. 020059. [CrossRef]
182. El-Hawa, M.A. Patulin production of *Penicillium digitatum* in citrus fruits as affected by Gamma Irradiation Combined with Heat. *Delta J. Sci.* **1993**, *17*, 172–182.
183. Aziz, N.H.; Moussa, L.A.; Far, F.M. reduction of fungi and mycotoxins formation in seeds by gamma-radiation. *J. Food Saf.* **2004**, *24*, 109–127. [CrossRef]
184. Abdelaal, S.S. Effects of Gamma radiation, temperature and water activity on the production of alternaria mycotoxins. *Egypt. J. Microbiol.* **1998**, *32*, 379–396.
185. Ibarz, R.; Garvín, A.; Falguera, V.; Pagán, J.; Garza, S.; Ibarz, A. Modelling of patulin photo-degradation by a UV multi-wavelength emitting lamp. *Food Res. Int.* **2014**, *66*, 158–166. [CrossRef]
186. Ibarz, R.; Garvín, A.; Azuara, E.; Ibarz, A. Modelling of ochratoxin A photo-degradation by a UV multi-wavelength emitting lamp. *LWT* **2015**, *61*, 385–392. [CrossRef]
187. Zou, Z.Y.; Huang, F.; Li, H.J. Removal of deoxynivalenol and T-2 Toxin by ultraviolet irradiation. *Food Sci.* **2015**, *36*, 7–11. [CrossRef]
188. Diao, E.; Wang, J.; Li, X.; Wang, X.; Gao, D. Patulin degradation in apple juice using ozone detoxification equipment and its effects on quality. *J. Food Process. Preserv.* **2018**, *42*, e13645. [CrossRef]
189. Yang, K.; Li, K.; Pan, L.; Luo, X.; Xing, J.; Wang, J.; Wang, L.; Wang, R.; Zhai, Y.; Chen, Z. Effect of Ozone and Electron Beam Irradiation on Degradation of Zearalenone and Ochratoxin A. *Toxins* **2020**, *12*, 138. [CrossRef] [PubMed]
190. Liu, W.; Zhao, Q.; Wang, T.; Duan, X.; Li, C.; Lei, X. Degradation of Organic Pollutants Using Atmospheric Pressure Glow Discharge Plasma. *Plasma Chem. Plasma Process.* **2016**, *36*, 1011–1020. [CrossRef]
191. Pu, L.; Bi, Y.; Long, H.; Xue, H.; Lu, J.; Zong, Y.; Kankam, F. Glow Discharge Plasma Efficiently Degrades T-2 Toxin in Aqueous Solution and Patulin in Apple Juice. *Adv. Tech. Biol. Med.* **2017**, *5*, 1–7. [CrossRef]
192. Drusch, S.; Kopka, S.; Kaeding, J. Stability of patulin in a juice-like aqueous model system in the presence of ascorbic acid. *Food Chem.* **2007**, *100*, 192–197. [CrossRef]

193. Liuzzi, V.C.; Fanelli, F.; Tristezza, M.; Haidukowski, M.; Picardi, E.; Manzari, C.; Lionetti, C.; Grieco, F.; Logrieco, A.F.; Thon, M.R.; et al. Transcriptional Analysis of Acinetobacter sp. neg1 Capable of Degrading Ochratoxin A. *Front. Microbiol.* **2017**, *7*, 2162. [CrossRef] [PubMed]

194. Wei, C.; Yu, L.; Qiao, N.; Wang, S.; Tian, F.; Zhao, J.; Zhang, H.; Zhai, Q.; Chen, W. The characteristics of patulin detoxification by *Lactobacillus plantarum* 13M5. *Food Chem. Toxicol.* **2020**, *146*, 111787. [CrossRef]

195. Hatab, S.; Yue, T.; Mohamad, O. Reduction of Patulin in Aqueous Solution by Lactic Acid Bacteria. *J. Food Sci.* **2012**, *77*, M238–M241. [CrossRef]

196. Fuchs, S.; Sontag, G.; Stidl, R.; Ehrlich, V.; Kundi, M.; Knasmüller, S. Detoxification of patulin and ochratoxin A, two abundant mycotoxins, by lactic acid bacteria. *Food Chem. Toxicol.* **2008**, *46*, 1398–1407. [CrossRef]

197. Schatzmayr, G.; Zehner, F.; Täubel, M.; Schatzmayr, D.; Klimitsch, A.; Loibner, A.P.; Binder, E.M. Microbiologicals for deactivating mycotoxins. *Mol. Nutr. Food Res.* **2006**, *50*, 543–551. [CrossRef]

198. Eriksen, G.S.; Pettersson, H.; Lundh, T. Comparative cytotoxicity of deoxynivalenol, nivalenol, their acetylated derivatives and de-epoxy metabolites. *Food Chem. Toxicol.* **2004**, *42*, 619–624. [CrossRef]

199. Young, J.C.; Zhou, T.; Yu, H.; Zhu, H.; Gong, J. Degradation of trichothecene mycotoxins by chicken intestinal microbes. *Food Chem. Toxicol.* **2007**, *45*, 136–143. [CrossRef] [PubMed]

200. He, J.; Zhou, T.; Young, J.C.; Boland, G.J.; Scott, P.M. Chemical and biological transformations for detoxification of trichothecene mycotoxins in human and animal food chains: A review. *Trends Food Sci. Technol.* **2010**, *21*, 67–76. [CrossRef]

201. Xue, H.L.; Bi, Y.; Wang, Y.; Ge, Y.H.; Li, Y.C. Advances in toxicity, detection and detoxification of trichothecenes. *Food Sci.* **2013**, *34*, 350–355. (In Chinese) [CrossRef]

202. Zhu, R.; Feussner, K.; Wu, T.; Yan, F.; Karlovsky, P.; Zheng, X. Detoxification of mycotoxin patulin by the yeast *Rhodosporidium paludigenum*. *Food Chem.* **2015**, *179*, 1–5. [CrossRef] [PubMed]

203. Li, X.; Tang, H.; Yang, C.; Meng, X.; Liu, B. Detoxification of mycotoxin patulin by the yeast *Rhodotorula mucilaginosa*. *Food Control* **2018**, *96*, 47–52. [CrossRef]

204. Tang, H.; Li, X.H.; Zhang, F.; Meng, X.H.; Liu, B.J. Biodegradation of the mycotoxin patulin in apple juice by orotate phosphoribosyltransferase from *Rhodotorula mucilaginosa*. *Food Control* **2019**, *100*, 158–164. [CrossRef]

205. Xing, M.; Chen, Y.; Li, B.; Tian, S. Characterization of a short-chain dehydrogenase/reductase and its function in patulin biodegradation in apple juice. *Food Chem.* **2021**, *348*, 129046. [CrossRef]

206. Zhang, X.Y.; Gu, X.Y.; Zhao, L.; Zheng, X.F.; Yang, H.J.; Li, J.; Zhang, H.Y. Biocontrol of postharvest disease of grapes and OTA accumulation by *Yarrowia lipolytica* Y$_{-2}$. *J. Chin. Inst. Food Sci. Technol.* **2020**, *20*, 201–206. (In Chinese) [CrossRef]

207. Zhang, X.; Yang, H.; Apaliya, M.; Zhao, L.; Gu, X.; Zheng, X.; Hu, W.; Zhang, H. The mechanisms involved in ochratoxin A elimination by *Yarrowia lipolytica* Y-2. *Ann. Appl. Biol.* **2018**, *173*, 164–174. [CrossRef]

208. Gómez-Albarrán, C.; Melguizo, C.; Patiño, B.; Vázquez, C.; Gil-Serna, J. Diversity of Mycobiota in Spanish Grape Berries and Selection of *Hanseniaspora uvarum* U1 to Prevent Mycotoxin Contamination. *Toxins* **2021**, *13*, 649. [CrossRef]

209. Tang, H.; Peng, X.; Li, X.; Meng, X.; Liu, B. Biodegradation of mycotoxin patulin in apple juice by calcium carbonate immobilized porcine pancreatic lipase. *Food Control* **2018**, *88*, 69–74. [CrossRef]

210. Liu, B.; Peng, X.; Meng, X. Effective Biodegradation of Mycotoxin Patulin by Porcine Pancreatic Lipase. *Front. Microbiol.* **2018**, *9*, 615. [CrossRef] [PubMed]

Article

Bioaccessibility Study of Aflatoxin B$_1$ and Ochratoxin A in Bread Enriched with Fermented Milk Whey and/or Pumpkin

Laura Escrivá, Fojan Agahi, Pilar Vila-Donat *, Jordi Mañes, Giuseppe Meca and Lara Manyes

Laboratory of Food Chemistry and Toxicology, Faculty of Pharmacy, University of Valencia, 46100 Burjassot, València, Spain; Laura.Escriva@uv.es (L.E.); fojan@alumni.uv.es (F.A.); jordi.manes@uv.es (J.M.); giuseppe.meca@uv.es (G.M.); lara.manyes@uv.es (L.M.)
* Correspondence: pilar.vila@uv.es

Abstract: The presence of mycotoxins in cereals and cereal products remains a significant issue. The use of natural ingredients such as pumpkin and whey, which contain bioactive compounds, could be a strategy to reduce the use of conventional chemical preservatives. The aim of the present work was to study the bioaccessibility of aflatoxin B$_1$ (AFB1) and ochratoxin (OTA) in bread, as well as to evaluate the effect of milk whey (with and without lactic acid bacteria fermentation) and pumpkin on reducing mycotoxins bioaccessibility. Different bread typologies were prepared and subjected to an in vitro digestion model. Gastric and intestinal extracts were analyzed by HPLC–MS/qTOF and mycotoxins bioaccessibility was calculated. All the tested ingredients but one significantly reduced mycotoxin intestinal bioaccessibility. Pumpkin powder demonstrated to be the most effective ingredient showing significant reductions of AFB1 and OTA bioaccessibility up to 74% and 34%, respectively. Whey, fermented whey, and the combination of pumpkin-fermented whey showed intestinal bioaccessibility reductions between 57–68% for AFB1, and between 11–20% for OTA. These results pointed to pumpkin and milk whey as potential bioactive ingredients that may have promising applications in the bakery industry.

Keywords: bioaccessibility; aflatoxin B1; ochratoxin A; bread; pumpkin; whey; lactic acid bacteria

Key Contribution: The enrichment of bread with bioactive ingredients such as pumpkin and whey significantly reduced AFB1 and OTA bioaccessibility. Pumpkin seems to be the most efficient ingredient. The addition of pumpkin and whey in the bread making process could be a strategy to reduce the absorbable fraction of mycotoxins at the intestinal level.

Citation: Escrivá, L.; Agahi, F.; Vila-Donat, P.; Mañes, J.; Meca, G.; Manyes, L. Bioaccessibility Study of Aflatoxin B$_1$ and Ochratoxin A in Bread Enriched with Fermented Milk Whey and/or Pumpkin. *Toxins* **2022**, *14*, 6. https://doi.org/10.3390/toxins14010006

Received: 25 November 2021
Accepted: 20 December 2021
Published: 22 December 2021

Publisher's Note: MDPI stays neutral with regard to jurisdictional claims in published maps and institutional affiliations.

1. Introduction

The most dangerous group of mycotoxins found in food are aflatoxins (AFs), produced by *Aspergillus* species. Among them, aflatoxin B$_1$ (AFB1) is considered to be the most toxic mutagenic, teratogenic, and carcinogenic mycotoxin, which is classified as a group 1A (carcinogenic to humans) by the International Agency for Research on Cancer (IARC), and the most toxic compound by the European Commission [1,2]. AFs are commonly found in foodstuffs such as cereals and cereal-by products, corn, nuts, peanuts, coconut, dried fruits, and beer. On the other hand, ochratoxin A (OTA), produced by *Aspergillus* and *Penicillium* species, is also found in a wide variety of foodstuffs, mainly cereals and cereal-based products. The nephrotoxicity, immunotoxicity, mutagenicity, and neurotoxicity effects of OTA in humans has been proven by numerous studies [3]. Moreover, OTA is classified as group 2B, "possibly carcinogenic to humans" [1]. Maximum levels (MLs) of AFB1 and OTA have been established in different food products with values up to 4 and 5 μg/kg, respectively, for cereals and cereal-based products [2].

Despite the regulation of the MLs for both mycotoxins by the EU, several studies have demonstrated their presence in different cereals such as wheat and maize-derived products (bread, pasta, semolina, bulgur, cookie, infant foods etc.) reaching in worst case scenarios levels up to 150 μg/kg, exceeding the legal limits [4–7]. Bread and bakery products are the main foodstuff consumed around the world and are particularly important as a source of carbohydrates, proteins, and vitamins B and E [8]. However, it has been revealed that the baking process does not contribute to the reduction of mycotoxins levels. In fact, the effect of fermentation in bread only reduces OTA and AFB1 by 7% and 6% [9]. The high stability of OTA was confirmed as no significant change in its content was observed after fermentation and bread making process [10,11].

Different strategies have been developed to prevent or reduce mycotoxigenic fungal growth on food and feed by adding non-nutritional adsorbing agents [12–15].

Recently, there has been increasing interest in whether the absorption of mycotoxins in food could be reduced by microorganisms in the gastrointestinal tract. In light of this fact, one of the most used strategies in the reduction of mycotoxins bioaccessibility during the gastrointestinal digestion is the use of biocontrol agents. Certain lactic acid bacteria (LAB) strains were found to be able to remove different mycotoxins from foodstuffs by binding to their cell wall or by degradation with their enzymes in vitro [16–19]. In addition, the use of natural ingredients rich in bioactive compounds with antifungal properties (such as mustard flour or milk whey) are being studied nowadays as bio-preservatives in bread as response to the increasing consumers demand against the use of chemical additives [20–23].

Pumpkin, which belongs to the *Cucurbitaceae* family, is rich in carotenoids (i.e., lycopene, α- and β-carotene, lutein, and zeaxanthin) that play an important role in protecting cells from oxidation and cellular damage, preventing the incidence of human diseases such as mutagenic processes, cardiovascular diseases, osteoporosis, and diabetes [24]. Recent studies have shown that carotenoids-rich food such as pumpkin could counteract the toxic effects produced by mycotoxins [25,26]. Moreover, carotenoids revealed the ability of reducing mycotoxins such as AFB1 in rat tissues [27].

Milk whey is a cheese by-product in the dairy industry that has a high nutritional value and represents an important source of bioactive compounds, such as antifungal peptides [28]. The application of hydrolyzed goat milk whey as a bread ingredient improved the shelf-life and reduced mycotoxigenic fungal growth *(P. verrucosum)*, as well as OTA production in pita bread [23]. Moreover, whey is also an optimum substrate for LAB fermentation and an excellent natural bio-preservative candidate in food production [29].

The aim of the present work was to evaluate the bioaccessibility of OTA and AFB1 when released from contaminated bread after a simulated in vitro human gastrointestinal digestion, as well as to study the effect on mycotoxins bioaccessibility of milk whey (with and without LAB fermentation), pumpkin, and the combination of both natural ingredients (pumpkin and whey) added to bread.

2. Results and Discussion

2.1. Bread Contamination and Analysis

Contaminated flour was analyzed in triplicated as previously explained. Barley flour showed OTA concentrations of 511 ± 84 mg/kg, while maize flour presented AFB1 levels of 6.2 ± 0.3 mg/kg. Based on those concentrations, contaminated breads with the single mycotoxin were prepared for each bread type; control (C), whey (W), fermented whey, (FW), pumpkin (P), fermented whey-pumpkin (FW-P) by adding 1.2 g of OTA contaminated flour or 10 g of AFB1 contaminated flour, aiming to obtain bread concentrations around 1000 μg/kg of OTA and 100 μg/kg of AFB1. Differences in both mycotoxins' concentration are attributed to the natural contamination of cereals by-products that is usually much higher in the case of OTA than AFB1 [30,31].

Mycotoxin concentrations in the final bread were then analyzed by HPLC–MS/qTOF showing some differences among the bread typologies. As it is shown in Table 1, AFB1 bread concentrations ranged between 78–164 μg/kg, while OTA concentrations were between 1184–1540 μg/kg.

Table 1. Aflatoxin B1 (AFB1) and ochratoxin (OTA) concentration (μg/kg) in the prepared breads. Average results and standard deviation of triplicate samples (*n* = 3).

		Bread Concentration
		(μg/kg)
AFB1	C-AFB1	78 ± 3
	W-AFB1	92 ± 12
	FW-AFB1	85 ± 6
	P-AFB1	148 ± 5
	FW-P AFB1	164 ± 16
OTA	C-OTA	1184 ± 118
	W-OTA	1258 ± 220
	FW-OTA	1173 ± 78
	P-OTA	1336 ± 202
	FW-P-OTA	1540 ± 291

Control bread + AFB1 (C-AFB1); bread + whey + AFB1 (W-AFB1); bread + fermented whey + AFB1 (FW-AFB1); bread + pumpkin + AFB1 (P-AFB1); bread + fermented whey + pumpkin +AFB1 (FW-P-AFB1); control bread + OTA (C-OTA); bread + whey + OTA (W-OTA); bread + fermented whey + OTA (FW-OTA); bread + pumpkin + OTA (P-OTA); bread + fermented whey + pumpkin + OTA (FW-P-OTA).

Several factors may affect the final bread composition (and in consequence, mycotoxin concentration) including nutrient concentration and changes in weight mainly due to water loss from the baking process. Depending on the ingredients of the initial dough, the baking process may be different, obtaining breads with different physicochemical properties. For example, whey addition as bread ingredient influenced rheological properties of wheat flour doughs reducing water absorption and increasing arrival time and dough development time [32]. On the other hand, fresh pumpkin that contains 80–96% moisture content, 4.6–6.5% sugars, 0.6–1.8% protein, 0.0–0.2% lipids, and 0.5–1.3% fiber, as well as other minor bioactive substances such as carotenoids, lose high amount of water in the lyophilization process, therefore nutrients are concentrated and bread enriched with dried pumpkin additive is expected to be richer in fiber, carotenoids, and other phytochemicals than the fresh vegetable. Moreover, pumpkin powder introduced as nutritional supplement was found to produce very large and unexpected increases in the loaf volume, as well as increase the organoleptic acceptability of wheat with comparatively poor bread making properties [33].

2.2. AFB1 Bioaccessibility

AFB1 release from bread during the simulated gastrointestinal digestion, as well as its bioaccessibility, was evaluated in control and enriched breads by the analysis of gastric and intestinal extracts. As it is shown in Figure 1, AFB1 concentration in gastric and intestinal extract from control bread (C-AFB1) was 4.7 ± 1.4 μg/L and 9.4 ± 1.2 μg/L, respectively, indicating a progressive mycotoxin release. However, gastric AFB1 concentration in enriched breads were between 0.6 ± 0.5 μg/L (P-AFB1) and 3.7 ± 0.9 μg/L (FW-AFB1), while in intestinal extracts it ranged between 3.4 ± 1.1 (W-AFB1), and 6.0 ± 0.4 μg/L (P-FW-AFB1). Enriched breads showed, in all cases, AFB1 concentrations lower than control bread for both gastric and intestinal phases. Significant differences from the control ($p < 0.05$) were observed for P-AFB1 and P-FW-AFB1 in the gastric extract, as well as in all enriched breads in the case of the intestinal concentrations.

The lower values obtained from AFB1 after gastric digestion may be due to the binding with long-chain carbohydrates, abundant in bread, which are hydrolyzed in duodenal digestion. In both gastric and intestinal extracts, the highest concentration was reached in the non-enriched bread (C-AFB1) giving a first indication that the natural added ingredients may have an effect on reducing the accessible fraction (bioaccessibility) of mycotoxins, hence reducing the amount of mycotoxin that could exert its effects at the gastrointestinal level.

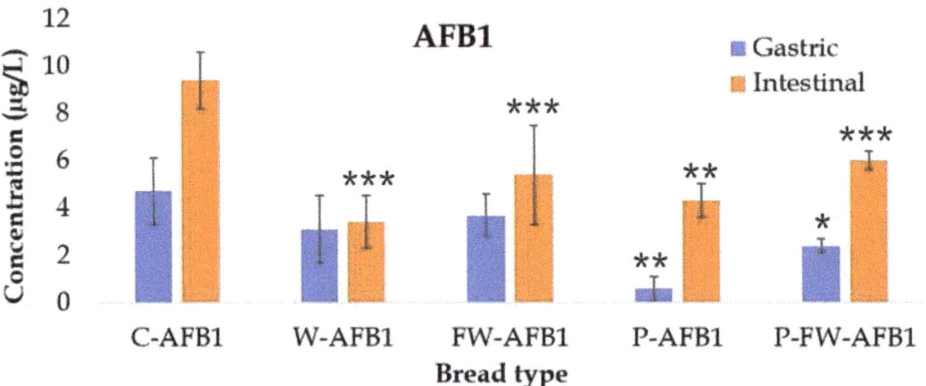

Figure 1. Gastric and intestinal concentration (µg/L) of aflatoxin B_1 (AFB1) during the in vitro simulated digestion (n = 5). Significant differences from the control indicated as $p < 0.05$ (*), $p < 0.01$ (**), $p < 0.001$ (***). Control bread + AFB1 (C-AFB1); bread + whey + AFB1 (W-AFB1); bread + fermented whey + AFB1 (FW-AFB1); bread + pumpkin + AFB1 (P-AFB1); bread + fermented whey + pumpkin +AFB1 (FW-P-AFB1).

The same trend was confirmed by the bioaccessibility results, where the highest values were achieved for non-enriched bread (C-AFB1) with gastric and intestinal bioaccessibility of 61 ± 19% and 114 ± 9%, respectively. Bioaccessibility values higher than 100% have been previously reported for mycotoxins in processed cereal-based food samples, and it could be attributed to possible interactions established between food matrix, mycotoxins, and digestive fluids [34,35]. Table 2 shows AFB1 gastric and intestinal bioaccessibility (%) from bread with and without enrichment ingredients. As it is shown, all enriched breads (W, FW, P, and FW-P) managed to reduce AFB1 bioaccessibility in some extent, and breads enriched with pumpkin showed the lowest bioaccessibility (4 ± 2 % and 29 ± 5 %) in gastric ($p \leq 0.05$) and intestinal ($p \leq 0.0001$) extracts, respectively, followed by bread enriched with the combination of fermented whey and pumpkin (15 ± 2% and 37 ± 3%, respectively). Since several studies have suggested the use of LAB fermentation, in order to bind AFB1 and reduce its gastrointestinal bioaccessibility [17,18,23,36,37] milk whey additive with and without previous LAB fermentation was examined in order to determine whether there could be any reduction on AFB1 bioaccessibility due to the bioactive compounds generated during fermentation. As shown in Table 2, enrichment treatments with W and FW showed lower AFB1 intestinal bioaccessibility (41–52%) compared to control bread (114%) (Table 2).

Table 2. Gastric and intestinal bioaccessibility (%) of AFB1 in bread (C), and enriched breads with milk whey (W), fermented milk whey (FW), pumpkin (P), and fermented milk whey + pumpkin (P-FW). Significantly different from the control, $p \leq 0.05$ (*), $p \leq 0.0001$ (***). Mean ± standard deviation (n = 5).

Bread Type	Gastric Bioaccessibility (%)	Intestinal Bioaccessibility (%)
C-AFB1	61 ± 19	114 ± 9
W-AFB1	34 ± 15	41 ± 13 ***
FW-AFB1	43 ± 11	52 ± 5 ***
P-AFB1	4 ± 2 *	29 ± 5 ***
P-FW-AFB1	15 ± 2 *	37 ± 3 ***

Control bread + AFB1 (C-AFB1); bread + whey + AFB1 (W-AFB1); bread + fermented whey + AFB1 (FW-AFB1); bread + pumpkin + AFB1 (P-AFB1); bread + fermented whey + pumpkin +AFB1 (FW-P-AFB1).

Data about AFs bioaccessibility is valuable to mitigate chronic hazards, mainly regarding food and feed that are frequently consumed; also, mycotoxins are thermal-resistant compounds, making it difficult to mitigate their presence in food and feed by conventional processes that apply moderated conditions such as cooking, extrusion, and baking. Studies concerning the bioaccessibility of AFB1 are scarce. Similarly to the obtained results, AFB1 bioaccessibility was between 83–108% for peanut slurry [35], between 85 and 98%, for various spiked food matrices (peanut, pistachio, hazelnut, dried figs, paprika, wheat, and maize) [38]; and 85% in fish feed [39]. Conversely, lower values were obtained for AFB1 bioaccessibility in spiked loaf bread during the stomach and the duodenal digestion reaching 54% and 26%, respectively [17]. It should be noted that food matrix, contamination level, compound, and type of contamination (spiked versus naturally contaminated) and even the gastrointestinal digestion model used can interfere with mycotoxin bioaccessibility performances as well [37].

There are no available studies assessing pumpkin (or carotenoids-rich food) and whey on AFB1 bioaccessibility reduction. However, the protective effect of pumpkin extract against AFB1 toxicity and *Aspergillus flavus* induced lung damage was reported in rats, effect that may be related to the antioxidant constituents of pumpkin extract [40].

Along the same line, other authors demonstrated that the antioxidant compounds of grape seed meal counteracted AFB1 toxicity in piglet mesenteric lymph nodes, showing a reduction of AFB1-induced oxidative stress by increasing the activity of glutathione peroxidase (GPx) and superoxide dismutase (SOD) and decreasing lipid peroxidation [41]. Supplementing the diet of ducks with antioxidants such as curcumin increased antioxidant capacity, inhibiting lipid and protein oxidation in their meat [42]. Moreover, the protective effects of lycopene against the toxic effects of AFB1 in kidney and heart were evaluated in rats [27], and the cellular and molecular mechanisms by which the antioxidants β-carotene and lycopene inhibit AFB1-induced toxic changes were also investigated in several cell lines, including Hep-G2 (Hep G2) cells [43].

On the other hand, Ferrer et al. [44] investigated the influence of some food ingredients (including milk whey, β-lactoglobulin, and calcium caseinate) and probiotic microorganisms on the bioaccessibility of deoxynivalenol (DON), zearalenone (ZEN), beauvericin (BEA), and enniatins (ENs A, A1, B, B1) in wheat crispy breads concluding that the addition of prebiotics and bioactive microorganisms decreased the bioaccessibility of mycotoxins, with a concentration-dependent behavior [44].

2.3. OTA Bioaccessibility

Regarding OTA gastrointestinal digestion, it was observed that concentrations in the gastric compartment were considerably lower than those detected in the intestinal stage (Figure 2), indicating that OTA release from bread mainly occurs at the intestinal digestion step with an important effect of medium pH. OTA consists of a dihydroisocoumarin moiety linked through its 7-carboxyl group by an amide linkage to L-phenylalanine. The pKa values are in the ranges of 4.2–4.4 and 7.0–7.3, respectively, for the carboxyl group of the phenylalanine moiety and the phenolic hydroxyl group of the isocoumarin part. This indicates that, in aqueous solutions near pH 7, both the monoanion (OTA$-$) and the dianion (OTA$^{2}-$) are present, whereas the molecular toxin is prevalent in acid solutions (pH < 3). This would cause the mycotoxin to be released with pH medium changes from gastric to intestinal conditions. In fact, numerous studies in which OTA adsorption (by adsorbing agents) was investigated by in vitro gastrointestinal digestions showed that OTA adsorption was significantly affected by the pH at acid gastric conditions [13,45].

Gastric and intestinal OTA concentrations in control bread (C-OTA) were 9.1 $\pm$ 0.1 and 103.8 $\pm$ 11.3 g/L, respectively. Gastric OTA concentrations in enriched breads were significantly lower (p < 0.05) in the case of whey and pumpkin-fermented whey additives, while lower intestinal OTA concentrations were observed in fermented whey and pumpkin enriched breads (Figure 2). OTA gastric concentrations in enriched breads ranged between 1.8 $\pm$ 0.6 μg/L (W) and 33.3 $\pm$ 0.9 μg/L (P-FW), while intestinal digests showed concentrations from 77.5 $\pm$ 7.0 (P) to 119.7 $\pm$ 7.7 μg/L (P-FW). The lowest OTA intestinal concentration was found for pumpkin treatment, however, in the gastric extract the lowest OTA levels were found for whey, followed by pumpkin enrichments (Figure 2).

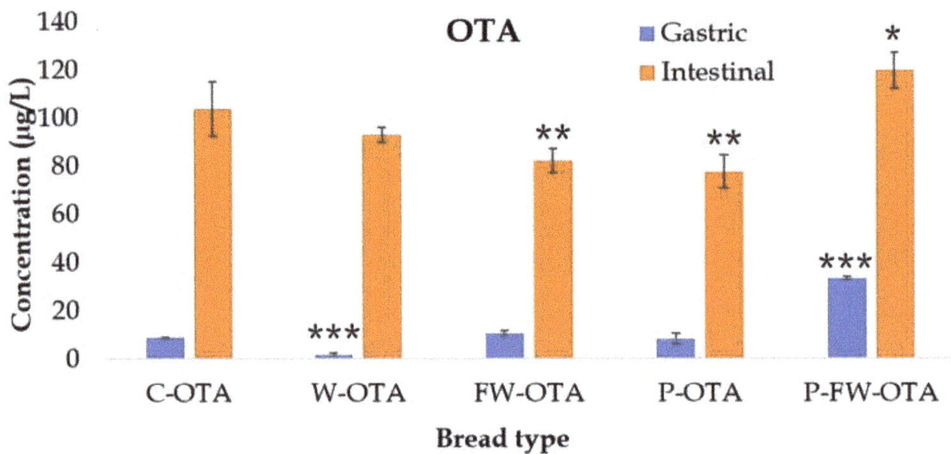

Figure 2. Gastric and intestinal concentration (µg/L) of ochratoxin A (OTA) during the in vitro simulated digestion ($n = 5$). Significant differences from the control indicated as $p < 0.05$ (*), $p < 0.01$ (**), $p < 0.001$ (***). Control bread + OTA (C-OTA); bread + whey + OTA (W-OTA); bread + fermented whey + OTA (FW-OTA); bread + pumpkin + OTA (P-OTA); bread + fermented whey + pumpkin + OTA (FW-P-OTA).

OTA bioaccesibility was examined with and without enriched ingredients, showing gastric and intestinal bioaccessibility in the control bread of 7.7 ± 0.1% and 88 ± 10%, respectively (Table 3). This difference may be explained since OTA can be released when the medium pH increases up to 7 (intestinal conditions), due to the physical-chemical properties of this molecule, as previously discussed. Previous works demonstrated high OTA bioaccessibility (100%) in naturally contaminated buckwheat [35] and lower bioaccessibility in naturally (22%) and artificially contaminated infant food (29–32%) [37], suggesting that bioaccessibility depends on several factors, such as food product, contamination level, compound, and type of contamination [34,38,46].

Table 3. Gastric and intestinal bioaccessibility (%) of OTA in bread (C), and enriched breads with milk whey (W), fermented milk whey (FW), pumpkin (P), and fermented milk whey + pumpkin (P-FW). Significantly different from the control, $p \leq 0.05$ (*), $p \leq 0.001$ (**), $p \leq 0.0001$ (***). Mean ± standard deviation ($n = 5$).

Bread Type	Gastric Bioaccessibility (%)	Intestinal Bioaccessibility (%)
C-OTA	7.7 ± 0.1	88 ± 10
W-OTA	1.4 ± 0.5 **	74 ± 2 *
FW-OTA	9.0 ± 1.1	70 ± 4 *
P-OTA	6.1 ± 1.8	58 ± 5 ***
P-FW-OTA	21.6 ± 0.6 ***	78 ± 5

Control bread + OTA (C-OTA); bread + whey + OTA (W-OTA); bread + fermented whey + OTA (FW-OTA); bread + pumpkin + OTA (P-OTA); bread + fermented whey + pumpkin + OTA (FW-P-OTA).

With regard to the enriched breads, gastric bioaccessibility values found in the present study were between 1.4 ± 0.5% (W-OTA) and 21.6% (P-FW-OTA), being statistically lower than the control in the case of whey- and pumpkin-enriched breads. Intestinal bioaccessibility of enriched breads was in all cases lower than in the control, ranging from 58 ± 5% (P-OTA) to 74 ± 2% (W-OTA), with significant difference in all enrichments but the combination pumpkin-fermented-whey (P-FW-OTA). Similarly, as for AFB1, the lowest OTA intestinal bioaccessibility was found for pumpkin, while the lowest OTA gastric bioaccessibility was found for whey, followed by pumpkin enrichments (Table 3).

There are no available studies assessing the effect of pumpkin and whey on reducing OTA bioaccessibility in vitro. However, the detoxification potential of whey powder against OTA harmful effects was investigated in broilers showing significant reduction of the hematobiochemical parameters raised by OTA treatment, as well as reduction in OTA residues detected in several organs including kidney, suggesting the potential application of whey ingredient in broiler feeds to reduce the negative effects of OTA in animals as efficiently as commercial mycotoxin binders [47]. Moreover, whey supplementation showed a vital role in maintaining the integrity of liver and kidney functions in fish exposed to OTA, ameliorating animals' performance, histopathological alterations, and biochemical parameters, and demonstrating its protective role against OTA toxicity especially with low dose [48,49].

2.4. Effect of Bread Enrichment with Bioactive Ingredients on Mycotoxin Reduction

This is the first work in which these natural ingredients have been added to bread formulation, alone and in combination, to study AFB1 and OTA bioaccessibility and their efficacy in reducing mycotoxins bioaccessibility. In the present work, pumpkin extract demonstrated to be the most effective treatment applied in bread, with significant intestinal bioaccessibility reductions ($p < 0.0001$) of both mycotoxins. AFB1 bioaccessibility decreased from 114% in control bread, up to 29% in bread enriched with pumpkin (P) (Table 2). Moreover, pumpkin enrichment revealed the major effect on reducing OTA intestinal bioaccessibility with significant reductions ($p < 0.0001$) from 88% in control bread up to 58% in bread enriched with pumpkin (P) (Table 3). This means AFB1 and OTA reductions of 74% and 34%, respectively, when pumpkin is present in bread formulation (Table 4).

Table 4. Intestinal bioaccessibility reduction (%) of AFB1 and OTA in bread enriched with milk whey (W), fermented milk whey (FW), pumpkin (P), and fermented milk whey + pumpkin (P-FW) compared to the control bread.

Bread Ingredient	Intestinal Bioaccessibility Reduction (%)	
	AFB1	OTA
W bread	64 ± 11	16 ± 3
FW bread	57 ± 4	20 ± 5
P bread	74 ± 4	34 ± 6
P-FW bread	68 ± 2	11 ± 6

W bread (bread enriched with whey), FW bread (bread enriched with fermented whey), P bread (bread enriched with pumpkin), and P-FW bread (bread enriched with pumpkin+ fermented whey).

Whey additive also showed significant reductions ($p < 0.0001$) of AFB1 bioaccessibility, from 114% in control bread, up to 41% and 52% in bread enriched with whey (W) and fermented whey (FW), respectively (Table 2). However, in the case of OTA, bioaccessibility values decreased from 88% in control bread up to 74% (W) and 70% (FW) in enriched breads (Table 3). Therefore, AFB1 and OTA intestinal bioaccessibility reductions between 57–64% and 16–20%, respectively, were achieved compared to control breads without preservatives (Table 4). However, no high differences were observed when both ingredients were combined (FW-P), reaching AFB1 and OTA bioaccessibility reductions of 68% and 11% for each mycotoxin, respectively (Table 4).

3. Conclusions

In this study, mycotoxins (AFB1 and OTA) bioaccessibility in bread enriched with milk whey and pumpkin has been studied by an in vitro digestion model. The obtained results indicate that almost all enrichment treatments significantly reduced AFB1 and OTA bioaccessibility compared to the control bread without preservatives. Interesting reductions were observed for milk whey; however, pumpkin seems to be the most efficient ingredient on reducing mycotoxin bioaccessibility. Hence, the addition of pumpkin and whey in the bread making process could be a strategy to reduce the absorbable fraction of mycotoxins at the intestinal level. Therefore, they can be potential candidates as bioactive ingredients for bread formulation.

One of the main challenges in the food technology field is to develop new biopreservatives that would support detoxification in a broad spectrum of food matrices. These ingredients could be used along with other biocontrol agents to counteract AFB1 and OTA in bakery products.

In addition, these natural ingredients introduced in bread dough at levels (1%) that can be applied in the industrial bread-making process could produce increases in the loaf volume, as well as improve the organoleptic acceptability and breads' shelf-life, potentially enriching their nutritional value as previously shown in the literature. Therefore, their potential application at the industry level is very feasible and should be further studied considering their organoleptic and nutritional qualities, but a larger sample size should be included.

The determination of mycotoxins bioaccessibility by in vitro methods offers an appealing alternative to human and animal studies avoiding the use of more complex cell culture techniques or the use of animals in expensive in vivo experiments, complying with the principle of the "3Rs": reducing, reusing, and recycling resources. However, despite the wide applications of these static in vitro digestion models, limitations such as oversimplify the digestive physiology and failing to mimic the dynamic aspects of the digestive process should be considered when interpreting the data.

4. Material and Methods

4.1. Chemicals, Reagents and Biological Strains

AFB1 and OTA standard solutions (purity > 99%) were purchased from Sigma-Aldrich (St. Louis, MO, USA). Potassium chloride (KCl), potassium thiocyanate (KSCN), sodium dihydrogen phosphate (NaH_2PO4), sodium sulfate (Na_2SO_4), sodium chloride (NaCl), sodium hydrogen carbonate (NaHCO3), urea ($CO(NH_2)_2$), α-amylase (930 U mg^{-1} A3403), hydrochloric acid (HCl), sodium hydroxide (NaOH), formic acid (HCOOH), pepsin A (674 U mg^{-1} P7000), pancreatin (762 U mg^{-1} P1750), bile salts (B8631), and phosphate buffer saline (PBS, pH 7.5) were purchased from Sigma-Aldrich (Madrid, Spain). Methanol and ethyl acetate were supplied by Fisher Scientific (Madrid, Spain). Deionized water was purchased from a Milli-Q water purification system (Millipore, Bedford, MA, USA). LAB used in this study (*L. plantarum CECT 220*) was obtained from the Spanish Type Culture Collection, CECT, Science Park of the University of Valencia (Paterna, València, Spain). The cultures were kept at −80 °C in glycerol 25% until their use.

4.2. Flour Contamination

Barley and maize flour were naturally contaminated by the fungal species-producers of AFB1 and OTA *Aspergillus steynii 20510* (obtained from CECT) and *A. flavus ITEM 8111* (obtained from the Agro-Food Microbial Culture Collection of The Institute of Sciences and Food Production (ISPA, Bari, Italy)), respectively. To that aim, 300–350 g of maize or barley were introduced in 1 L glass jars previously autoclaved. Then, cereals were contaminated with 15–20 mL of spores and mycelium suspension in peptone water with Tween 80 (0.1% both) of the corresponding fungal strain. Glass jars were then left at room temperature in darkness for one month. After that, cereals were autoclaved to remove the fungal contamination and samples were ground to flour until complete homogenization. Mycotoxins in contaminated flour were quantified by high performance liquid chromatography coupled to time-of-flight mass spectrometry (HPLC–MS/qTOF) after a solid-liquid extraction, as explained below in Section 4.5.

4.3. Bread Natural Ingredients

Milk whey strained from goat's milk coagulated by commercial rennet (starter culture R-604) was obtained from the ALCLIPOR society, S.A.L. (Benassal, Spain). Milk whey ingredient was studied with and without LAB fermentation. For milk whey fermentation, 4 mL of LAB suspension at concentration of 10^8 CFU/mL were added to 40 mL milk whey, previously pasteurized in accordance with standardized guidelines [50], and samples were incubated 72 h at 37 °C to allow LAB fermentation. Fermented and non-fermented whey were then lyophilized to obtain a homogeneous powder.

Pumpkin was obtained from a commercial supermarket in Valencia (Spain). Pumpkin powder was prepared by pealing and cutting the fresh vegetable (skin and seeds previously removed) followed by lyophilization and grinding to obtain a homogeneous powder. Both ingredients were analyzed to confirm the absence of mycotoxins, and stored at −20 °C until their use.

4.4. Bread Preparation and Baking

Fifteen breads were prepared combining the bioactive ingredients with and without the studied mycotoxins, OTA and AFB1. Control bread preparation was performed by the following recipe: 300 g of wheat flour, 165 mL of mineral water (37 °C), 20 g of yeast for bakery products (Levital, Spain), 10 g of sucrose, and 6.5 g of NaCl. After mixing all the ingredients, doughs were homogenized in a bakery machine (Silver Crest) for 5 min, shaped in molds (100 g), covered with a damp cloth, and left 1 h to ferment at room temperature. After that, breads were covered with silver foil and baked at 200 °C for 45 min in a Memmert ULE 500 muffle furnace (Madrid, Spain). Finally, breads were unmolded and cooled at room temperature (1 h). Enriched breads were then prepared by slight modifications on the control recipe obtaining a) control bread (C), b) 1% milk whey bread (W), c) 1% of fermented milk whey bread (FW), d) 1% of lyophilized pumpkin bread (P), and e) 1% of fermented milk whey + 1% of lyophilized pumpkin bread (FW-P). Then, contaminated breads with OTA and AFB1 were prepared for the five bread conditions (C, W, FW, P, FW-P) substituting a fraction of wheat flour by 1.2 g of barley flour contaminated with OTA and/or 10 g of maize flour contaminated with AFB1. Fifteen breads were obtained, three for each enrichment type, as shown in Table 5.

4.5. Bread Analysis

For bread analysis the method described by Saladino et al. [5] was followed. Briefly, 5 g of bread were accurately weighed (precision 0.1 mg), transferred to centrifuge tubes (50 mL), and 25 mL of methanol were added. Samples were crushed in Ultraturrax (T 18 digital ULTRA-TURRAX®, Staufen, Germany) for 5 min and centrifuged at 4500 rpm for 10 min (Centrifuge 5810R, Eppendorff, Germany). Supernatant was collected in new centrifuge tubes and evaporated until complete dryness using a rotavapor (BUCHI Rotavapor R-200, Postfach, Switzerland) and turbovap (TurboVap LV Evaporator, Zymark, Hopkinton, MA, USA). Samples were then reconstituted in 1 mL of methanol, filtered with a 0.22-μm filter (Phenomenex, Madrid, Spain) and injected for HPLC–MS/qTOF analysis. All the analyses were performed by triplicate ($n = 3$).

4.6. HPLC–MS/qTOF Conditions

Chromatographic analysis was performed on an Agilent 1200 (Agilent Technologies, Santa Clara, CA, USA), consisting of an auto sampler, vacuum degasser, and binary pump. Analyte separation was carried out using a Gemini C18 column (50 mm × 2 mm, 110 Å and particle size 3 μm; Phenomenex, (Phenomenex, Palo Alto, CA, USA)). The mobile phases consisted of water (solvent A) and acetonitrile (solvent B) both with 0.1% of formic acid with an elution flow rate of 0.3 mL/min and an elution gradient as follows: 0 min, 5% B; 30 min, 95% B; 35 min, 5% B. Total analysis run was achieved in 25 min and the injection volume was 5 μL. For mass spectrometry analyses, a MS/qTOF (6540 Agilent Ultra High-Definition Accurate Mass, Agilent Technologies, Santa Clara, CA, USA), coupled to an Agilent Dual Jet Stream electrospray ionization (Dual AJS ESI) interface operating in positive ion mode was used. Optimized mass spectrometry parameters included: fragment voltage 175 V; capillary voltage 3.5 kV; collision energy 10, 20 and 40 eV, nebulizer pressure 30 psi; drying gas flow (N_2) 8 L/min, temperature 350 °C. Data analysis was performed by MassHunter Qualitative Analysis Software B.08.00. (Agilent Technologies, Santa Clara, CA, USA).

Table 5. Bread conditions prepared in the present study. Whey concentration (1%), fermented whey concentration (1%), pumpkin concentration (1%).

Bread Type
Control Bread
Bread (C)   Bread + AFB1 (C-AFB1)   Bread + OTA (C-OTA)
Whey bread
Bread + whey (W)   Bread + whey + AFB1 (W-AFB1)   Bread + whey + OTA (W-OTA)
Fermented whey bread
Bread + fermented whey (FW)   Bread + fermented whey + AFB1 (FW-AFB1)   Bread + fermented whey + OTA (FW-OTA)
Pumpkin bread
Bread + pumpkin (P)   Bread + pumpkin + AFB1 (P-AFB1)   Bread + pumpkin + OTA (P-OTA)
Fermented whey-Pumpkin bread
Bread + fermented whey + pumpkin (FW-P)   Bread + fermented whey + pumpkin +AFB1 (FW-P-AFB1)   Bread + fermented whey+ pumpkin + OTA (FW-P-OTA)

4.7. In Vitro Static Digestion Model

To reproduce the complete process of the human digestion, an in vitro digestion model was applied based on Manzini et al. (2015) [51], with some modifications. Briefly, 10 g of bread were placed in sterilized plastic bags (500 mL), mixed with 6 mL of artificial saliva and 80 mL of Milli-Q water (37 °C) (Millipore, Bedford, MA, USA). To replicate the oral phase, an IUL Stomacher (IUL S.A, Barcelona, Spain) was applied for 30 s simulating mastication process. Saliva was prepared the day before (and adjusted to pH = 6.8) by mixing 10 mL of KCl (89.6 g/L), 10 mL of KSCN (20 g/L), 10 mL of NaH_2PO_4 (88.8 g/L), 10 mL of Na_2SO_4 (57 g/L), 1.7 mL of NaCl (175.3 g/L), 20 mL $NaHCO_3$ (84.7 g/L), 8 mL of urea (25 g/L), 290 mg of α-amylase, and 430 mL of distilled water. After simulation of the oral phase, the content was transferred to a topaz Erlenmeyer flask to continue with the gastric phase. The mixture was acidified to pH = 2 with 6N HCl solution. Then 0.5 g of pepsin solution (1 g in 25 mL of 0.1N HCl) and 14 mL of Milli-Q water (37 °C) were added to complete the final volume to 100 mL. Samples were incubated 2h at 37 °C under darkness and slight agitation (100 rpm) using an orbital shaker (Infors AG CH-4103, Bottmingen, Switzerland). After that, gastric aliquots (15 mL) were kept for analysis and pancreatic digestion was reproduced by adding 1.25 g bile salts/pancreatin mixture (0.1 g pancreatin and 0.625 g of bile salts dissolved in 25 mL of 0.1N $NaHCO_3$), at pH = 6.5 (0.5N $NaHCO_3$). Extracts were incubated as previously (2 h at 37 °C in darkness and slight agitation), pH was finally adjusted to 7.2 (0.5N NaOH), samples were centrifuged (4500 rpm for 10 min at 4°C), and supernatant was collected obtaining the intestinal digested extracts. All the digestions were performed with five replicates ($n = 5$).

4.8. Gastrointestinal Extracts Analysis and Bioaccessibility

Gastric aliquots and intestinal extracts were filtered (0.22 µm filter) and directly injected in the HPLC–MS/qTOF for mycotoxins determination. Standard calibration curves were prepared in methanol (1–1000 µg/L) from OTA and AFB1 standards (1000 mg/L). For quantitation purposes, matrix-matched calibration curves were prepared for all bread conditions (C, W, FW, P, FW-P) by spiking blank digested extracts with OTA and AFB1 at the same concentrations.

Mycotoxins bioaccessibility (%) was calculated as the percentage of mycotoxins from the initial digested bread that were detected in the digested extracts. The mycotoxin quantity (µg) in 10g of bread (A) was calculated from bread concentration (µg/kg) by conversion factors ($\times 10/1000$). The mycotoxin quantity (µg) in 100 mL of digest (B) was calculated from digest concentration (µg/L) by conversion factors ($\times 100/1000$). To calculate the bioaccessibility percentage, both quantities were related as $B/A \times 100$.

Combining all the calculations, the bioaccessibility could be directly calculated as indicated in Equation (1):

$$\text{Bioaccesibility} = \frac{\text{digest concentration} \left(\frac{\mu g}{L}\right) \times 1000}{\text{bread concentration} \left(\frac{\mu g}{kg}\right)} \tag{1}$$

Statistical analysis was performed by a Student's repeated measures t-test ($n \geq 3$) to analyze all the results considered as significant, p values < 0.05.

Author Contributions: Conceptualization, G.M. and L.M.; methodology, L.E.; validation, L.E.; formal analysis, F.A.; investigation, F.A. and L.E.; data curation, L.E. and P.V.-D.; writing—original draft preparation, F.A. and L.E.; writing—review and editing, P.V.-D.; visualization, L.E.; supervision, G.M. and L.M.; project administration, L.M. and G.M.; funding acquisition, J.M. and G.M. All authors have read and agreed to the published version of the manuscript.

Funding: This research was funded by [Ministry of Science and Innovation] grant number [PID2019-108070RB-100].

Institutional Review Board Statement: Not applicable.

Informed Consent Statement: Not applicable.

Data Availability Statement: Data are available upon request; please contact the contributing authors.

Conflicts of Interest: There are no conflict of interest to declare.

References

1. International Agency for Research on Cancer (IARC). *Monographs. Some Naturally Occurring Substances: Food Items and Constituents, Heterocyclic Aromatic Amines and Mycotoxins*; IARC: Lyon, France, 1993; Volume 56.
2. European Commission (EC). No 1881/2006 of 19 December 2006 setting maximum levels for certain contaminants in foodstuffs. *Off. J. Eur. Union* **2006**, *364*, 5–24.
3. EFSA Panel on Contaminants in the Food Chain (CONTAM). Scientific opinion on the risk assessment of ochratoxin A in food. *EFSA J.* **2020**, *18*, 6113. [CrossRef]
4. Serrano, A.B.; Font, G.; Ruiz, M.J.; Ferrer, E. Co-occurrence and risk assessment of mycotoxins in food and diet from Mediterranean area. *Food Chem.* **2012**, *135*, 423–429. [CrossRef] [PubMed]
5. Saladino, F.; Quiles, J.M.; Mañes, J.; Fernández-Franzón, M.; Bittencourt, F.L.; Meca, G. Dietary exposure to mycotoxins through the consumption of commercial bread loaf in Valencia, Spain. *LWT—Food Sci. Tech.* **2017**, *75*, 697–701. [CrossRef]
6. Coppa, S.C.C.F.; Khaneghah, A.M.; Alvito, P.; Assunção, R.; Martins, C.; Eş, I.; Gonçalves, B.L.; Valganon de Neeff, D.; Sant'Ana, A.S.; Corassin, C.H.; et al. The occurrence of mycotoxins in breast milk, fruit products and cereal-based infant formula: A review. *Trends Food Sci. Techol.* **2019**, *92*, 81–93. [CrossRef]
7. Palumbo, R.; Crisci, A.; Venâncio, A.; Cortiñas Abrahantes, J.; Dorne, J.-L.; Battilani, P.; Toscano, P. Occurrence and co-occurrence of mycotoxins in cereal-based feed and food. *Microorganisms* **2020**, *8*, 74. [CrossRef]

8.   Maietti, A.; Tedeschi, P.; Catani, M.; Stevanin, C.; Pasti, L.; Cavazzini, A.; Marchetti, N. Nutrient composition and antioxidant performances of bread-making products. *Foods* **2021**, *10*, 938. [CrossRef]

9.   Bol, E.K.; Araujo, L.; Veras, F.V.; Welke, J.E. Estimated exposure to zearalenone, ochratoxin A and aflatoxin B1 through the consume of bakery products and pasta considering effects of food processing. *Food Chem. Toxicol.* **2016**, *89*, 85–91. [CrossRef]

10.   Bryła, M.; Ksieniewicz-Woźniak, E.; Stępniewska, S.; Modrzewska, M.; Waśkiewicz, A.; Szymczyk, K.; Szafrańska, A. Transformation of ochratoxin A during bread-making processes. *Food Control* **2021**, *125*, 107950. [CrossRef]

11.   Vidal, A.; Morales, H.; Sanchis, V.; Ramos, A.J.; Marín, S. Stability of DON and OTA during the bread making process and determination of process and performance criteria. *Food Control* **2014**, *40*, 234–242. [CrossRef]

12.   Vila-Donat, P.; Marín, S.; Sanchis, V.; Ramos, A.J. A review of the mycotoxin adsorbing agents, with an emphasis on their multi-binding capacity, for animal feed decontamination. *Food Chem. Toxicol.* **2018**, *114*, 246–259. [CrossRef] [PubMed]

13.   Vila-Donat, P.; Marín, S.; Sanchis, V.; Ramos, A.J. New mycotoxin adsorbents based on tri-octahedral bentonites for animal feed. *Anim. Feed Sci. Technol.* **2019**, *255*, 114228. [CrossRef]

14.   Vila-Donat, P.; Marín, S.; Sanchis, V.; Ramos, A.J. Tri-octahedral bentonites as potential technological feed additive for Fusarium mycotoxin reduction. *Food Addit. Contam. Part A Chem. Anal. Control* **2020**, *37*, 1374–1387. [CrossRef]

15.   Awuchi, C.G.; Ondari, E.N.; Ogbonna, C.U.; Upadhyay, A.K.; Baran, K.; Okpala, C.O.R.; Korzeniowska, M.; Guiné, R.P.F. Mycotoxins affecting animals, foods, humans, and plants: Types, occurrence, toxicities, action mechanisms, prevention, and detoxification strategies—A revisit. *Foods* **2021**, *10*, 1279. [CrossRef]

16.   Serrano-Niño, J.C.; Cavazos-Garduño, A.; Hernandez-Mendoza, A.; Applegate, B.; Ferruzzi, M.G.; San Martin-González, M.F.; García, H.S. Assessment of probiotic strains ability to reduce the bioaccessibility of aflatoxin M1 in artificially contaminated milk using an in vitro digestive model. *Food Control* **2013**, *31*, 202–207. [CrossRef]

17.   Saladino, F.; Posarelli, E.; Luz, C.; Luciano, F.B.; Rodriguez-Estrada, M.T.; Mañes, J.; Meca, G. Influence of probiotic microorganisms on aflatoxins B1 and B2 bioaccessibility evaluated with a simulated gastrointestinal digestion. *J. Food Composit. Anal.* **2018**, *68*, 128–132. [CrossRef]

18.   Nazareth, T.D.M.; Luz, C.; Torrijos, R.; Quiles, J.M.; Luciano, F.B.; Mañes, J.; Meca, G. Potential application of lactic acid bacteria to reduce aflatoxin B1 and fumonisin B1 occurrence on corn kernels and corn ears. *Toxins* **2019**, *12*, 21. [CrossRef]

19.   Illueca, F.; Vila-Donat, P.; Calpe, J.; Luz, C.; Meca, G.; Quiles, J.M. Antifungal activity of biocontrol agents in vitro and potential application to reduce mycotoxins (aflatoxin B1 and ochratoxin A). *Toxins* **2021**, *13*, 752. [CrossRef] [PubMed]

20.   Quiles, J.M.; Manyes, L.; Luciano, F.B.; Mañes, J.; Meca, G. Effect of the oriental and yellow mustard flours as natural preservative against aflatoxins B1, B2, G1 and G2 production in wheat tortillas. *J. Food Sci. Technol.* **2015**, *52*, 8315–8321. [CrossRef]

21.   Quiles, J.M.; Torrijos, R.; Luciano, F.B.; Mañes, J.; Meca, G. Aflatoxins and A. flavus reduction in loaf bread through the use of natural ingredients. *Molecules* **2018**, *4*, 1638. [CrossRef]

22.   Torrijos, R.; Nazareth, T.M.; Pérez, J.; Mañes, J.; Meca, G. Development of a bioactive sauce based on oriental mustard flour with antifungal properties for pita bread shelf life improvement. *Molecules* **2019**, *24*, 1019. [CrossRef]

23.   Luz, C.; Izzo, L.; Ritieni, A.; Mañes, J.; Meca, G. Antifungal and antimycotoxigenic activity of hydrolyzed goat whey on *Penicillium* spp: An application as biopreservation agent in pita bread. *LWT* **2020**, *118*, 108717. [CrossRef]

24.   Bergantin, C.; Maietti, A.; Tedeschi, P.; Font, G.; Manyes, L.; Marchetti, N. HPLC-UV/Vis-APCI-MS/MS determination of major carotenoids and their bioaccessibility from "Delica" (*Cucurbita maxima*) and "Violina" (*Cucurbita moschata*) pumpkins as food traceability markers. *Molecules* **2018**, *23*, 2791. [CrossRef] [PubMed]

25.   Alonso-Garrido, M.; Tedeschi, P.; Maietti, A.; Font, G.; Marchetti, N.; Manyes, L. Mitochondrial transcriptional study of the effect of aflatoxins, enniatins and carotenoids in vitro in a blood brain barrier model. *Food Chem. Toxicol.* **2020**, *137*, 111077. [CrossRef]

26.   Alonso-Garrido, M.; Frangiamone, M.; Font, G.; Cimbalo, A.; Manyes, L. In vitro blood brain barrier exposure to mycotoxins and carotenoids pumpkin extract alters mitochondrial gene expression and oxidative stress. *Food Chem. Toxicol.* **2020**, *153*, 112261. [CrossRef]

27.   Yilmaz, S.; Kaya, E.; Karaca, A.; Karatas, O. Aflatoxin B1 induced renal and cardiac damage in rats: Protective effect of lycopene. *Res. Vet. Sci.* **2018**, *119*, 268–275. [CrossRef] [PubMed]

28.   Chatzipaschali, A.A.; Stamatis, A.G. Biotechnological utilization with a focus on anaerobic treatment of cheese whey: Current status and prospects. *Energies* **2012**, *5*, 3492–3525. [CrossRef]

29.   Escrivá, L.; Manyes, L.; Vila-Donat, P.; Font, G.; Meca, G.; Lozano, M. Bioaccessibility and bioavailability of bioactive compounds from yellow mustard flour and milk whey fermented with lactic acid bacteria. *Food Funct.* **2021**, *12*, 11250. [CrossRef]

30.   Coppa, S.C.C.F.; Cirelli, A.C.; Gonçalves, B.L.; Barnabé, E.M.B.; Khaneghah, A.M.; Corassin, C.H.; Oliveira, C.A.F. Dietary exposure assessment and risk characterization of mycotoxins in lactating women: Case study of São Paulo state, Brazil. *Food Res. Int.* **2020**, *134*, 109272. [CrossRef] [PubMed]

31.   Zinedine, A.; Brera, C.; Faid, M.; Benlemlih, M.; Miraglia, M. Ochratoxin A and fusarium toxins in cereals from Morocco. *Cah. Agric.* **2017**, *16*, 11–15.

32.   Nogueira, A.; Oliveira, R.; Steel, C. Protein enrichment of wheat flour doughs: Empirical rheology using protein hydrolysates. *Food Sci. Technol.* **2020**, *4*, 97–105. [CrossRef]

33.   Rakcejeva, T.; Galoburda, R.; Cude, L.; Strautniece, E. Use of dried pumpkins in wheat bread production. *Procedia Food Sci.* **2011**, *1*, 441–447. [CrossRef]

34. Assunção, R.; Martins, C.; Dupont, D.; Alvito, P. Patulin and ochratoxin A co-occurrence and their bioaccessibility in processed cereal-based foods: A contribution for Portuguese children risk assessment. *Food Chem. Toxicol.* **2016**, *96*, 205–214. [CrossRef] [PubMed]

35. Versantvoort, C.H.M.; Oomen, A.G.; Van de Kamp, E.; Rompelberg, C.J.M.; Sips, A.J.A.M. Applicability of an in vitro digestion model in assessing the bioaccessibility of mycotoxins from food. *Food Chem. Toxicol.* **2005**, *43*, 31–40. [CrossRef] [PubMed]

36. Gratz, S.; Mykkanen, H.; Ouwehand, A.C.; Juvonen, R.; Salminen, S.; El-Nezami, H. Intestinal mucus alters the ability of probiotic bacteria to bind aflatoxin B1 in vitro. *Appl. Environ. Microbiol.* **2004**, *70*, 6306–6308. [CrossRef] [PubMed]

37. Kabak, B.; Brandon, F.A.E.; Var, I.; Blokland, M.; Sips, J.A. Effects of probiotic bacteria on the bioaccessibility of aflatoxin B1 and ochratoxin A using an in vitro digestion model under fed conditions. *J. Environ. Sci. Health Part B* **2009**, *44*, 472–480. [CrossRef] [PubMed]

38. Kabak, B.; Ozbey, F. Assessment of the bioaccessibility of aflatoxins from various food matrices using an in vitro digestion model, and the efcacy of probiotic bacteria in reducing bioaccessibility. *J. Food Compost. Anal.* **2012**, *27*, 21–31. [CrossRef]

39. Nogueira, W.V.; Oliveira, F.K.; Sibaja, K.V.M.; Garcia, S.O.; Kupski, L.; Souza, M.M.; Tesser, M.B.; Garda-Buffon, J. Occurrence and bioacessibility of mycotoxins in fish feed. *Food Addit. Contam. Part B* **2020**, *13*, 244–251. [CrossRef]

40. Saddiq, A.A.N.; Awedh, M.H. Pumpkin (*Cucurbita moschata*) against *Aspergillus flavus* and aflatoxin B1 induced lung cyto-morphological damage in rats. *Pak. J. Pharm. Sci.* **2019**, *32*, 575–579.

41. Marin, D.E.; Bulgaru, C.V.; Anghel, C.A.; Pistol, G.C.; Dore, M.I.; Palade, M.L.; Taranu, I. Grape seed waste counteracts aflatoxin B1 toxicity in piglet mesenteric lymph nodes. *Toxins* **2020**, *12*, 800. [CrossRef]

42. Jin, S.; Yang, H.; Liu, F.; Pang, Q.; Shan, A.; Feng, X. Effect of dietary curcumin supplementation on duck growth performance, antioxidant capacity and breast meat quality. *Foods* **2021**, *10*, 2981. [CrossRef]

43. Reddy, L.; Odhav, B.; Bhoola, K. Aflatoxin B1-induced toxicity in HepG2 cells inhibited by carotenoids: Morphology, apoptosis and DNA damage. *Biol. Chem.* **2006**, *387*, 87–93. [CrossRef]

44. Ferrer, E.; Manyes, L.; Mañes, J.; Meca, G. Influence of prebiotics, probiotics and protein ingredients on mycotoxins bioaccessibility. *Food Funct.* **2015**, *6*, 987–994. [CrossRef]

45. Avantaggiato, G.; Greco, G.; Damascelli, A.; Solfrizzo, M.; Visconti, A. Assessment of multi-mycotoxin adsorption efficacy of grape pomace. *J. Agric. Food Chem.* **2014**, *62*, 497–507. [CrossRef]

46. González-Arias, C.A.; Piquer-Garcia, I.; Marin, S.; Sanchis, V.; Ramos, A.J. Bioaccessibility of ochratoxin A from red wine in an in vitro dynamic gastrointestinal model. *World Mycotoxin J.* **2015**, *8*, 107–112. [CrossRef]

47. Mujahid, H.; Hashmi, A.S.M.Z.; Tayyab, M.; Shehzad, W. Protective effect of yeast sludge and whey powder against ochratoxicosis in broiler chicks. *Pak. Vet. J.* **2019**, *39*, 588–592. [CrossRef]

48. Mansour, T.A.; Safinaz, G.; Soliman, M.K.; Eglal, A.; Srour, T.M.; Mona, S.Z.; Shahinaz, M.H. Ameliorate the drastic effect of ochratoxin A by using yeast and whey in cultured *Oreochromus niloticus* in Egypt. *Life Sci. J.* **2011**, *8*, 68–81.

49. Mansour, T.A.; Omar, A.E.; Soliman, K.M.; Snour, M.T.; Nour, M.A. The antagonistic effect of whey on ochratoxin a toxicity on the growth performance, feed utilization, liver and kidney functions of Nile tilapia (*Oreochromis niloticus*). *Middle East. J. Appl. Sci.* **2015**, *5*, 176–183.

50. Izzo, L.; Luz, C.; Ritieni, A.; Mañes, J.; Meca, G. Whey fermented by using *Lactobacillus plantarum* strains: A promising approach to increase the shelf life of pita bread. *J. Dairy Sci.* **2020**, *103*, 5906–5915. [CrossRef] [PubMed]

51. Manzini, M.; Rodriguez-Estrada, M.T.; Meca, G.; Mañes, J. Reduction of beauvericin and enniatins bioaccessibility by prebiotic compounds: Evaluated in static and dynamic simulated gastrointestinal digestion. *Food Control* **2015**, *47*, 203–211. [CrossRef]

Article

Fusarium Mycotoxins in Maize Field Soils: Method Validation and Implications for Sampling Strategy

Kilian G. J. Kenngott [1], Julius Albert [1], Friederike Meyer-Wolfarth [2], Gabriele E. Schaumann [1] and Katherine Muñoz [3,*]

[1] Group of Environmental and Soil Chemistry, Institute for Environmental Sciences (iES) Landau, University of Koblenz-Landau, Fortstraße 7, 76829 Landau, Germany; kenngott@uni-landau.de (K.G.J.K.); albert.j@uni-landau.de (J.A.); schaumann@uni-landau.de (G.E.S.)

[2] Julius Kühn Institute (JKI), Federal Research Centre for Cultivated Plants, Institute for Plant Protection in Field Crops and Grassland, Messeweg 11/12, 38104 Braunschweig, Germany; friederike.meyer-wolfarth@julius-kuehn.de

[3] Group of Organic and Ecological Chemistry, Institute for Environmental Sciences (iES) Landau, University of Koblenz-Landau, Fortstraße 7, 76829 Landau, Germany

* Correspondence: munoz@uni-landau.de

Citation: Kenngott, K.G.J.; Albert, J.; Meyer-Wolfarth, F.; Schaumann, G.E.; Muñoz, K. *Fusarium* Mycotoxins in Maize Field Soils: Method Validation and Implications for Sampling Strategy. *Toxins* **2022**, *14*, 130. https://doi.org/10.3390/toxins14020130

Received: 4 December 2021
Accepted: 25 January 2022
Published: 9 February 2022

Publisher's Note: MDPI stays neutral with regard to jurisdictional claims in published maps and institutional affiliations.

Abstract: While mycotoxins are generally regarded as food contamination issues, there is growing interest in mycotoxins as environmental pollutants. The main sources of trichothecene and zearalenone mycotoxins in the environment are mainly attributed to *Fusarium* infested fields, where mycotoxins can wash off in infested plants or harvest residues. Subsequently, mycotoxins inevitably enter the soil. In this context, investigations into the effects, fate, and transport are still needed. However, there is a lack of analytical methods used to determine *Fusarium* toxins in soil matrices. We aimed to validate an analytical method capable of determining the toxins nivalenol (NIV), deoxynivalenol (DON), 15-acetyl-deoxynivalenol (15-AcDON), and zearalenone (ZEN), at environmentally relevant concentrations, in five contrasting agricultural soils. Soils were spiked at three levels (3, 9 and 15 ng g^{-1}), extracted by solid-liquid extraction assisted with ultrasonication, using a generic solvent composition of acetonitrile:water 84:16 (v:v) and measured by LC–HRMS. Method validation was successful for NIV, DON, and 15-AcDON with mean recoveries > 93% and RSD$_r$ < 10%. ZEN failed the validation criteria. The validated method was applied to eight conventionally managed maize field soils during harvest season, to provide a first insight into DON, NIV, and 15-AcDON levels. Mycotoxins were present in two out of eight sampled maize fields. Soil mycotoxin concentrations ranged from 0.53 to 19.4 ng g^{-1} and 0.8 to 2.2 ng g^{-1} for DON and NIV, respectively. Additionally, we found indication that "hot-spot" concentrations were restricted to small scales (<5 cm) with implications for field scale soil monitoring strategies.

Keywords: mycotoxins; *Fusarium*; soil; deoxynivalenol; nivalenol; environment; monitoring strategies; validated method; maize field

Key Contribution: Validation of a method for the quantitative analysis of DON, NIV and 15-ADON in agricultural soils and considerations for sampling strategies aimed to soil monitoring.

1. Introduction

Fusarium is a genus of filamentous fungi, of which some species are known to be pathogenic to various plants, including cereal crops. These pathogenic species cause several serious plant diseases, such as stalk rot, which is globally the main reason for crop loss in maize production [1]. Furthermore, they are able to produce toxic secondary metabolites of the mycotoxin families, Type B trichothecene and zearalenone [2]. Among these mycotoxin groups, deoxynivalenol (DON), nivalenol (NIV), and zearalenone (ZEN) are of particular interest due to the frequent occurrence and the concentration levels in harvest samples and freshwater streams [3–6]. Type B trichothecenes, such as DON and NIV, are toxic to humans and animals [7]. ZEN is an endocrine disrupting substance [8], which may have estrogenic effects on fish when transported into rivers [9]. For these reasons, concentrations of mycotoxins in food and feed are regulated in many countries for the safe commercialization of food and feed commodities [10,11], which is why there are many analytical methods for various food matrices [12,13].

Mycotoxins are usually regarded as food contamination problems but there is growing interest in mycotoxins as soil and aquatic pollutants of emerging concern. Some studies have detected mycotoxins in soil, surface water, and sewage sludge [4,14–19]. Diffusive sources of DON, NIV, and ZEN in the environment are mainly attributed to *Fusarium* infested fields, where mycotoxins can be washed off the infested plants entering soil and drainage water [4,20]. Another source can be harvesting residues leaching out mycotoxins on the fields [21,22]. In both cases, mycotoxins will initially enter the soil, where the effects, fate, and transport are scarcely investigated.

To assess mycotoxin occurrence and its fate in soil, suitable quantification methods are imperative. However, the available methods from food analyses cannot be easily applied to soil due to the complexity and heterogeneity of the soil matrices and the much lower expected concentration levels in soil (a lower $ng\,g^{-1}$ scale) compared to a food and feed ($\mu g\,g^{-1}$ scale, [6]). The most commonly applied analytical method for mycotoxin analysis is liquid chromatography with mass spectrometry (LC–MS) [13,23,24]. However, LC–MS measurement of mycotoxins is sensitive to matrix effects [25,26], which is caused by ions and organic acids contained in a sample, leading to signal enhancement or suppression. The potential load of such a matrix in soil samples is mainly described by physicochemical parameters, such as soil pH, soil organic carbon, and clay content. These parameters can be highly variable even at small field scales [27,28], leading to the necessity of the matrix effect assessment of individual samples. Therefore, extraction and exact quantification of mycotoxins in soils is challenging [29]. The usefulness of isotopically-labeled internal standards has been reported on in several studies [20,30,31]. However, it remains unclear whether this tool is suitable to compensate matrix effects of different soils, leading to a reduced number of matrix-matched calibrations.

Some methods were already reported to be effective for extraction and quantification of trichothecenes from sandy loam, loamy silty sand, and sandy loamy silt of a strawberry field [17,18]. Mortensen et al. [32] established an extraction method for ZEN using three different agricultural soils with varying texture and carbon content, but used spiking concentrations (30–$60\,ng\,g^{-1}$) that were clearly higher compared to natural levels (0.6–$1.1\,ng\,g^{-1}$, [19]; 0–$7.5\,ng\,g^{-1}$, [16]). In this study, we validate an extraction and quantification method with the aim of being suitable for selected Type B trichothecenes (NIV, DON, 15-AcDON) and ZEN in various agricultural soils at environmentally realistic concentrations.

Another challenge in soil mycotoxin analysis is the sampling strategy. Natural infestations of plants may not be homogeneously distributed over the field [33]. Randomly taken soil cores that are pooled together may lead to dilution and consequently no detection of mycotoxins or underestimation of mycotoxin concentrations. Furthermore, small scale concentration patterns may also hamper mycotoxin detection. Mycotoxins that are produced above ground in the plant can be washed off by the rain and rinse down the maize plant stem to the soil surface. Furthermore, *Fusarium* is able to grow on harvest residues left on top of the soil. Sub-soil mycotoxin production is yet not investigated, but it seems to be possible [19]. Both cases would lead to a "hot-spot" concentration close to individual infested plants or harvest residues, i.e., maize stubble. Therefore, the sampling strategy on the filed scale is crucial for reliable mycotoxin detection. Knowledge about these hot-spot soil mycotoxin concentrations is necessary to assess the potential dose-dependent effect on the soil microbiome or meso organisms. Additionally, plant uptake of mycotoxins has been shown for ZEN [34], ochratoxin A [35], DON [36], and was shown to be dose-dependent for aflatoxins [37]. The species *F. verticillioides* and *F. graminearum* are frequently found on maize in Germany [38]. These species are able to produce various type A and type B trichothecenes, of which DON and NIV were found most frequently (>60%) in the maize sampled in Germany [38,39]. Additionally, these mycotoxins were also frequently found in freshwater streams of basins with widespread maize production [4]. Therefore, these mycotoxins are of particular interest for environmental soil sampling.

We hypothesize that *Fusarium* toxins predominately enter the soil via washing off plants, leading to elevated soil concentrations (hot-spots) in the soil. These hot-spots can be identified by screening the pooled samples for increased levels. We applied the validated method on conventionally managed maize field soils that were not artificially infested with *Fusarium*, to provide the first insight into natural mycotoxin levels and to derive sampling strategies for future research.

2. Materials and Methods

2.1. Chemicals

HPLC grade acetonitrile used for extraction was supplied by Carl Roth (Karlsruhe, Germany). LC–MS grade methanol and formic acid for mobile phase and standard preparations were supplied by Fisher Scientific (Schwerte, Germany). Ultrapure water was used throughout the whole study (Thermolyne EASYpure II, EnviroFALK PharmaWaterSystems, Leverkusen, Germany). Non-labeled standards of NIV, DON, 15-AcDON, ZEN, and isotopically labeled standards of $^{13}C_{15}$-DON and $^{13}C_{15}$-NIV, all dissolved in acetonitrile, were supplied by Biopure^{TM}, Romer Labs (Butzbach, Germany).

Non-labeled standards were combined to one standard mix stock solution containing $1000\,ng\,mL^{-1}$, which was further diluted with methanol to spiking stocks containing 150, 450 and $750\,ng\,mL^{-1}$ of each compound, respectively. All calibrations were prepared from the spiking stocks, prior to the measurement. Isotopically-labeled standards were mixed and diluted with methanol to a spiking stock containing $250\,ng\,mL^{-1}$ of each compound, respectively. The same isotopically labeled standard stock was used throughout the whole study and was spiked to each individual sample and calibration prior to measurement.

2.2. Soil Samples for Method Validation

Five soils were selected representing a broad spectrum of agricultural soils, to ensure robustness of the analytical procedure. Clay and soil organic carbon content are assumed to be the most influential parameters for method performance. Due to the great surface and number of sorption sites, these soil fractions bear the greatest potential in retaining the mycotoxins [40]. Furthermore, these soil fractions contain the majority of extractable organic and inorganic molecules, which may cause matrix effects during the measurement. The soils used for method validation were the reference soils LUFA2.4, LUFA6S supplied from LUFA Speyer (Speyer, Germany), RefeSol01A and RefeSol02A supplied by Fraunhofer IME (Schmallenberg, Germany), and one natural maize field soil, supplied by JKI (Braunschweig, Germany). The reference soils were air-dried and sieved (2 mm). Selected soils can be classified as sandy loam (LUFA2.4), clayey loam (LUFA6S), silty loam (RefeSol01A), and clayey silt (RefeSol02A), see Table 1 for detailed physicochemical parameters. The non-reference JKI soil was sampled in September 2019 from four non-infested sub-plots within one maize field located in Braunschweig (Germany), where 10 samples of approximately 40 mL of soil volume were taken from the topsoil (12 cm) of each sub-plot using an auger. Samples were frozen and sent to Landau (Germany), where they were pooled, freeze-dried, sieved using a 2 mm mesh size stainless steel sieve, and homogenized.

Table 1. Physicochemical soil properties of reference soils (LUFA2.4, LUFA6S, RefeSol01A, RefeSol02A) and fresh soil samples. Soil texture classes are abbreviated with T/t = clay/clayey, U/u = silt/silty, S/s = sand/sandy, L/l = loam/loamy, 2 = slight, 3 = moderate, e.g., slight silty sand = Su2.

Soil	Sand %	Silt %	Clay %	Soil Texture	Carbon Content %	Soil pH
LUFA2.4	32.1	41.6	26.3	Lt2	1.78	7.4
LUFA6S	23.8	35.3	40.9	Lt3	1.99	7.2
RefeSol01A	74	19.8	6.2	Sl2	0.89	5.3
RefeSol02A	5.7	78.3	16	Ut3	1.04	6.6
JKI soil	63.5	32.5	4	Su3	0.75	Na
Field 1	92.5	5	2.5	Ss	1.42	6.6
Field 2	63.5	32.5	4	Su3	0.68	5.9
Field 3	31	65	4	Us	0.89	7.1
Field 4	80	17.5	2.5	Su2	0.54	5.7
Field 5	19	57.5	23.5	Lu	1.41	7.4
Field 6	19	57.5	23.5	Lu	2.22	7.2
Field 7	22.5	22.5	55	Tl	2.4	7.5
Field 8	Na	17.5	82.5	Tt	1.2	6.8

2.3. Environmental Monitoring Strategy

To validate the applicability of the proposed method, mycotoxins were investigated in eight conventionally managed maize field soils located around Geinsheim, Germany. Sampling was done in September/October 2020 during harvest season when plants were dried for harvest (BBCH-scale 97). Since fields were used for forage maize production, none were irrigated for several weeks before sampling. Sampling of conventionally managed maize fields was designed to detect potential concentration gradients and to suggest sampling strategies for future monitoring studies. Since many fields potentially do not contain any mycotoxins, analysis of all individual samples may lead to a large number of negative results. To reduce the number of samples, samples pooled by plot and position were analyzed, resulting in eight composite samples per field. When elevated levels of mycotoxins were detected in the pooled samples, individual samples of the respective plot and position were analyzed to assess small-scale concentration patterns (10 samples per plot and position, 3 replicates per sample).

A sampling scheme is presented in Figure 1. Field sizes ranged between 2700 and 30,300 m^2 (median 7700 m^2). On each field, four sub-plots (30 m^2) were defined at equally distributed positions within the field. To detect potential concentration differences depending on the sampling position distance to the plant, we sampled on two positions, i.e., between plants ("plant") and inter-row ("inter"). The mean distances between the plants and plant rows were 20 ± 10 cm and 76 ± 7 cm, respectively. Each position was sampled five times (1 m distance) in two rows (approximately 6 m distance) on each sub-plot, resulting in ten samples per position and sub-plot. Sampling was done by horizontally pushing a cylindrical plastic container (5 cm diameter) into the soil, thereby scratch-sampling a topsoil volume of about 80 mL. The first scratch sample of each sampling point was collected in a plastic bag, resulting in a pooled sample for each position and sub-plot. The second scratch sample was kept in the respective plastic cup, closed and stored at −20 °C in the lab until further preparation.

2.4. Soil Characterization

The soils were characterized by soil texture, total soil carbon, soil pH, and water content when field fresh samples were available. Reference soils were characterized by the supplier. Soil pH and water content of non-reference soils were measured in plot-wise pooled samples. Soil texture was determined according to DIN 19682-2. Total soil carbon content was measured by dry combustion in tin foils using an elemental analyzer (vario MICRO cube, Elementar Analysensysteme GmbH, Langenselbold, Germany). Water content was measured gravimetrically as mass loss after 48 h of freeze-drying. Soil pH was measured electrochemically in a 0.01 mol L^{-1} CaCl$_2$ solution (1:5 ratio soil:solution). Physicochemical properties of the soils are shown in Table 1.

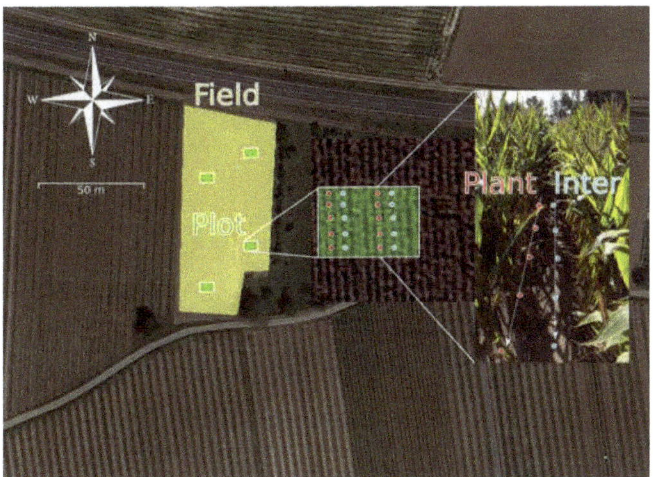

Figure 1. Sampling scheme for naturally infected maize field soils. Four plots were selected on each of the eight fields. Each plot was sampled on two positions, "plant" and "inter" row, represented by red and blue dots, respectively. Each position was sampled five times in two rows of each Plot. Map source: © Google Earth Pro.

2.5. Soil Spiking and Extraction

Fractions of 5 g air-dried reference soils were weighed in 50 mL centrifuge tubes and spiked with 100 µL spiking standards or pure methanol to obtain three soil concentration levels at 3, 9 and 15 ng g^{-1} and blank control. Ten replicates were prepared for each reference soil, concentration level, and blank control. Additionally, the method was tested with the JKI maize field soil samples, which were spiked similarly at the levels 3 and 9 ng g^{-1}. Extraction of the JKI soil samples was repeated three times on different days to assess within-field and inter-day variability. Spiking levels were chosen to be close to the expected limit of detection and in the same order of magnitude as natural soil concentrations (i.e., 2–28 ng g^{-1} reported in [17,18]). After spiking, all samples were vortexed for 10 s and left open under the fume hood for 30 min to ensure homogeneous distribution of the sample and methanol evaporation.

Samples were extracted using 15 mL acetonitrile:water (84:16 v:v) as it is a commonly used solvent mixture for multi mycotoxin extraction in different food and feed matrices [13]. The soils were mixed with the extraction solvent for 30 min at 180 rpm in an orbital shaker. After 15 min in an ultrasonic bath, samples were centrifuged for 15 min at 3500 rpm. Aliquots of 4 mL of each extract were evaporated to dryness at 40 °C under a gentle nitrogen stream. The samples were reconstituted with 0.8 mL 1:9 methanol:water mixture by vortexing for 10 s and ultrasonication for 10 min. Concentrated extracts were filtered through 0.2 µm PET syringe filters. Aliquots of 90 µL of the filtered extracts were mixed with 10 µL isotopically labeled standard mixture before measurement.

2.6. LC–HRMS Analysis

Separation and quantification of mycotoxins was carried out by a liquid chromatography high resolution mass spectrometry (LC–HRMS) Thermo Scientific® system equipped with an Accela Quaternary® pump, an Exactive® Orbitrap MS detector (Thermo Fisher

Scientific, Waltham, MA, USA), and a Hypersil GOLD™ C18 column with dimensions 100×2.1 mm, and 1.9 µm particle size (Thermo Fisher Scientific, Waltham, USA). The mobile phase consisted of methanol (Solvent A) and water (Solvent B) both conditioned with 0.1% formic acid, in the following gradient program: 0–2 min 19% Solvent A; 2–6 min 19–100% Solvent A; 6–11 min 100% Solvent A; 11–12 min 100–19% Solvent A; 12–15 min 19% Solvent A. Flow rate was constant at $0.2\,\text{mL min}^{-1}$. Injection volume was 10 µL. Electron spray ionization was performed in positive and negative mode simultaneously at a capillary temperature of 275 °C. Electronic settings were as follows: spray voltage 4 kV, capillary voltage 25 V, tube lens voltage 75 V, and skimmer voltage 14 V.

The target masses of all mycotoxins with the respective adduct and retention time used for HRMS detection are shown in Table 2. NIV, DON, and the respective isotopically labeled standards were measured in negative modes, as the formic acid adduct. ZEN was measured in a negative mode, as the [M-H]$^-$ adduct. Parameters for NIV, DON, and ZEN matched well with data reported in the Thermo Scientific application note 51,961 by Zachariášová [41], using a very similar technical setup.

The masses 361.1257 *m/z* and 356.1705 *m/z* measured in positive mode, were found suitable for 15-AcDON detection and were attributed to the respective sodium and ammonium adduct. The sodium adduct has been described in literature several times [42–44], and was used as a qualifier, while the ammonium adduct was used as a quantifier due to the slightly higher signal intensities.

Matrix matched calibrations for method validation were measured at concentration levels 0.8, 2, 4, 8, 20 and $80\,\text{ng mL}^{-1}$. Field soil samples were quantified in a narrower range to adapt to environmentally realistic concentrations (0.75, 1.5, 3.75, 7.5, 15, 22.5 and $30\,\text{ng mL}^{-1}$) in methanol:water (20:80, v:v). Matrix matched calibration and internal standard correction were applied for recovery experiment quantification. Internal standard correction was applied for environmental sample quantification.

Table 2. Overview of the retention time, molecular mass, and the exact masses of the most intensive ions used for quantification.

Analyte	Retention Time (min)	Molecular Weight (Da)	Adduct	Target Mass (*m/z*)
NIV	2.2	312.32	[M + COOH]$^-$	357.1195
DON	3.4	296.16	[M + COOH]$^-$	341.1242
15-AcDON	6.45	338.35	[M + Na]$^+$	361.1257
			[M + NH$_4$]$^+$	356.1705
ZEN	8.4	318.36	[M – H]$^-$	317.1389
^{13}C$_{15}$-NIV	2.2	327.32	[M + COOH]$^-$	372.1701
^{13}C$_{15}$-DON	3.4	311.21	[M + COOH]$^-$	356.1750

2.7. Method Validation Criteria

In line with the Eurachem guide [45], method validation criteria were selectivity, linearity of the working range, matrix effect, trueness (bias), precision, method limit of detection (LOD), and quantification (LOQ). Selectivity was ensured by identifying characteristic *m/z* ratios, comparison of *m/z* ratios with literature and comparison of chromatograms with internal and external standard measurements. Linearity of the working range was assessed by visual inspection of the calibration curve and calculation of the adjusted coefficient of determination R^2_{adj} after linear regression. Assumption of normality and homoscedasticity were evaluated with QQ and residual vs. fitted plots [46]. A weighted least squares linear regression (WLS) was performed according to Almeida et al. [47] when assumption of homoscedasticity was not met. Briefly, the increase in variation, i.e., heteroscedasticity, may often be described as a function of concentration. Therefore, by selecting an appropriate weighing factor, the WLS aims to account for greater influence of greater concentrations on the regression and minimize the sum of relative errors instead of the sum of squares. As suggested by the authors, the following weighing factors were tested: $1/x^{0.5}$, $1/x$, $1/x^2$,

$1/y^{0.5}$, $1/y$ and $1/y^2$ with x as the nominal concentration and y as the peak area. The matrix effect was assessed as the signal suppression enhancement ratio (SSE) and calculated as the deviation of matrix-matched calibration slopes from the solvent calibration slope, expressed in percent [29]. Additionally, the signals of NIV and DON were related to the respective internal standard signal. These standardized signals were used for SSE recalculation to evaluate the use of internal standards to eliminate the matrix effect, i.e., minimizing SSE. Trueness (in terms of mean spike recovery) and precision (in terms of relative standard deviation of recoveries, RSD_r) were calculated as described in Magnusson and Örnemark [45]. According to the Commission of the European Communities [48], the acceptable recovery range for DON at concentrations in the range 100–500 ng g^{-1} in foodstuffs is 60–110% with $RSD_r < 20\%$. Since method performance criteria for lower concentrations or other similar mycotoxins, such as NIV and 15-AcDON, are not defined, we use the same as for DON. The criteria for ZEN at concentrations below 50 ng g^{-1} are defined as 60–120% recovery and $RSD_r < 20\%$

The method LOD and LOQ were calculated based on the lowest recovery spiking level of 3 ng g^{-1} (the near instrumental limit of quantification). Standard error of the 10 replicates was multiplied by 3 and 10 to calculate LOD and LOQ, respectively [45].

2.8. Data Evaluation

All data evaluations and presentations were performed using the software R [49], including the package 'data.table' and 'ggplot2' [50,51]. Assumption of normality and homoscedasticity of linear model residuals were evaluated with QQ and residual vs. fitted plots [46]. When linear models for calibration showed heteroscedasticity of residuals, the best weighting factor for WLS calibration was selected using the function *weight_select* and WLS was performed using the function *calibration* from the package 'envalysis' [52]. To reduce the matrix effects, peak areas of NIV and DON were divided by the peak area of the respective isotopically labeled internal standard. To assess within-field and inter-day variability in the JKI soil extraction, the effects of predictor variables "extraction batch" and "sub-plot" on the recovery of individual mycotoxins were tested using a two-way analysis of variance models (F-ANOVA). The effect of the predictor variable "spiking level" on mean recoveries of respective mycotoxins was tested via the linear mixed effect models (LME) using the function *lmer* from the package 'lmerTest' [53]. To account for the variability caused by different reference soils, the factor "soils" was included in the models as a random effect. The effect of predictor variables "clay" and "soil organic carbon content" on mean recoveries was tested via LME, including "spiking level" as a random effect to account for potentially caused variation. Degrees of freedom and F-statistics were performed via the Kenward–Roger approximation, according to [54], using the package 'lmerTest' and 'pbkrtest' [53,54].

3. Results

3.1. Soil Characterization

Soil physicochemical properties are shown in Table 1. Soil types of field samples varied greatly from sand (Field 1) to clay (Field 8). The lowest soil pH was found in Fields 2 and 4, with values around 5.8. All other field soil pH values ranged from 6.6 to 7.5. Total carbon content in field samples ranged from 0.68 to 2.4.

3.2. Linearity and Matrix Effect

The R^2_{adj} for all matrix-matched and solvent calibrations was well above 0.99, with one exception for 15-AcDON with 0.987. A total of 3 out of 16 calibrations for reference soils had heteroscedastic residues and WLS was applied.

SSE was highly variable between soils and compounds (Figure A1). For all matrices, a minor SSE of $-10 \pm 5\%$ was observed for 15-AcDON. SSE of DON and NIV ranged from -4 to 68 % and -10 to 88 %, respectively. This was clearly reduced by the use of isotopically labeled internal standards to SSE of $-4 \pm 6\%$ for DON and $6 \pm 5\%$ for NIV. Greatest variation in SSE between soils was observed for ZEN with $90 \pm 80\%$.

No systematic effect of clay or soil organic carbon content on matrix effect was observed.

3.3. Trueness and Precision

Recovery of NIV, DON, and 15-AcDON was well above the required limit of 60% for all reference soils (Figure 2, Table A1). Mean recovery of NIV and DON was within the required range of 60–110% in 94% of all reference soil–concentration level combinations. There was only one exception for the lowest spiking level in LUFA2.4 soil, with mean recoveries of 112.2% and 111.6% for NIV and DON, respectively. Mean recoveries of 15-AcDON were always within the required range. Mean recoveries of ZEN ranged from 14 to 45% and, thus, never met the requirements. ZEN was therefore excluded from further evaluations.

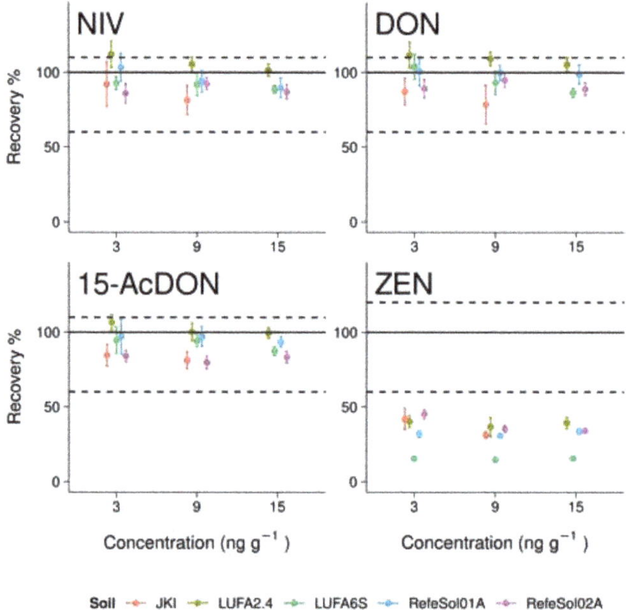

Figure 2. Recovery of nivalenol (NIV), deoxynivalenol (DON), 15-acetyl deoxynivalenol (15-AcDON), and zearalenol (ZEN) from different reference soils (LUFA Speyer and IME Fraunhofer) and one maize field soil (JKI). The dashed lines refer to the required performance criteria, according to Commission Regulation (EC) no. 401/2006.

There was a slight but significant trend towards lower recoveries at higher spiking concentrations for NIV ($p = 0.035$, F-test, df = 7), DON ($p = 0.059$, F-test, df = 7), and 15-AcDON ($p = 0.029$, F-test, df = 7). The variation of recovery significantly increased with decreasing spiking levels for NIV ($p = 0.046$, F-test, df = 7) and DON ($p = 0.01$, F-test, df = 7). The RSD_r was below the required limit of 20% for NIV, DON, and 15-AcDON in all reference soils and even below 10% for the majority with one exception for 15-AcDON.

NIV, DON, and 15-AcDON met all validation criteria in the recovery experiment using JKI soil. A large variation of recovery was observed for almost all compound and concentration combinations, which was, on average, 3.14% larger compared to reference soils. Individual sub-plot soils were spiked to account for within-field variations and extracted in three batches on different days. A visual inspection and data analysis showed that NIV recovery was particularly clustered by the extraction batch ($p < 0.001$, F-ANOVA, df = 2) instead of the sub-plot ($p = 0.146$, F-ANOVA, df = 3), which indicates inter-day variability (Figure A2). However, there was no statistically significant effect on other mycotoxins.

Mean recoveries of NIV, DON, and 15-AcDON were not systematically related to soil organic carbon and clay content, respectively. Therefore, no significant effects of the respective contents on recovery were observed.

3.4. Limit of Detection and Quantification

The overall method LOD and LOQ obtained from the reference soils ranged from 0.11 to 0.33 $ng\,g^{-1}$ and 0.36 to 1.10 $ng\,g^{-1}$, respectively (Table A1). Highest mean LOQ was found for DON ($0.8 \pm 0.1 ng\,g^{-1}$) followed by 15-AcDON ($0.7 \pm 0.3\,ng\,g^{-1}$) and NIV ($0.7 \pm 0.2\,ng\,g^{-1}$). The instrumental LOD and LOQ, measured in the mobile phase, ranged within 0.4–1.1 $ng\,mL^{-1}$ and 1.3–3.7 $ng\,mL^{-1}$, respectively. These parameters were clearly higher in matrix-matched and internal standard-corrected calibrations for recovery experiments with LOD and LOQ, ranging from 0.3 to 3.6 $ng\,mL^{-1}$ and 1.1 to 12.7 $ng\,mL^{-1}$, respectively. Translated into soil concentrations, matrix-matched calibration LOD and LOQ corresponds to 0.2–2.4 $ng\,g^{-1}$ and 0.7–8.4 $ng\,g^{-1}$, which is above the method LOD and LOQ. Instrumental limits of detection and quantification during method validation were estimated based on the parameters obtained from the calibration curve. However, this method was found to be too conservative, since signals can be visually identified at much lower concentrations. Therefore, instrumental LOD and LOQ in environmental sample assessments were calculated as the signal standard deviation of repetitive measurements of the smallest calibration standard (0.75 $ng\,mL^{-1}$, $n > 7$) multiplied with 3 and 10, respectively. This resulted in LOD and LOQ below the smallest calibration standard, which was further defined as the instrumental LOQ (0.75 $ng\,mL^{-1}$, 0.5 $ng\,g^{-1}$ soil).

3.5. Maize Field Soil Samples

We report all data that were above the instrumental LOQ, i.e., 0.5 $ng\,g^{-1}$ soil, and mention the number of samples that were above the mean method LOQ (DON: 0.8 $ng\,g^{-1}$ soil; NIV: 0.7 $ng\,g^{-1}$ soil; Figure 3). Samples containing concentrations above the instrumental LOQ are further referred to as positive samples. We found 7.8% (5 out of 64) positive samples in the initial pooled sample testing. The positive samples originated from 2 of the 8 investigated fields (Field 3 and Field 4), with 4 positive samples from Field 3. Four of the five positive samples were collected between plant rows ("Inter"). However, due to the small number of positive samples, statistical evaluation of position effect was not possible.

DON was detected in the samples "Field 3 Plot 2 Inter" (0.7 $ng\,g^{-1}$) and "Field 4 Plot 4 Plant" (0.5 $ng\,g^{-1}$) at levels above the instrumental LOQ but below mean method LOQ. NIV exceeded instrumental and mean method LOQ in the samples "Field 3 Plot 1 Inter" (0.8 $ng\,g^{-1}$), "Field 3 Plot 2 Inter" (1.5 $ng\,g^{-1}$), and "Field 3 Plot 3 Inter" (0.7 $ng\,g^{-1}$). No sample contained 15-AcDON at levels above instrumental LOQ.

Since the pooled sample "Field 3 Plot 2 Inter" had the highest contents of both DON and NIV, it was selected for measurement of the individual samples to detect potential within plot concentration patterns, i.e., "hot-spots" or continuous levels (Figure 4). Three samples showed concentrations of DON that were above the instrumental LOQ, of which, two were above the mean method LOQ (Sample 3, $1.4 \pm 0.3\,ng\,g^{-1}$; Sample 4, $0.59 \pm 0.08\,ng\,g^{-1}$; Sample 5, $19 \pm 2\,ng\,g^{-1}$). Overall mean concentration of DON in "Field 3 Plot 2 Inter" based on individual measurements was 2.3 $ng\,g^{-1}$, which is almost four-fold higher compared to the expected mean concentration based on the previously measured pooled sample. Concentrations of NIV were above the mean method LOQ in three of four positive samples: Sample 4, $1.29 \pm 0.06\,ng\,g^{-1}$; Sample 5, $0.8 \pm 0.1\,ng\,g^{-1}$; Sample 6, $0.6 \pm 0.1\,ng\,g^{-1}$; Sample 9, $1.0 \pm 0.2\,ng\,g^{-1}$. The overall mean concentration of NIV was 0.4 $ng\,g^{-1}$, which is one third of the expected concentration.

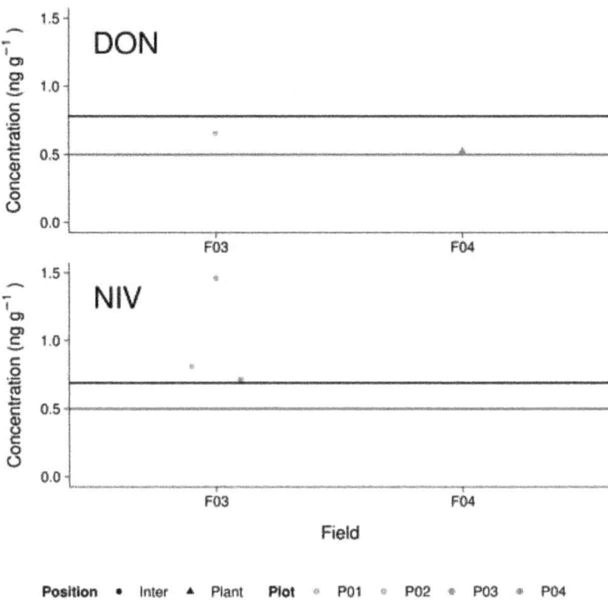

Figure 3. Levels of deoxynivalenol (DON) and nivalenol (NIV) found in pooled maize field soils. Red lines refer to the instrumental limit of detection, black lines refer to the mean extraction method limit of detection.

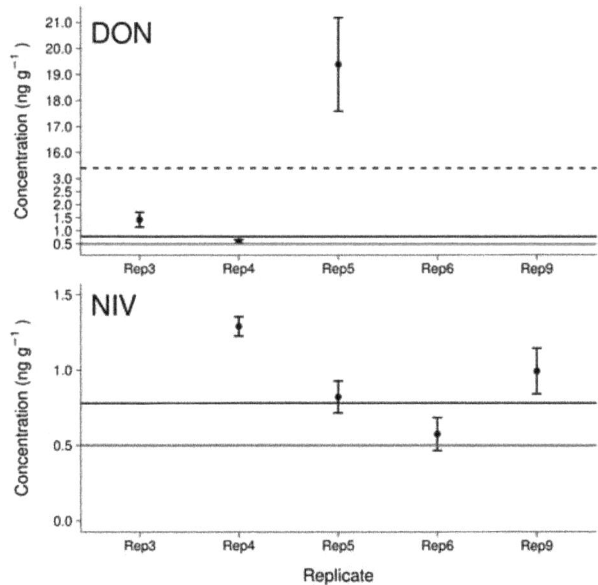

Figure 4. Levels of deoxynivalenol (DON) and nivalenol (NIV) found in single replicates of maize field soil samples. The red lines refer to the instrumental limit of detection, the black lines refer to the mean extraction method limit of detection. The dashed line highlights a break in the y-axis.

4. Discussion

4.1. Linearity and Matrix Effect

Linearity of the relevant working range was acceptable for the whole method validation process. No signal decrease for increasing concentrations was observed during visual inspection of the data, but WLS calibration was necessary in some cases where the measurement variation increased with concentration. In this study, linearity was tested in a comparably small range of 0.8–80 ng mL^{-1} but based on available literature [17–19], soil mycotoxin concentrations can be expected to be mostly in the lower ng g^{-1} range, and higher calibration concentrations are not of particular interest. Additionally, wider concentration ranges have already been tested elsewhere with good linearity [41].

Similar to our results, strong matrix effects have already been observed for other matrix-rich environmental samples, such as wastewater treatment plant effluent or agricultural drainage water, even when samples were cleaned via solid phase extraction [20,25]. In line with our findings, Schenzel et al. [25] observed a predominantly positive matrix effect for Type B trichothecenes, around 20–60%, but negative matrix effects were also shown for DON in aqueous environmental samples (−31 to 20%, [30]; −18 to −14%, [20]), and ZEN in soil (−8%) and sewage sludge (−49%, [16]). In a recent, very similar study, Kappenberg and Juraschek [55] showed only signal suppression ranging from −17 to −10% and −18 to −11% for DON and ZEN in various soils, respectively. The high variation in matrix effects observed in this study could not be attributed to the investigated physicochemical parameters of the soils. Similarly, the differences in Kappenberg and Juraschek [55] do not appear to be related to soil organic carbon content or soil texture. This is contrary to the expectations, since clay content and soil organic carbon represent the most important fraction for sorption sites and, thereby, the potential matrix load. Additionally, soil organic carbon content has been shown to be significant for matrix effects on steroid hormones and penicillin G, measured in LC–ESI–MS/MS [56]. However, soil organic carbon and clay content may be "too general" sum parameters to describe physicochemical properties of soil and a more detailed analysis of co-extracted molecules, such as organic acids and inorganic ions may provide better explanatory variables for matrix effects.

The high variation in matrix effects between soils and compounds observed here shows the importance of isotopically labeled internal standards in the analysis of mycotoxins in diverse soil matrices. Even the structurally very similar NIV and DON deviated in matrix effects by 15 ± 7%. Correction with internal standards compensated the matrix effect for DON an NIV to less than 10% and, thus, to an acceptable range [57]. The usefulness of isotopically labeled internal standards has also been pointed out in other studies [20,30,31]. Additionally, the compound specific matrix effects show that there is a need for several analyte-specific internal standards. The remaining variation of our results after internal standard corrections can be explained with inter-day variability of laboratory work, which was also observed for batch-wise JKI soil extraction.

Assessing the matrix effect is of particular interest for analysis of soil samples since soils can have very different physicochemical properties, even on the field scale [27,28] and, therefore, matrices may also be different. Application of matrix-matched calibrations, as suggested by Zachariasova et al. [58], would be necessary for almost every sample where differences in matrices cannot be excluded. This would lead to an inflating number of measurements and costs. Our results show that it is possible to quantify different soil samples with the same external calibration curve, as long as proper isotopically labeled internal standards are included and all measurements belong to the same measurement batch.

4.2. Trueness and Precision

Recovery and repeatability were acceptable for NIV, DON, and 15-AcDON, with two exceptions at the lowest spiking level of LUFA2.4 for NIV and DON, respectively. In general, within and between soil variability appeared to increase with a decreasing spiking level while recovery decreased with spiking level. One reason could be the mismatch in internal standard concentration (25 ng mL^{-1}) and the lowest spiking level concentration

(4.5 ng mL^{-1}) in the measurement. The ratio of signals may not be stable, leading to an overestimation of small concentrations. Furthermore, the soils were spiked level-by-level from small to high concentrations, and a systematic error cannot be excluded. Analysis of JKI soil recovery indicates that within-field variability may be less important than inter-day variability. Therefore, the recovery differences between reference soils may be rather attributed to the batch-wise extraction of respective reference soil samples than physico-chemical soil differences. With an overall RSD$_r$ of 10% for NIV, DON, and 15-AcDON, the variation was still in the required range [48].

The recovery and RSD$_r$ of DON observed at the intermediate spiking level are very similar to the results that were recently published by Kappenberg and Juraschek [55] (concentration: 10 ng g^{-1}, recovery: 97–100%, RSD$_r$: 6–9%) applying a similar extraction approach (79:20:1 acetonitrile/water/glacial acetic acid, ultrasonic bath for 1 h). However, the differences between the methods had a great effect on ZEN recovery, which was successfully extracted from the soil with satisfactory recovery and RSD$_r$ by Kappenberg and Juraschek [55]. One reason may be the addition of glacial acetic acid, which has been reported to potentially improve the extraction process by breaking the interactions between mycotoxins and sample constituents (i.e., sugars and proteins) [59]. Furthermore, ZEN is stronger adsorbed to soil organic carbon [40], and a prolonged ultrasonic bath may be necessary for acceptable recovery rates.

While ZEN does not seem to be extractable by this method without modification, the method proposed in this study successfully validated two additional mycotoxins NIV and 15-AcDON. Furthermore, a recent study by Albert et al. [29], with a very similar acetonitrile:water approach, proved that this method is also suitable for analysis of aflatoxins in soil. This may be of particular interest, since climate change may cause a spread of *Aspergillus* to European regions where *Fusarium* is currently predominant [60,61].

Our results show that the applied method is able to extract trace levels of NIV, DON, and 15-AcDON from soil with satisfactory recovery and repeatability. We recommend choosing the internal standard concentration close to the mean of the expected concentrations. Extraction and measurement should be randomly distributed between batches to account for inter-day variability.

4.3. Limit of Detection and Quantification

Estimates of the method LOD and LOQ were mostly below 1 ng g^{-1}, which indicates that the method potentially allows detection of NIV, DON, and 15-AcDON below the lowest spiking level of 3 ng g^{-1}. This is even below the instrumental limits estimated from the matrix-matched calibration curve, which were above 3 ng g^{-1} in some cases and indicates that quantification was rather limited by calibration and instrumental performance than the extraction method. However, the measurements of the lowest levels were accepted here, since signals could be clearly identified and quantification matched with expected concentrations (Figures A3 and A4). The comparatively high instrumental limits during method validation are attributed to the LOD and LOQ estimation method, which was based on the calibration curve. Therefore, the calibration was adapted for the second part of the environmental sample analysis. The calibration range was narrowed, the number of calibration levels increased, and the smallest calibration standard was measured repetitively. This allowed LOD and LOQ estimation based on "artificial" background noise close to the instrumental LOD, which were subsequently below the smallest calibration standard, which was further used as LOQ. The resulting instrumental and method LOD and LOQ are still in a similar range. To further develop this method, improved instrumentation with higher sensitivity is needed. One approach could be the usage of atmospheric pressure chemical ionization (APCI), which showed 12- and 10-fold higher signal, for NIV and DON, respectively [58], and 7.4-fold higher signals for DON and 15-AcDON in Jensen et al. [62].

4.4. Environmental Samples and Implications for Further Investigations

Our results show the presence of mycotoxins in 2 out of 8 field soils. Soil mycotoxin concentrations ranged from 0.53 to 19.38 ng g^{-1} and 0.8 to 2.2 ng g^{-1} for DON and NIV, respectively.

Occurrence of mycotoxins seemed to be more frequent between planting rows than individual plants. However, due to the low number of positive samples, there was no clear difference between the sampling positions, i.e., between plants and planting rows. If mycotoxins are washed off the plants during rain events, the rain water may distribute more homogeneously around the plants, leading to more widespread and homogeneous soil mycotoxin levels, but it is unclear whether mycotoxins are on the maize plant surface and accessible to rain. Although all above-ground maize plant parts can contain trichothecenes, the highest contents were found in rudimentary ears [39], which are physically protected from rain by husks. Therefore, washing off may not be the main route of mycotoxins from maize plants to the soil. Another path may be the translocation within the plant to the roots when above-ground parts are infected. Unfortunately, Schollenberger et al. [39] did not investigate natural maize root contents, but since trichothecenes were found in all parts of the plants, it seems likely that they are also present in plant roots. Furthermore, *Fusarium* can cause root rot, which can lead to trichothecene concentrations in the range 0.4–1.6 µg g^{-1} [63] and 0.3–22.6 µg g^{-1} [64] in root dry mass from wheat and maize, respectively. Consequently, the levels observed in this study may also originate from in situ biosynthesis. Further studies should identify the main introduction routes of mycotoxins from maize plants to soil and the potential spatiotemporal distribution from the rhizosphere to bulk soil. Fate and transport of mycotoxins within soil is scarcely investigated and the analytical method proposed in this study with a low LOQ may help to determine the mycotoxin pathways.

We observed a mismatch between concentrations in the pooled sample and mean of individually analyzed samples of DON and NIV, respectively. The mean DON concentration was higher than expected while NIV was lower. This indicates that soil mycotoxin levels may be heterogeneous on very small scales of 10 cm (two times the diameter of the sampling containers) and "hot-spot" concentrations may be restricted to areas smaller than 10 cm within soils. As mentioned before, infested roots may have a great contribution to soil trichothecene concentration. If the edge of an infested plant root zone was sampled, it is possible that the two samples at the same position had different contents of fine root material. Additionally, soil microbial communities may vary even on cm scales [28], which may explain why the pooled sample concentration is different from the mean of individual samples.

The apparent small scale heterogeneity has implications for soil sampling strategies. When aiming at a representative mean level on the field scale, which is necessary for field emission estimates, many individual samples collected from a fine mesh of sampling points are necessary. However, the pooling of many samples causes dilution of the concentrations. The method proposed in this study may help to overcome this issue, due to the low limits of quantification. Further method developments can improve the analysis, as for example using an APCI to achieve lower instrument limits [58]. Additionally, method limits can still be improved, for example by increasing the extract concentration in the evaporation step. A greater concentration step also increases matrix concentrations and may require further clean-up steps, such as immunoaffinity columns or solid phase extraction. The addition of acetic acid to the extraction solvent and a longer ultrasonic bath step allows to include ZEN in the analysis [55].

Based on our results, extrapolation to a mean field mycotoxin concentration is not possible or linked to a high level of uncertainty. In this study design, we aimed to identify "hot-spot" concentrations in the soil with a minimum number of measurements. The screening approach, with initial analysis of pooled sub-sets of samples, clearly reduced the number of potential hot-spot samples. Identifying the soil concentration range of trichothecenes is of ecological relevance, since the fate and effect of trichothecenes in the environment is still largely unknown and current studies report contrasting results. While no degradation of DON was observed 28 day after direct application to soil [55], there was a clear reduction within 21 day when applied as contaminated harvesting residues [22,65]. This indicatesthat the way in which mycotoxins are introduced to soil also affects their degradation. However, while degradation of pure DON in soil has not been proven yet, DON-degrading microbial cultures have been extracted from soil in several studies [66–70], which indicates that some soils may have a mitigating effect on soil mycotoxin pollution. These cultures were able to degrade DON within 48 h to 2 weeks with 3-epi-DON, de-epoxy-DON and 3-keto-DON being the main degradation products. These substances, as well as masked mycotoxins and temporal changes, were not assessed in this study. Therefore, the number of fields containing mycotoxins and related substance soils may be even higher than 25% found here (2 out of 8 fields). Further investigations should assess degradation rates and main degradation products in soil to estimate the potential of soil to mitigate environmental mycotoxin pollution.

5. Conclusions

In this study, we successfully validated an extraction and measurement method for the simultaneous measurement of NIV, DON, and 15-AcDON in agriculturally managed soils. The method was applied to provide first insights into natural mycotoxin levels in maize field soils. Mycotoxins were detected at varying levels, down to trace amounts, which shows that reliable methods are imperative for analysis of environmental soil samples. Furthermore, we showed that mycotoxin levels in soil can be highly variable, even on small scales. We therefore recommend a fine mesh of sampling points for mean field concentration estimates or "hot-spot" detection. An initial pooled sample analysis combined with a low method limit of detection can clearly reduce the number of measurements.

Author Contributions: Conceptualization, K.G.J.K., J.A. and K.M.; methodology, K.G.J.K. and J.A.; software, K.G.J.K.; formal analysis, K.G.J.K.; investigation, K.G.J.K. and J.A.; resources, K.G.J.K., J.A. and K.M.; writing—original draft preparation, K.G.J.K.; writing—review and editing, K.G.J.K., J.A., F.M.-W., G.E.S. and K.M.; visualization, K.G.J.K.; supervision, G.E.S. and K.M.; project administration, K.M.; funding acquisition, K.M. All authors have read and agreed to the published version of the manuscript.

Funding: The project is funded by the Federal Ministry of Food and Agriculture BLE under the reference AflaZ 2816PROC14 and by the Ministry for Education, Sciences, Further Education, and Culture of the State of Rhineland-Palatinate (MBWWK) in the frame of the Interdisciplinary Research Group for Environmental Studies (IFG-Umwelt) of University Koblenz-Landau.

Institutional Review Board Statement: Not applicable.

Informed Consent Statement: Not applicable.

Data Availability Statement: The data presented in this study are available upon request from the corresponding author.

Acknowledgments: We thank Werena und Ricardino for sampling the JKI soil. We thank Camilla More, Niklaus Dahlke, Vivien Lenard, and Aaron Kintzi for their help in the field- and laboratory work.

Conflicts of Interest: The authors declare no conflict of interest. The funders had no role in the design of the study; in the collection, analyses, or interpretation of data; in the writing of the manuscript, or in the decision to publish the results.

Abbreviations

The following abbreviations are used in this manuscript:

JKI	Julius Kühn Institute
NIV	nivalenol
DON	deoxynivalenol
15-AcDON	15-acetyl-deoxynivalenol

ZEN	zearalenone
LC	liquid chromatography
HRMS	high-resolution mass spectrometry
RSD_r	relative standard deviation
HPLC	high pressure liquid chromatography
PET	polyethylene
LOD	limit of detection
LOQ	limit of quantification
QQ plot	quantile–quantile plot
WLS	weighted least squares
F-ANOVA	analysis of variance based on F-test
LME	linear mixed effect
APCI	atmospheric pressure chemical ionization
EU	European Union
EC	European Commission

Appendix A

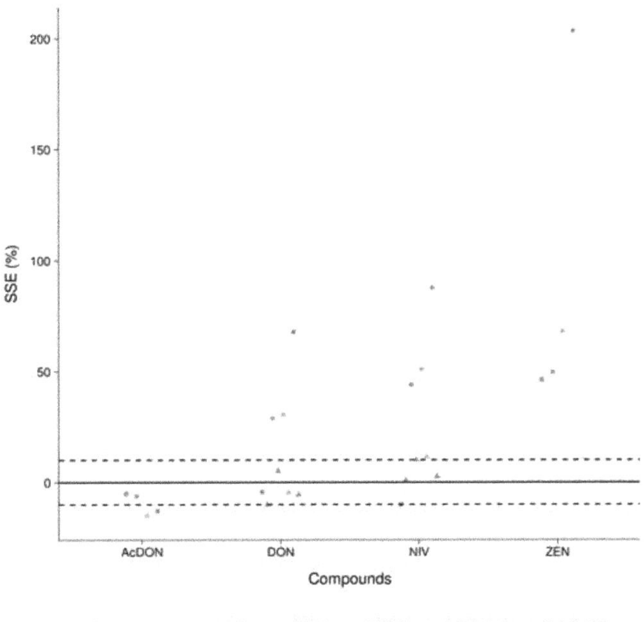

Figure A1. Comparison of the matrix-matched calibration slope with the calibration slope measured in pure solvent. The matrix effect is calculated as the difference between slopes divided by the solvent slope and expressed in percent (relative deviation). Different shapes (IS) indicate whether signals were standardized with an internal standard (triangle) or not (dot). Colors refer to the different matrices extracted from reference soils. Dashed lines denote the 10% interval around zero signal suppression/enhancement (straight line).

Table A1. Mean recovery (standard deviation in parentheses), method limit of detection (LOD), and method limit of quantification (LOQ) of deoxynivalenol (DON), nivalenol (NIV), 15-acetyl-deoxynivalenol (15-AcDON), and zearalenone (ZEN) extracted from four standard reference soils spiked with 3–15 ng g^{-1} soil dry weight. LOD and LOQ were calculated according to the Eurachem guideline, based on standard deviation of ten independent sample extract measurements with a nominal concentration of 3 ng g^{-1}. The LOD and LOQ of ZEN were not determined (n.d.).

Compound	Soil	Recovery			LOD	LOQ
		3 ng g^{-1}	9 ng g^{-1}	15 ng g^{-1}	ng g^{-1}	ng g^{-1}
NIV	LUFA2.4	112 (9)	106 (5)	102 (4)	0.15	0.5
	LUFA6S	93 (4)	92 (8)	89 (3)	0.25	0.82
	RefeSol01A	104 (9)	94 (7)	90 (7)	0.25	0.84
	RefeSol02A	86 (6)	93 (4)	87 (5)	0.26	0.86
DON	LUFA2.4	112 (9)	109 (5)	105 (5)	0.24	0.79
	LUFA6S	104 (8)	93 (8)	87 (3)	0.12	0.41
	RefeSol01A	100 (10)	100 (5)	99 (6)	0.33	1.1
	RefeSol02A	89 (6)	95 (5)	89 (4)	0.28	0.92
15-AcDON	LUFA2.4	107 (5)	100 (6)	100 (3)	0.26	0.88
	LUFA6S	95 (9)	94 (4)	88 (3)	0.11	0.36
	RefeSol01A	100 (10)	97 (7)	94 (3)	0.18	0.58
	RefeSol02A	84 (4)	80 (4)	84 (4)	0.18	0.62
ZEN	LUFA2.4	40 (4)	37 (6)	39 (4)	n.d.	n.d.
	LUFA6S	16 (1)	15 (2)	16 (1)	n.d.	n.d.
	RefeSol01A	32 (2)	31 (1)	34 (2)	n.d.	n.d.
	RefeSol02A	45 (3)	35 (3)	34 (2)	n.d.	n.d.

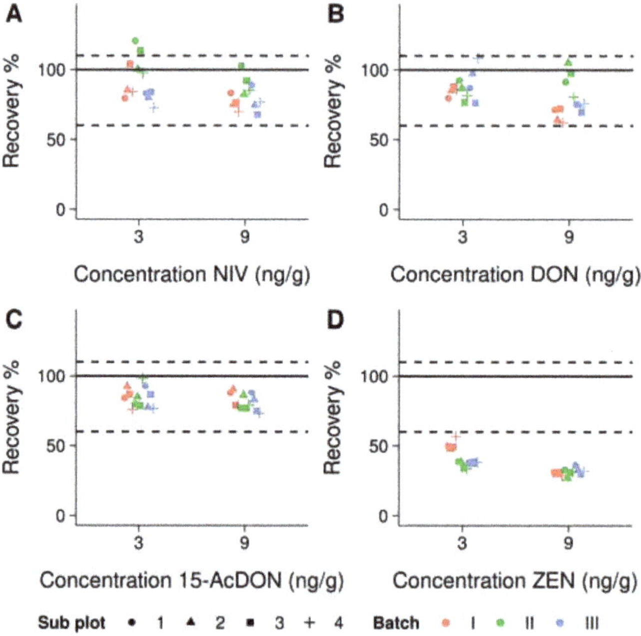

Figure A2. Recovery of nivalenol (NIV), deoxynivalenol (DON), 15-acetyl deoxynivalenol (15-AcDON), and zearalenol (ZEN) from a maize field soil provided by JKI Braunschweig, Germany. The dashed line refers to the required performance criteria according to Commission Regulation (EC) no. 401/2006 for DON 100–500 ng g^{-1}. Point shapes refer to the different sub plots where the samples were taken. Point colors refer to the respective extraction batch, processed on different days.

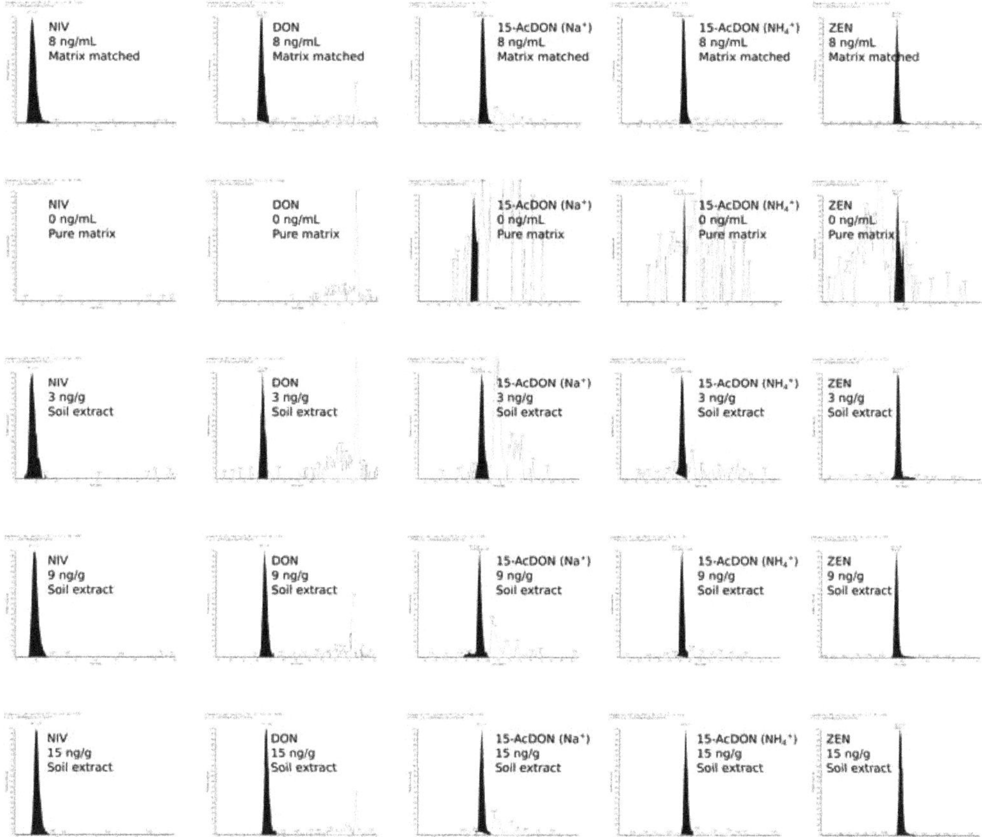

Figure A3. Example chromatograms of nivalenol (NIV), deoxynivalenol (DON), 15-acetyl deoxynivalenol (15-AcDON), and zearalenol (ZEN) measured in the matrix-matched calibration, pure matrix, and in soil extracts from RefeSol02A spiked at the levels 3, 9 and 15 ng g^{-1} soil.

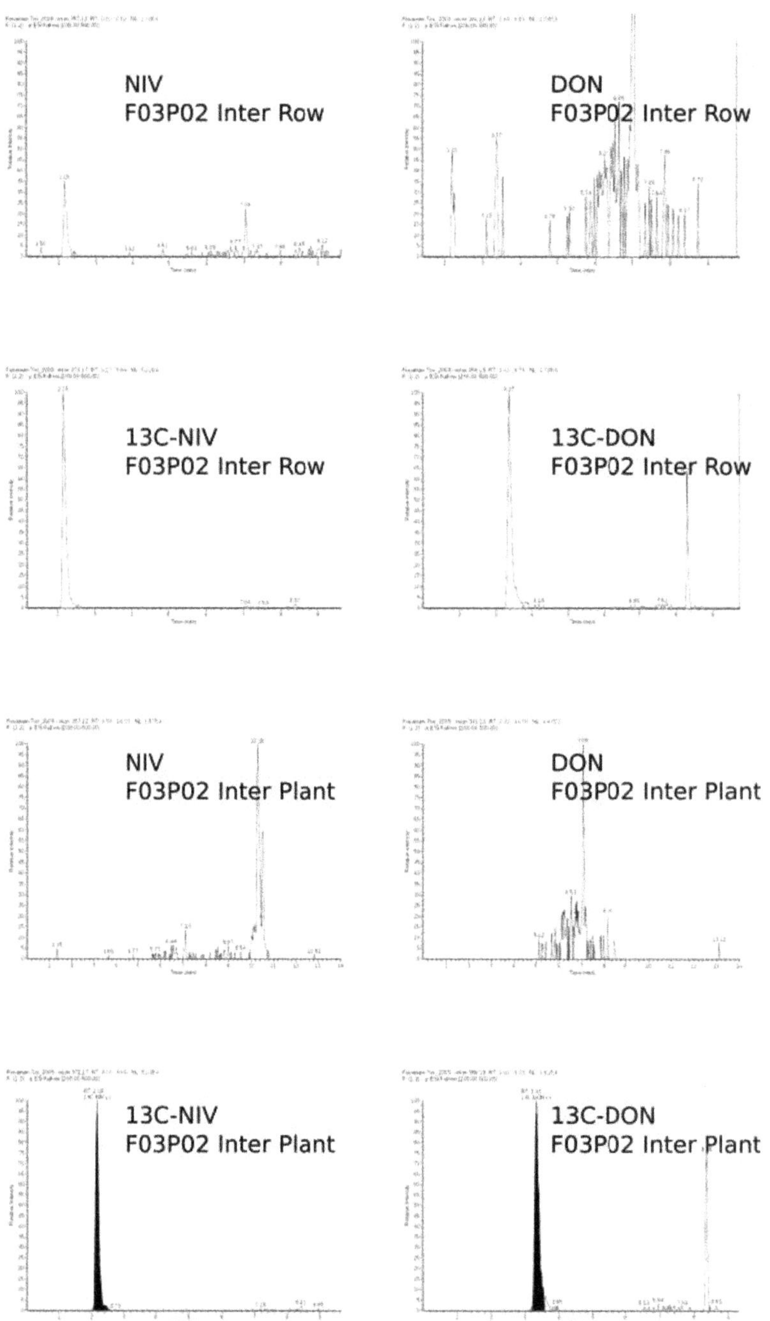

Figure A4. Example chromatograms of nivalenol (NIV, retention time 2.15 min), deoxynivalenol (DON, retention time 3.37 min), and respective stable isotope standards (13C-NIV, 13C-DON) measured in the maize field soil. Samples were taken from the same field and sub plot, but at different positions, i.e., between plants (inter plant) and between rows (inter row).

References

1. Savary, S.; Willocquet, L.; Pethybridge, S.J.; Esker, P.; McRoberts, N.; Nelson, A. The Global Burden of Pathogens and Pests on Major Food Crops. *Nat. Ecol. Evol.* **2019**, *3*, 430–439. [CrossRef] [PubMed]
2. Elmholt, S. Mycotoxins in the Soil Environment. In *Secondary Metabolites in Soil Ecology*; Springer: Berlin/Heidelberg, Germany, 2008; Volume 14, pp. 167–203. [CrossRef]
3. Bryła, M.; Ksieniewicz-Woźniak, E.; Yoshinari, T.; Waśkiewicz, A.; Szymczyk, K. Contamination of Wheat Cultivated in Various Regions of Poland during 2017 and 2018 Agricultural Seasons with Selected Trichothecenes and Their Modified Forms. *Toxins* **2019**, *11*, 88. [CrossRef] [PubMed]
4. Kolpin, D.W.; Schenzel, J.; Meyer, M.T.; Phillips, P.J.; Hubbard, L.E.; Scott, T.M.; Bucheli, T.D. Mycotoxins: Diffuse and Point Source Contributions of Natural Contaminants of Emerging Concern to Streams. *Sci. Total Environ.* **2014**, *470*, 669–676. [CrossRef] [PubMed]
5. Vogelgsang, S.; Musa, T.; Bänziger, I.; Kägi, A.; Bucheli, T.D.; Wettstein, F.E.; Pasquali, M.; Forrer, H.R. Fusarium Mycotoxins in Swiss Wheat: A Survey of Growers' Samples between 2007 and 2014 Shows Strong Year and Minor Geographic Effects. *Toxins* **2017**, *9*, 246. [CrossRef]
6. Knutsen, H.K.; Alexander, J.; Barregård, L.; Bignami, M.; Brüschweiler, B.; Ceccatelli, S.; Cottrill, B.; Dinovi, M.; Grasl-Kraupp, B.; Hogstrand, C.; et al. Risks to Human and Animal Health Related to the Presence of Deoxynivalenol and Its Acetylated and Modified Forms in Food and Feed. *EFSA J.* **2017**, *15*, e04718. [CrossRef] [PubMed]
7. Escrivá, L.; Font, G.; Manyes, L. In Vivo Toxicity Studies of Fusarium Mycotoxins in the Last Decade: A Review. *Food Chem. Toxicol.* **2015**, *78*, 185–206. [CrossRef]
8. Kowalska, K.; Habrowska-Górczyńska, D.E.; Piastowska-Ciesielska, A.W. Zearalenone as an Endocrine Disruptor in Humans. *Environ. Toxicol. Pharmacol.* **2016**, *48*, 141–149. [CrossRef]
9. Schwartz, P.; Thorpe, K.L.; Bucheli, T.D.; Wettstein, F.E.; Burkhardt-Holm, P. Short-Term Exposure to the Environmentally Relevant Estrogenic Mycotoxin Zearalenone Impairs Reproduction in Fish. *Sci. Total Environ.* **2010**, *409*, 326–333. [CrossRef]
10. Commission of the European Communities. Commission Regulation (EC) No 1881/2006 of 19 December 2006 Setting Maximum Levels for Certain Contaminants in Foodstuffs (Text with EEA Relevance). *Off. J. Eur. Union* **2006**, *364*, 5–24.
11. U.S. FDA. *Guidance for Industry and FDA: Advisory Levels for Deoxynivalenol (DON) in Finished Wheat Products for Human Consumption and Grains and Grain By-Products Used for Animal Feed*; U.S. FDA: Silver Spring, MD, USA, 2010.
12. Kralj Cigić, I.; Prosen, H. An Overview of Conventional and Emerging Analytical Methods for the Determination of Mycotoxins. *Int. J. Mol. Sci.* **2009**, *10*, 62–115. [CrossRef]
13. Pereira, V.L.; Fernandes, J.O.; Cunha, S.C. Mycotoxins in Cereals and Related Foodstuffs: A Review on Occurrence and Recent Methods of Analysis. *Trends Food Sci. Technol.* **2014**, *36*, 96–136. [CrossRef]
14. Schenzel, J.; Hungerbühler, K.; Bucheli, T.D. Mycotoxins in the Environment: II. Occurrence and Origin in Swiss River Waters. *Environ. Sci. Technol.* **2012**, *46*, 13076–13084. [CrossRef] [PubMed]
15. Bucheli, T.D. Phytotoxins: Environmental Micropollutants of Concern? *Environ. Sci. Technol.* **2014**, *48*, 13027–13033. [CrossRef] [PubMed]
16. Hartmann, N.; Erbs, M.; Wettstein, F.E.; Hoerger, C.C.; Schwarzenbach, R.P.; Bucheli, T.D. Quantification of Zearalenone in Various Solid Agroenvironmental Samples Using D6-Zearalenone as the Internal Standard. *J. Agric. Food Chem.* **2008**, *56*, 2926–2932. [CrossRef] [PubMed]
17. Muñoz, K.; Schmidt-Heydt, M.; Stoll, D.; Diehl, D.; Ziegler, J.; Geisen, R.; Schaumann, G.E. Effect of Plastic Mulching on Mycotoxin Occurrence and Mycobiome Abundance in Soil Samples from Asparagus Crops. *Mycotoxin Res.* **2015**, *31*, 191–201. [CrossRef]
18. Muñoz, K.; Buchmann, C.; Meyer, M.; Schmidt-Heydt, M.; Steinmetz, Z.; Diehl, D.; Thiele-Bruhn, S.; Schaumann, G.E. Physico-chemical and Microbial Soil Quality Indicators as Affected by the Agricultural Management System in Strawberry Cultivation Using Straw or Black Polyethylene Mulching. *Appl. Soil Ecol.* **2017**, *113*, 36–44. [CrossRef]
19. Meyer, M.; Diehl, D.; Schaumann, G.E.; Muñoz, K. Agricultural Mulching and Fungicides—Impacts on Fungal Biomass, Mycotoxin Occurrence, and Soil Organic Matter Decomposition. *Environ. Sci. Pollut. Res.* **2021**, *28*, 36535–36550. [CrossRef]
20. Bucheli, T.D.; Wettstein, F.E.; Hartmann, N.; Erbs, M.; Vogelgsang, S.; Forrer, H.R.; Schwarzenbach, R.P. Fusarium Mycotoxins: Overlooked Aquatic Micropollutants? *J. Agric. Food Chem.* **2008**, *56*, 1029–1034. [CrossRef]
21. Schenzel, J.; Forrer, H.R.; Vogelgsang, S.; Hungerbühler, K.; Bucheli, T.D. Mycotoxins in the Environment: I. Production and Emission from an Agricultural Test Field. *Environ. Sci. Technol.* **2012**, *46*, 13067–13075. [CrossRef]
22. Meyer-Wolfarth, F.; Oldenburg, E.; Meiners, T.; Muñoz, K.; Schrader, S. Effects of Temperature and Soil Fauna on the Reduction and Leaching of Deoxynivalenol and Zearalenone from Fusarium Graminearum-Infected Maize Stubbles. *Mycotoxin Res.* **2021**, *37*, 249–263. [CrossRef]
23. Chen, P.; Xiang, B.; Shi, H.; Yu, P.; Song, Y.; Li, S. Recent Advances on Type A Trichothecenes in Food and Feed: Analysis, Prevalence, Toxicity, and Decontamination Techniques. *Food Control* **2020**, *118*, 107371. [CrossRef]
24. Malachová, A.; Stránská, M.; Václavíková, M.; Elliott, C.T.; Black, C.; Meneely, J.; Hajšlová, J.; Ezekiel, C.N.; Schuhmacher, R.; Krska, R. Advanced LC–MS-based Methods to Study the Co-Occurrence and Metabolization of Multiple Mycotoxins in Cereals and Cereal-Based Food. *Anal. Bioanal. Chem.* **2018**, *410*, 801–825. [CrossRef] [PubMed]

25. Schenzel, J.; Schwarzenbach, R.P.; Bucheli, T.D. Multi-Residue Screening Method To Quantify Mycotoxins in Aqueous Environmental Samples. *J. Agric. Food Chem.* **2010**, *58*, 11207–11217. [CrossRef]

26. De Santis, B.; Debegnach, F.; Gregori, E.; Russo, S.; Marchegiani, F.; Moracci, G.; Brera, C. Development of a LC–MS/MS Method for the Multi-Mycotoxin Determination in Composite Cereal-Based Samples. *Toxins* **2017**, *9*, 169. [CrossRef] [PubMed]

27. Patzold, S.; Mertens, F.M.; Bornemann, L.; Koleczek, B.; Franke, J.; Feilhauer, H.; Welp, G. Soil Heterogeneity at the Field Scale: A Challenge for Precision Crop Protection. *Precis. Agric.* **2008**, *9*, 367–390. [CrossRef]

28. Franklin, R.B.; Mills, A.L. Importance of Spatially Structured Environmental Heterogeneity in Controlling Microbial Community Composition at Small Spatial Scales in an Agricultural Field. *Soil Biol. Biochem.* **2009**, *41*, 1833–1840. [CrossRef]

29. Albert, J.; More, C.A.; Dahlke, N.R.P.; Steinmetz, Z.; Schaumann, G.E.; Muñoz, K. Validation of a Simple and Reliable Method for the Determination of Aflatoxins in Soil and Food Matrices. *ACS Omega* **2021**, *6*, 18684–18693. [CrossRef]

30. Wettstein, F.E.; Bucheli, T.D. Poor Elimination Rates in Waste Water Treatment Plants Lead to Continuous Emission of Deoxynivalenol into the Aquatic Environment. *Water Res.* **2010**, *44*, 4137–4142. [CrossRef]

31. Fabregat-Cabello, N.; Zomer, P.; Sancho, J.; Roig-Navarro, A.; Mol, H. Comparison of Approaches to Deal with Matrix Effects in LC–MS/MS Based Determinations of Mycotoxins in Food and Feed. *World Mycotoxin J.* **2016**, *9*, 149–161. [CrossRef]

32. Mortensen, G.K.; Strobel, B.W.; Hansen, H.C.B. Determination of Zearalenone and Ochratoxin A in Soil. *Anal. Bioanal. Chem.* **2003**, *376*, 98–101. [CrossRef]

33. Klein, T.A.; Burgess, L.W.; Ellison, F.W. The Incidence and Spatial Patterns of Wheat Plants Infected by Fusarium Graminearum Group 1 and the Effect of Crown Rot on Yield. *Aust. J. Agric. Res.* **1991**, *42*, 399–407. [CrossRef]

34. Rolli, E.; Righetti, L.; Galaverna, G.; Suman, M.; Dall'Asta, C.; Bruni, R. Zearalenone Uptake and Biotransformation in Micropropagated Triticum Durum Desf. Plants: A Xenobolomic Approach. *J. Agric. Food Chem.* **2018**, *66*, 1523–1532. [CrossRef] [PubMed]

35. Mantle, P.G. Uptake of Radiolabelled Ochratoxin A from Soil by Coffee Plants. *Phytochemistry* **2000**, *53*, 377–378. [CrossRef]

36. Righetti, L.; Damiani, T.; Rolli, E.; Galaverna, G.; Suman, M.; Bruni, R.; Dall'Asta, C. Exploiting the Potential of Micropropagated Durum Wheat Organs as Modified Mycotoxin Biofactories: The Case of Deoxynivalenol. *Phytochemistry* **2020**, *170*, 112194. [CrossRef] [PubMed]

37. Snigdha, M.; Hariprasad, P.; Venkateswaran, G. Mechanism of Aflatoxin Uptake in Roots of Intact Groundnut (*Arachis hypogaea* L.) Seedlings. *Environ. Sci. Pollut. Res.* **2013**, *20*, 8502–8510. [CrossRef]

38. Goertz, A.; Zuehlke, S.; Spiteller, M.; Steiner, U.; Dehne, H.W.; Waalwijk, C.; de Vries, I.; Oerke, E.C. Fusarium Species and Mycotoxin Profiles on Commercial Maize Hybrids in Germany. *Eur. J. Plant Pathol.* **2010**, *128*, 101–111. [CrossRef]

39. Schollenberger, M.; Müller, H.M.; Ernst, K.; Sondermann, S.; Liebscher, M.; Schlecker, C.; Wischer, G.; Drochner, W.; Hartung, K.; Piepho, H.P. Occurrence and Distribution of 13 Trichothecene Toxins in Naturally Contaminated Maize Plants in Germany. *Toxins* **2012**, *4*, 778–787. [CrossRef] [PubMed]

40. Schenzel, J.; Goss, K.U.; Schwarzenbach, R.P.; Bucheli, T.D.; Droge, S.T.J. Experimentally Determined Soil Organic Matter–Water Sorption Coefficients for Different Classes of Natural Toxins and Comparison with Estimated Numbers. *Environ. Sci. Technol.* **2012**, *46*, 6118–6126. [CrossRef]

41. Zachariášová, M. Multi-Mycotoxin Screening and Quantitation Using UHPLC, High Resolution and Accurate Mass. 2011. Available online: https://www.thermofisher.com/document-connect/document-connect.html?url=https://assets.thermofisher.com/TFS-Assets/LSG/Application-Notes/Article-Food-Science-Standout-plus-AN-51961-Multi-mycotoxin-Screening-Quantitation.pdf (accessed on 17 December 2018).

42. McCullough, B.J.; Hopley, C.J. A Validated LC–HRMS Method for the Detection of T-2 and HT-2 Toxins in Cereals. *J. Assoc. Public Anal.* **2017**, *45*, 41–73.

43. Tolosa, J.; Graziani, G.; Gaspari, A.; Chianese, D.; Ferrer, E.; Mañes, J.; Ritieni, A. Multi-Mycotoxin Analysis in Durum Wheat Pasta by Liquid Chromatography Coupled to Quadrupole Orbitrap Mass Spectrometry. *Toxins* **2017**, *9*, 59. [CrossRef]

44. Slobodchikova, I.; Sivakumar, R.; Rahman, M.S.; Vuckovic, D. Characterization of Phase I and Glucuronide Phase II Metabolites of 17 Mycotoxins Using Liquid Chromatography—High-Resolution Mass Spectrometry. *Toxins* **2019**, *11*, 433. [CrossRef]

45. Magnusson, B.; Örnemark, U. (Eds.) *Eurachem Guide: The Fitness for Purpose of Analytical Methods—A Laboratory Guide to Method Validation and Related Topics*, 2nd ed.; 2014; ISBN 978-91-87461-59-0. Available online: www.eurachem.org (accessed on 17 December 2018).

46. Zuur, A.F.; Ieno, E.N.; Elphick, C.S. A Protocol for Data Exploration to Avoid Common Statistical Problems. *Methods Ecol. Evol.* **2010**, *1*, 3–14. [CrossRef]

47. Almeida, A.M.; Castel-Branco, M.M.; Falcão, A.C. Linear Regression for Calibration Lines Revisited: Weighting Schemes for Bioanalytical Methods. *J. Chromatogr. B* **2002**, *774*, 215–222. [CrossRef]

48. Commission of the European Communities. Commission Regulation (EC) No 401/2006 of 23 February 2006 Laying down the Methods of Sampling and Analysis for the Official Control of the Levels of Mycotoxins in Foodstuffs (Text with EEA Relevance). *Off. J. Eur. Union* **2006**, *70*, 12–34.

49. R Core Team. *R: A Language and Environment for Statistical Computing*; R Foundation for Statistical Computing: Vienna, Austria, 2018.

50. Dowle, M.; Srinivasan, A. Extension of 'data.Frame' [R Package Data.Table Version 1.14.0]. 2021. Available online: https://CRAN.R-project.org/package=data.table (accessed on 14 May 2021).

51. Wickham, H. *Ggplot2: Elegant Graphics for Data Analysis*, 2nd ed.; Springer International Publishing: Berlin/Heidelberg, Germany, 2016. [CrossRef]

52. Steinmetz, Z. Envalysis: Miscellaneous Functions for Environmental Analyses. 2021. Available online: https://cran.r-project.org/package=envalysis (accessed on 14 June 2021).

53. Kuznetsova, A.; Brockhoff, P.B.; Christensen, R.H.B. lmerTest Package: Tests in Linear Mixed Effects Models. *J. Stat. Softw.* **2017**, *82*, 1–26. [CrossRef]

54. Halekoh, U.; Højsgaard, S. A Kenward–Roger Approximation and Parametric Bootstrap Methods for Tests in Linear Mixed Models - The R Package Pbkrtest. *J. Stat. Softw.* **2014**, *59*, 1–32. [CrossRef]

55. Kappenberg, A.; Juraschek, L.M. Development of a LC–MS/MS Method for the Simultaneous Determination of the Mycotoxins Deoxynivalenol (DON) and Zearalenone (ZEA) in Soil Matrix. *Toxins* **2021**, *13*, 470. [CrossRef] [PubMed]

56. Salvia, M.V.; Cren-Olivé, C.; Vulliet, E. Statistical Evaluation of the Influence of Soil Properties on Recoveries and Matrix Effects during the Analysis of Pharmaceutical Compounds and Steroids by Quick, Easy, Cheap, Effective, Rugged and Safe Extraction Followed by Liquid Chromatography–Tandem Mass Spectrometry. *J. Chromatogr. A* **2013**, *1315*, 53–60. [CrossRef] [PubMed]

57. European Commission. SANCO/12571/2013: Guidance Document on Analytical Quality Control and Validation Procedures for Pesticides Residues Analysis in Food and Feed. 2013. Available online: https://www.eurl-pesticides.eu/docs/public/tmplt_article.asp?CntID=727 (accessed on 14 June 2021).

58. Zachariasova, M.; Cajka, T.; Godula, M.; Malachova, A.; Veprikova, Z.; Hajslova, J. Analysis of Multiple Mycotoxins in Beer Employing (Ultra)-High-Resolution Mass Spectrometry. *Rapid Commun. Mass Spectrom.* **2010**, *24*, 3357–3367. [CrossRef]

59. Rahmani, A.; Jinap, S.; Soleimany, F. Qualitative and Quantitative Analysis of Mycotoxins. *Compr. Rev. Food Sci. Food Saf.* **2009**, *8*, 202–251. [CrossRef] [PubMed]

60. Liu, C.; Van der Fels-Klerx, H.J. Quantitative Modeling of Climate Change Impacts on Mycotoxins in Cereals: A Review. *Toxins* **2021**, *13*, 276. [CrossRef]

61. Moretti, A.; Pascale, M.; Logrieco, A.F. Mycotoxin Risks under a Climate Change Scenario in Europe. *Trends Food Sci. Technol.* **2019**, *84*, 38–40. [CrossRef]

62. Jensen, T.; de Boevre, M.; Preußke, N.; de Saeger, S.; Birr, T.; Verreet, J.A.; Sönnichsen, F.D. Evaluation of High-Resolution Mass Spectrometry for the Quantitative Analysis of Mycotoxins in Complex Feed Matrices. *Toxins* **2019**, *11*, 531. [CrossRef]

63. Winter, M.; Samuels, P.L.; Dong, Y.; Dill-Macky, R. Trichothecene Production Is Detrimental to Early Root Colonization by Fusarium Culmorum and F. Graminearum in Fusarium Crown and Root Rot of Wheat. *Plant Pathol.* **2019**, *68*, 185–195. [CrossRef]

64. Asran, M.R.; Buchenauer, H. Pathogenicity of Fusarium Graminearum Isolates on Maize (*Zea mays* L.) Cultivars and Relation with Deoxynivalenol and Ergosterol Contents/Pathogenität von Fusarium Graminearum Isolaten an Mais- (*Zea mays* L.) Sorten Und Beziehung Zu Deoxynivalenol- Und Ergosterol-Gehalten. *Z. Pflanzenkrankh. Pflanzenschutz J. Plant Dis. Prot.* **2003**, *110*, 209–219.

65. Abid, M.; Fayolle, L.; Edel-Hermann, V.; Gautheron, N.; Héraud, C.; Leplat, J.; Steinberg, C. Fate of Deoxynivalenol (DON) and Impact on the Soil Microflora and Soil Fauna. *Appl. Soil Ecol.* **2021**, *162*, 103898. [CrossRef]

66. Ahad, R.; Zhou, T.; Lepp, D.; Pauls, K.P. Microbial Detoxification of Eleven Food and Feed Contaminating Trichothecene Mycotoxins. *BMC Biotechnol.* **2017**, *17*, 30. [CrossRef]

67. Islam, R.; Zhou, T.; Christopher Young, J.; Goodwin, P.H.; Peter Pauls, K. Aerobic and Anaerobic De-Epoxydation of Mycotoxin Deoxynivalenol by Bacteria Originating from Agricultural Soil. *World J. Microbiol. Biotechnol.* **2012**, *28*, 7–13. [CrossRef] [PubMed]

68. Vanhoutte, I.; De Mets, L.; De Boevre, M.; Uka, V.; Di Mavungu, J.; De Saeger, S.; De Gelder, L.; Audenaert, K. Microbial Detoxification of Deoxynivalenol (DON), Assessed via a Lemna Minor L. Bioassay, through Biotransformation to 3-Epi-DON and 3-Epi-DOM-1. *Toxins* **2017**, *9*, 63. [CrossRef] [PubMed]

69. He, J.W.; Hassan, Y.I.; Perilla, N.; Li, X.Z.; Boland, G.J.; Zhou, T. Bacterial Epimerization as a Route for Deoxynivalenol Detoxification: The Influence of Growth and Environmental Conditions. *Front. Microbiol.* **2016**, *7*, 572. [CrossRef]

70. Zhai, Y.; Zhong, L.; Gao, H.; Lu, Z.; Bie, X.; Zhao, H.; Zhang, C.; Lu, F. Detoxification of Deoxynivalenol by a Mixed Culture of Soil Bacteria With 3-Epi-Deoxynivalenol as the Main Intermediate. *Front. Microbiol.* **2019**, *10*, 2172. [CrossRef] [PubMed]

Article

Development of a Novel UPLC-MS/MS Method for the Simultaneous Determination of 16 Mycotoxins in Different Tea Categories

Haiyan Zhou [1], Zheng Yan [1], Song Yu [2], Aibo Wu [1] and Na Liu [1,*]

[1] SIBS-UGENT-SJTU Joint Laboratory of Mycotoxin Research, CAS Key Laboratory of Nutrition, Metabolism and Food Safety, Shanghai Institute of Nutrition and Health, University of Chinese Academy of Sciences, Chinese Academy of Sciences, Shanghai 200030, China; zhouhaiyan2018@sibs.ac.cn (H.Z.); zyan@sibs.ac.cn (Z.Y.); abwu@sibs.ac.cn (A.W.)

[2] Division of Chemical Toxicity and Safety Assessment, Shanghai Municipal Center for Disease Control and Prevention, Shanghai 200336, China; yusohar@163.com

* Correspondence: liuna@sibs.ac.cn; Tel.: +86-21-54-920-716

Abstract: The contamination of potential mycotoxins in tea production and consumption has always been a concern. However, the risk monitoring on multiple mycotoxins remains a challenge by existing methods due to the high cost and complex operation in tea matrices. This research has developed a simple ultra-performance liquid chromatography-tandem mass spectrometry strategy based on our homemade purification column, which can be applied in the detections of mycotoxins in complex tea matrices with high-effectively purifying and removing pigment capacity for 16 mycotoxins. The limits of detection and the limits of quantification were in the ranges of 0.015~15.00 and 0.03~30.00 $\mu g \cdot kg^{-1}$ for 16 mycotoxins, respectively. Recoveries from mycotoxin-fortified tea samples (0.13~1200 $\mu g \cdot kg^{-1}$) in different tea matrices ranged from 61.27 to 118.46%, with their relative standard deviations below 20%. Moreover, this method has been successfully applied to the analysis and investigation of the levels of 16 mycotoxins in major categories of tea and the monitoring of multiple mycotoxins in processed samples of ripened Pu-erh. In conclusion, the proposed strategy is simple, effective, time-saving, and low-cost for the determination of a large number of tea samples.

Keywords: mycotoxins; tea; ultra-performance liquid chromatography-tandem mass spectrometry; simultaneous determination; purification

Key Contribution: A universal UPLC-MS/MS method based on our homemade purification column with high-efficiency and broad-selectivity capacity has been developed and validated for seven groups of mycotoxins (aflatoxins of B1, B2, G1 and G2, ochratoxin A, zearalenone and its derivatives (α-ZEL, β-ZEL, α-ZAL, and β-ZAL), deoxynivalenol and its derivates (15-Ac DON, 3-Ac DON), T-2, neosolaniol, and citrinin) determination in complex tea matrices, which was applied to survey the occurrence of multi-mycotoxins in 105 actual tea samples. The results demonstrated that the proposed strategy is time-saving and low-cost in the detection of large numbers of samples.

Citation: Zhou, H.; Yan, Z.; Yu, S.; Wu, A.; Liu, N. Development of a Novel UPLC-MS/MS Method for the Simultaneous Determination of 16 Mycotoxins in Different Tea Categories. *Toxins* **2022**, *14*, 169. https://doi.org/10.3390/toxins14030169

Received: 24 January 2022
Accepted: 23 February 2022
Published: 24 February 2022

Publisher's Note: MDPI stays neutral with regard to jurisdictional claims in published maps and institutional affiliations.

1. Introduction

Tea, the leaves of *Camellia sinensis*, is divided into green, white, yellow, oolong, black, and dark tea, depending on the type of fermentation process [1], which results in the differences of health-promoting components [2]. In a word, tea has attracted increasing attention for its unique human health benefits [3–5].

Tea products take a long time to process, with stages including planting, primary and refined processing, packaging, storage, and transportation for final consumption. Mycotoxins, potential hazardous pollutants generally produced by toxigenic fungi, are of great concern during manufacturing [6–10]. Furthermore, it is known that post-fermented tea contains many microorganisms. *Aspergillus* spp., *Fusarium* spp. and *Penicillium* spp. are the most prevalent microbial groups present during the processing and storage of post-fermented tea [11,12]. Some fungi are important contributors to tea manufacture, promoting the formation of active ingredients and the unique aroma of tea during the pile fermentation period. However, there are concerns that some fungal species might be mycotoxin producers and cause mycotoxin accumulation under suitable environmental conditions [13,14]. Currently, mycotoxins in foodstuffs are receiving increasing attention for their toxicity, carcinogenicity, teratogenicity, mutagenicity, genotoxicity, and immunotoxicity to humans and other animals. Some studies have focused on the occurrence of mycotoxin pollutants in tea, containing zearalenone (ZEN), aflatoxins (AFs), and ochratoxin A (OTA) in dark tea [14,15] and AFs, deoxynivalenol (DON) in black, green, and other teas [16,17]. However, there are no international maximum regulation limits (MRLs) for mycotoxins in tea due to the limited contamination data and exposure assessments in different types of tea, especially for the new toxicities of common mycotoxins or emerging mycotoxins.

Concerning the presence of multiple mycotoxins in a single tea, several studies have focused on the quantitative determination of multiple mycotoxins in a few kinds of tea through high-performance liquid chromatography (HPLC) and liquid chromatography-tandem mass spectrometry (LC-MS/MS) [14,17–19]. Tea samples, complex matrices containing abundant caffeine and polyphenols, are considered to be analytically challenging in trace pollutant residue detection methods [20,21]. The sample pretreatment processes of existing quantitative methods for mycotoxins in complex tea matrices are complicated and expensive, such as dispersive liquid–liquid microextraction (DLLME), which is based on isotopes [18]; solid phase extraction (SPE), which is based on salting out; dispersive solid-phase extraction (D-SPE) during extraction [22]; and the successive use of NH_2-SPE and C_{18}-SPE [19], multiple immunoaffinity columns (IACs) [17], and MFCs (multifunctional columns) integrated with IACs [14] in the clean-up stage. It is still a challenge to establish simple pretreatment processes to eliminate matrix effects (MEs) with broad applicability to monitor the potential multiple mycotoxins in various tea matrices.

The present study aimed to develop and validate an ultra-performance LC-MS/MS method with a homemade purification column to determine 16 mycotoxins in four categories of tea (green, oolong, black, and dark tea). Additionally, the occurrence and contamination of multi-mycotoxin in various tea samples were investigated with the newly established and successfully validated [23,24] method.

2. Results and Discussion

2.1. UPLC-MS/MS Analysis Conditions

The target mycotoxins in this paper are small molecules with a wide polarity range (log Kow = −0.85~5.37). C_{18} reversed-phase liquid chromatography columns are the most commonly used columns for the separation of mycotoxin (Table 1). During the application of the method, $[M + Na]^+$ ions can lead to incomplete cleavage of parent ions and reduce the signal abundance of the target analyte [7]. When mobile phase A is ammonium acetate, it can avoid the formation of $[M + Na]^+$ in some mycotoxins. In other words, 5 mM ammonium acetate can enhance the response of 15-Ac DON, NEO, and T-2 by promoting the $[M + NH_4]^+$ ion intensities [25]. AFB_1, AFB_2, AFG_1, AFG_2 and OTA contain methoxy or carbonyl groups that can result in a high abundance of the $[M + H]^+$ peak in positive ionization mode. ZEN and its modified forms easily generated $[M − H]^-$ ions at m/z values of 317, 319 and 321. The $[M + CH_3COO]^-$ ions among the precursor ions of DON and 3-Ac DON had the highest ion abundance. Furthermore, the main form of the precursor ions of 15-Ac DON, NEO and T-2 was $[M + NH_4]^+$. The highest CIT ion abundance was in the $[M + OH]^-$ ion form. The signal acquisition information of various mycotoxins in the optimized SRM mode, including characteristic ion pairs and collision energy parameters, is shown in Table 2. According to the abundance of ions, we selected appropriate ions as quantitative ions. The total ion chromatography separation effect of the 16 mycotoxins at middle concentrations (Table S1) under the optimal UPLC-MS conditions is shown in Figure 1.

Table 1. Comparison of multi-mycotoxins in tea determination by HPLC or LC-MS/MS recently reported.

Detection (Mycotoxins)	Extraction//Purification (The Chromatographic Column)	Matrix (Recoveries)	LODs//LOQs $\mu g \cdot kg^{-1}$
UPLC-MS/MS (22 Mycotoxins) [19]	1% formic acid-EtOAc//NH2-SPE, C18-SPE (Acquity UPLC BEH C18)	Raw tea materials (China and Belgium) (91–107%)	2.1–122// 4.1–243
HPLC-MS/MS-IT (16 Mycotoxins) [18]	DLLME: NaCl, EtOAc, ACN, MeOH, CHCL3 (Gemini-NX column C18)	Black, red, green tea beverages (Spain) (65–127%)	0.05–10.0// 0.2–33.0
HPLC-FD (5 Mycotoxins) [17]	80% MeOH, NaCl//AFs-IAC, OTA-IAC (ZORBAX Eclipse XDB C18)	Black, green tea (Iran) (74.1–99.6%)	0.1–0.47// 0.4–1.23
HPLC (10 Mycotoxins) [14]	ACN or MeOH//water containing NaCl//MFC-IAC (C18 column Xbridge)	Dark tea (China) (76.8–95.6%)	0.018–34.4// Not Found
UPLC-MS/MS (7 Mycotoxins) [22]	1.0 mol/L ammonium acetate, 98% ACN-DMSO// MgSO4, C18 (Acquity HSS-T3 column)	Black tea (China) (75.20–124.4%)	Not Found// 5
UPLC-MS/MS (16 Mycotoxins) This work	1% formic acid -ACN//MWCNTs-COOH, HLB, SG (Agilent Extend C18)	Green, oolong, black, and dark tea (China) (61.27–118.46%)	0.015–15.00// 0.03–30.00

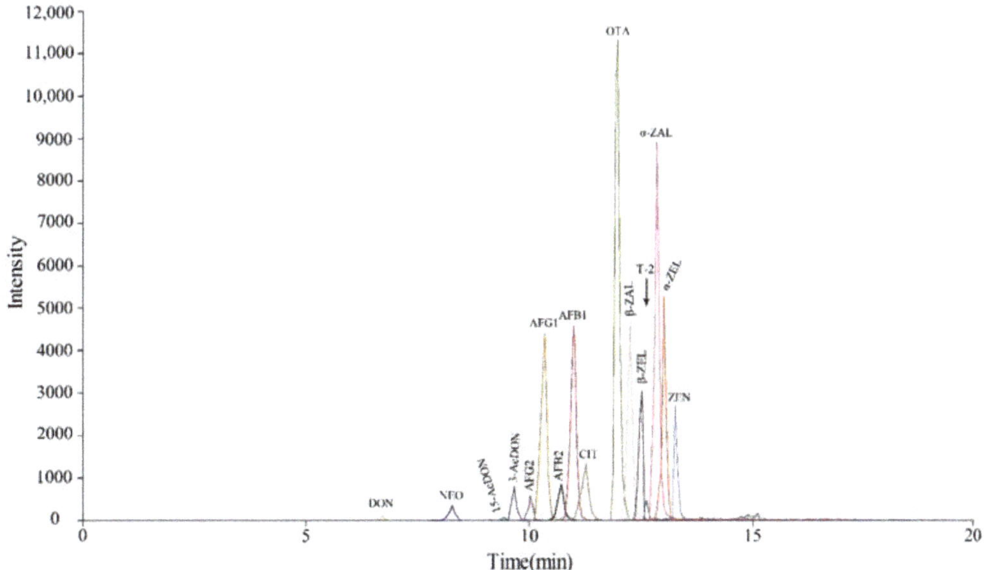

Figure 1. Total ion chromatograms of 16 mycotoxins at middle concentration under optimized chromatographic and mass spectrometry conditions.

Table 2. Retention time and MS parameters for the analysis of mycotoxins.

Analytes	Molecular Weight	TR (min)	Molecular Ion	ESI	Parent Ions (*m/z*)	Product Ions (*m/z*)	CE (eV)
AFB$_1$	312.27	10.82	[M + H]$^+$	ESI$^+$	313.110 313.120 313.130	241.100 [a] 285.200 213.100	44 23 46
AFB$_2$	314.29	10.55	[M + H]$^+$	ESI$^+$	315.110 315.120 315.130	287.200 [a] 259.300 243.200	30 33 40
AFG$_1$	328.27	10.18	[M + H]$^+$	ESI$^+$	329.110 329.120 329.130	199.000 243.300 200.100 [a]	58 32 41
AFG$_2$	330.29	9.86	[M + H]$^+$	ESI$^+$	331.110 331.120 331.130	189.100 [a] 245.100 314.200	42 34 25
ZEN	318.36	13.17	[M − H]$^-$	ESI$^-$	317.110 317.120 317.130	175.030 [a] 131.020 273.170	25 31 20
α-ZEL	320.38	12.90	[M − H]$^-$	ESI$^-$	319.110 319.120 319.130	275.200 [a] 160.000 301.200	21 33 23
β-ZEL	320.38	12.36	[M − H]$^-$	ESI$^-$	319.110 319.120 319.130	275.200 [a] 160.000 301.200	21 33 23
α-ZAL	322.40	12.74	[M − H]$^-$	ESI$^-$	321.110 321.120 321.130	277.200 [a] 303.200 259.200	23 22 25
β-ZAL	322.40	12.10	[M − H]$^-$	ESI$^-$	321.110 321.120 321.130	277.200 [a] 303.200 259.200	23 22 25
DON	296.32	6.55	[M + CH$_3$COO]$^-$	ESI$^-$	355.000 355.100	265.000 [a] 247.200	17 22
15-Ac DON	338.35	9.41	[M + NH$_4$]$^+$	ESI$^+$	356.000 356.100	137.000 [a] 321.000	5 13
3-Ac DON	338.35	9.46	[M + CH$_3$COO]$^-$	ESI$^-$	397.000 397.100	307.160 [a] 173.100	16 15
OTA	403.81	11.96	[M + H]$^+$	ESI$^+$	404.110 404.120 404.130	105.100 221.000 239.100 [a]	18 35 25
NEO	382.40	8.09	[M + NH$_4$]$^+$	ESI$^+$	400.100 400.120	185.100 [a] 215.100	16 14
T-2	466.52	12.51	[M + NH$_4$]$^+$	ESI$^+$	484.110 484.120 484.130	215.000 [a] 165.000 197.000	20 66 24
CIT	250.25	11.28	[M + OH]$^-$	ESI$^-$	267.110 267.120 267.130	221.000 177.000 249.000 [a]	20 28 21

RT: retention time; CE: collision energy; [a] Quantifying ions.

2.2. Improvement of Sample Pretreatment Processes

2.2.1. Optimization of Mycotoxins Extractant in Tea Samples

In our preliminary research, the extraction recoveries of the target analyte in tea gradually increased with increasing acetonitrile concentration. The ME of acetonitrile is the weakest among commonly used extraction agents. However, the extraction effect of mycotoxins sensitive to the polarity range was not ideal when only acetonitrile was used as the extraction agent. Therefore, the extraction recoveries of mycotoxins achieved by acetonitrile with different formic acid contents were considered and compared. The extraction recoveries of OTA, T-2 and NEO were close to 100%, with 5% formic acid in the extractant (Figure 2A). The recoveries of ZEN and its modified forms did not increase gradually with increasing acidity of acetonitrile but were inhibited to less than 60% at an acidity percentage greater than 5%, which may have been due to the influence of hydrogen ions on the response of negative ions. Recoveries of DON and its derivatives were better in acetonitrile with formic acid concentrations from 0 to 2%. The recoveries of AFs were better under extraction with acetonitrile containing formic acid. The extraction recoveries of CIT increased with increasing formic acid concentration, while some recoveries exceeded 120%. When the concentration of formic acid was 1%, the recovery was close to 100% ($p < 0.05$). To monitor the potential risk of multiple mycotoxins in tea simultaneously and improve the wide applicability compared with other established methods, acetonitrile with 1% formic acid was finally selected as the extraction agent to meet the recoveries and responses of the mycotoxins mentioned above (Figure 2A).

2.2.2. Evaluation of Purification Effects on Mycotoxins in Tea Samples

In recent years, multi-functional purification columns have been popular in LC-MS/MS preprocessing for their simple and fast impurity removal. However, the detection cost is expensive for large numbers of samples. We developed a homemade purification column with excellent purification effects for aflatoxins [26]. However, the purification effects for other mycotoxins are still needed to be further evaluated. In this study, the purification effects of homemade were compared with other commercial purification columns on purification for 16 mycotoxins in tea matrices. MWCNTs-COOH:HLB:SG (our homemade purification column) had distinct advantages in 16 mycotoxins purification in tea matrices. MWCNTs-COOH:HLB:SG has the good impurity removal ability among these columns, and it has a wider range of detection for mycotoxins (Figures 2B and 3C). The Cleanert MC purification column had better purification recoveries for 15-Ac DON, AFs, 3-Ac DON, T-2, NEO, and ZENs (61.24~119.58%). The recoveries of AFs, 3-Ac DON, T-2, NEO, and ZEN by the CNW 301 MFC purification column were 63.38~110.53%. The MFC 260 purification column had the best purification recovery for T-2 among the mycotoxins (85.13%). The purification recoveries of 3-Ac DON, DON, AFs, CIT, T-2, NEO, OTA, and ZENs on the Romerlab MycoSpin 400 were 64.95~116.93%. Our homemade purification column especially achieved ideal purification recoveries (74.26~119.98%) in tea matrices for 16 mycotoxins. Furthermore, the MEs of mycotoxins were improved through purification with our homemade purification column (Figure 3). Most current reports apply the successive use of NH_2-, C_{18}-SPE, multiple IACs, or MFCs-IACs to achieve clean-up in some tea matrices [14,17–19,22]. With our homemade purification column, the developed method does not require combining with other commercial columns for sample purification in complex tea matrices, which is mainly due to its high-efficiency and broad-selectivity capacity (Figures 2B and 3). Additionally, the cost price ratios of MWCNTs-COOH:HLB:SG (our homemade purification column), Cleanert MC, MFC 301, MFC 260, and MycoSpin 400 were 1:3.61:13.79:11.82:17.00.

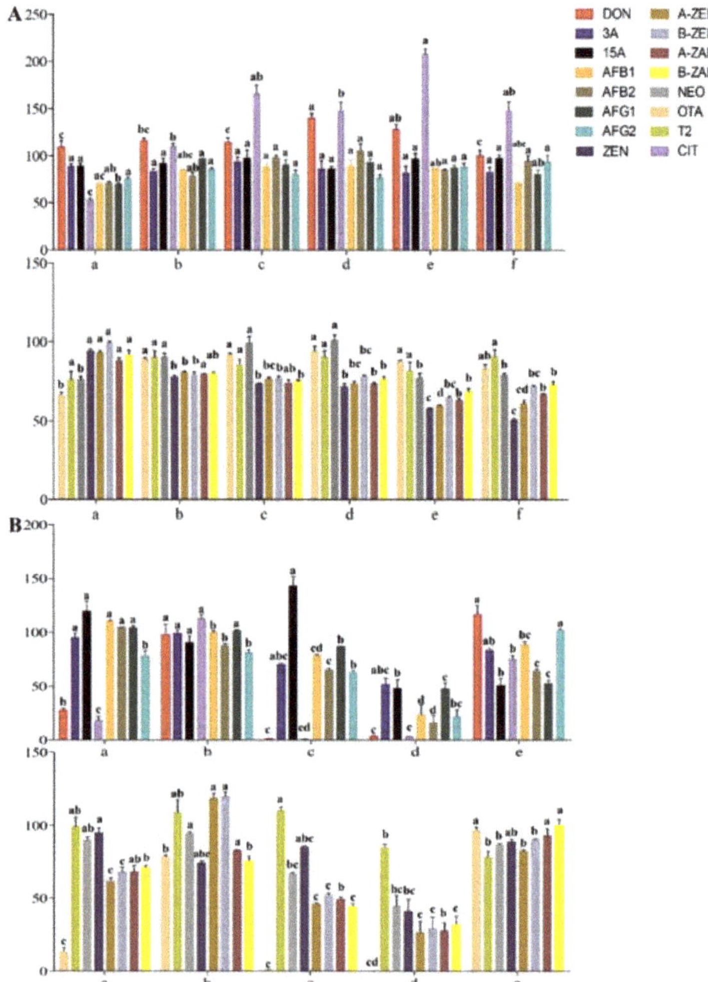

Figure 2. Optimization of extraction and purification methods: (**A**) The recoveries of mycotoxins in tea matrix by acetonitrile with different formic acid content; (a) acetonitrile; (b) acetonitrile with 1% formic acid; (c) acetonitrile with 2% formic acid; (d) acetonitrile with 5% formic acid; (e) acetonitrile with 10% formic acid; (f) acetonitrile with 15% formic acid. (**B**) The Recoveries of mycotoxins in tea matrix by different purification methods; (a) Cleanert MC; (b) MWCNTs-COOH:HLB:SG (1:7.5:7.5); (c) CNW 301 MFC; (d) MFC260; (e) Romerlab MycoSpin 400. Different low case letters above columns indicate statistical differences at $p < 0.05$; The bars ($n = 3$) represent mean $\pm$ SE.

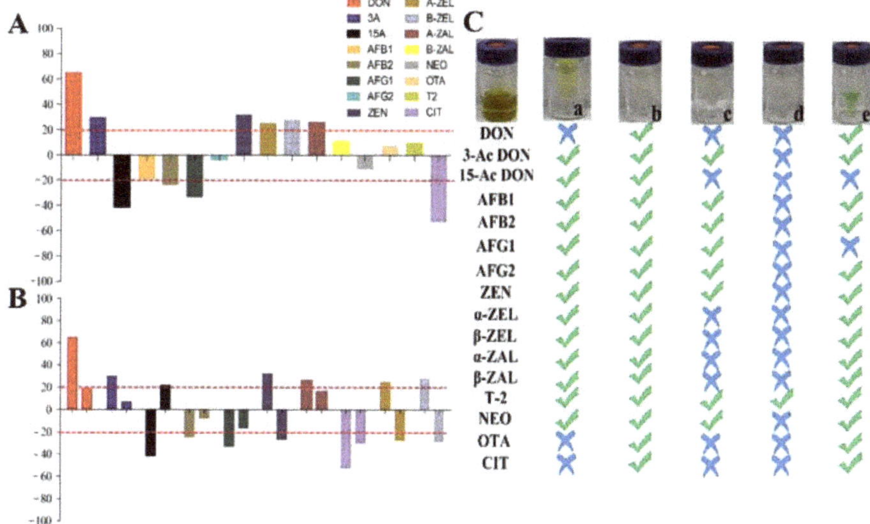

Figure 3. Matrix effects and purification effects of 16 mycotoxins after extraction and purification: (**A**) Matrix effects of 16 mycotoxins extracted by acetonitrile with 1% formic acid. (**B**) Matrix effects of 10 mycotoxins (matrix effects are unacceptable after extraction) before and after purification. (**C**) Purification effects for 16 mycotoxins with different clean-up columns; (a) Cleanert MC; (b) MWCNTs-COOH:HLB:SG (1:7.5:7.5); (c) CNW 301 MFC; (d) MFC260; (e) Romerlab MycoSpin 400; symbols "√" represent recoveries ranging from 60% to 120%, while symbols "×" represent recoveries <60% or >120%, respectively.

2.3. Method Validation

The LODs and LOQs of mycotoxins were in the ranges of 0.015~15.00 and 0.03~30.00 $\mu g \cdot kg^{-1}$ (Table S1), which are similar to those obtained with the method developed by Pallarés et al. [18]. There are no uniform limits for mycotoxins in tea. Custom Union countries have regulated a limit of AFB1 in raw tea of 5 $\mu g \cdot kg^{-1}$ [27]. Argentina set limits for AFB1 and AFs in herbal tea materials at 5 and 20 $\mu g \cdot kg^{-1}$, respectively [28]. In the European Union, the MRL for OTA in both roasted coffee beans and ground roasted coffee is 5.0 $\mu g \cdot kg^{-1}$; and the MRLs for DON and ZEN in cereals intended for direct human consumption are 750 $\mu g \cdot kg^{-1}$ and 75 $\mu g \cdot kg^{-1}$, respectively [24]. The LODs (or LOQs) of this method are far below the MRLs in foodstuffs described above. Recoveries of this method in different tea matrices ranged from 61.27 to 118.46% for mycotoxins. Furthermore, the intra- and inter-day relative standard deviations (RSDs) were below 20% (0.70~19.95%), which indicates that this strategy can quantitatively determine mycotoxins in tea (Table 3). In short, there are satisfactory data from the recoveries of extraction, purification (Figure 2), and the validation of the prosed method (Table 3). A comparison between our proposed method and previously reported methods was completed on mycotoxins quantitative detections (Table 1). Obviously, our method has the advantages of lower cost, wider applicability with satisfactory recovery rates and LOQs, and has the ability to eliminate interferences of more types of tea for multiple mycotoxins monitoring.

Table 3. Overview of the accuracy and precision of the developed LC-MS/MS method.

Targets	Spiked Recovery (%) RSD (%, $n = 6$)												Precision (RSD, %)	
	Green Tea			Oolong Tea			Black Tea			Dark Tea			RSDr	RSDR
	Low	Middle	High	Low	Middle	High	Low	Middle	High	Low	Middle	High	($n = 6$)	($n = 6$)
AFB1	92.90 13.75	101.72 7.81	65.51 4.55	86.60 7.76	95.47 5.84	68.48 3.49	90.99 5.97	94.08 2.67	63.27 3.17	92.65 8.06	99.67 10.45	74.88 4.00	94.26 4.75	108.08 3.36
AFB2	90.28 17.77	100.87 11.12	62.20 3.55	79.70 10.01	86.54 5.68	65.41 4.61	84.60 3.50	86.57 6.60	66.74 3.12	80.99 15.66	98.44 10.27	72.24 3.09	88.50 8.11	102.82 6.46
AFG1	107.01 10.72	109.72 8.55	64.29 4.12	115.83 4.17	105.93 2.58	69.55 4.75	104.52 8.46	104.10 6.92	65.24 4.37	114.44 6.04	101.74 4.38	77.21 5.69	107.91 4.36	109.54 4.79
AFG2	89.13 19.39	88.61 8.76	63.93 7.89	77.31 19.34	99.96 5.87	102.65 11.72	64.48 1.19	114.39 19.13	87.49 17.04	77.02 19.95	98.27 15.80	93.81 18.75	88.65 16.59	101.26 4.69
ZEN	76.58 8.19	91.47 10.06	63.55 4.40	78.48 8.57	88.46 10.16	71.93 7.57	71.46 5.96	76.65 3.50	61.81 2.68	69.62 8.35	73.45 3.53	60.79 0.70	78.27 9.35	85.55 8.99
α-ZEL	117.62 1.68	99.86 6.21	85.93 18.97	116.21 2.95	98.63 7.31	95.79 18.74	100.79 14.34	89.30 4.46	78.86 13.34	110.90 8.72	97.93 10.77	86.53 16.89	103.91 8.95	88.12 9.67
β-ZEL	88.37 7.34	69.75 11.08	94.55 17.71	101.14 11.50	67.84 8.84	103.51 12.59	77.20 3.94	62.80 3.18	91.34 11.48	85.42 7.88	64.25 6.63	101.90 14.62	77.10 16.43	78.42 8.29
α-ZAL	74.79 9.12	79.36 6.52	86.95 18.66	72.93 6.52	74.65 9.06	100.73 16.70	76.86 5.86	71.90 4.25	92.19 19.28	73.71 7.44	71.40 3.35	100.65 16.01	74.45 3.32	73.12 8.62
β-ZAL	67.43 6.79	69.41 9.61	89.46 19.17	62.71 4.03	67.92 8.25	101.03 14.71	64.14 9.31	61.27 1.56	86.52 18.90	63.42 9.36	63.26 5.08	97.79 15.91	64.95 4.19	70.39 12.55
DON	75.45 13.81	90.96 4.13	108.16 6.56	103.11 13.36	98.11 7.65	112.90 5.65	83.49 7.51	98.84 4.61	104.18 7.96	66.74 11.90	90.48 4.53	106.39 9.15	88.40 13.21	97.25 15.22
15AcDON	101.34 10.58	101.17 19.74	76.59 12.89	87.24 19.73	82.43 18.16	83.40 9.68	84.65 14.51	77.16 18.03	74.57 11.06	95.42 19.48	100.72 18.17	87.43 10.59	91.27 9.82	95.57 9.04
3AcDON	101.45 7.09	99.22 3.80	96.07 12.89	105.69 7.21	101.47 3.42	100.25 7.77	100.10 5.70	108.85 3.59	95.55 6.16	97.13 8.08	102.95 4.84	90.57 15.04	102.11 3.41	107.59 3.48
OTA	90.02 18.90	90.26 19.72	77.02 11.19	80.60 12.99	66.11 6.46	92.40 8.15	64.36 4.46	64.80 0.94	90.64 8.32	68.73 11.26	64.18 4.13	97.53 11.77	73.63 14.63	97.17 8.34
NEO	75.32 14.53	99.07 17.79	96.55 15.07	100.04 14.05	106.40 12.87	109.62 8.88	74.58 14.59	76.63 8.69	102.32 9.43	67.99 11.52	77.98 7.76	100.31 9.87	84.75 16.12	100.54 11.06
T-2	66.54 11.08	104.40 13.56	88.89 16.89	81.20 17.63	93.91 9.14	100.05 14.55	80.21 9.66	88.71 4.20	102.24 4.15	65.51 10.08	92.03 6.01	104.72 4.58	84.06 14.99	96.23 13.28
CIT	103.01 7.66	105.01 10.15	61.89 4.05	114.85 4.86	111.90 5.94	77.47 10.03	118.46 2.29	111.03 3.44	65.41 7.94	101.74 11.81	74.15 15.04	61.50 2.23	105.02 12.27	97.45 9.68

Low, Middle, and High represent the spiked low, middle, and high concentrations of mycotoxins respectively. RSD: repeatability for recoveries in each type of tea samples at each fortified concentration; RSDr: intraday precision (repeatability) in mycotoxin-fortified tea samples; RSDR: interday precision (reproducibility) in mycotoxin-fortified tea samples.

2.4. Occurrence of Mycotoxins in Tea Samples

The mycotoxins in the samples of different types of tea were detected to different degrees, with average concentrations ranging from 0.07 to 958.58 $\mu g \cdot kg^{-1}$ (Table 4). From green, oolong, and black tea to dark tea, the detection types and ranges of most mycotoxins increased, which may have been related to the differences among these teas in processing methods or biochemical components. However, the total positive rates of DON, AFs, α-ZEL, β-ZEL, OTA, and CIT were 6.25% (5/80), 1.25% (1/80), 6.25% (5/80), 10% (8/80), 5% (4/80), and 1.25% (1/80), respectively.

The co-occurrence rate of DON (10%) and its derivatives in tea samples was lower than the result obtained by Reinholds et al., possibly due to study differences in the samples analyzed [16,29]. Kiseleva et al. found that C. sinensis tea samples were the least contaminated among the herbal tea samples analyzed [30]. The average contents of DON in the 20 oolong tea samples in the present study exceeded the MRL of DON in cereals for direct consumption (750 $\mu g \cdot kg^{-1}$), with five tea samples having contents between 777.51 and 1181.48 $\mu g \cdot kg^{-1}$. Mannani et al. and Bogdanova et al. observed average content ranges similar to those in the present study: AFB_1 (LOD~1.04), AFB_2 (LOD~0.39), AFG_1 (0.07~0.62), and AFG_2 (2.97~8.98) [25,31,32]. Importantly, the average contents of AFB_1 and AFs in dark tea in this study did not exceed the Argentinean limits of 5 and 20 $\mu g \cdot kg^{-1}$, respectively. The occurrence of ZEN (2.5%) was similar to that detected in the Ye et al. study [14]. In addition, the average content of ZEN did not exceed the relevant MRL (75 $\mu g \cdot kg^{-1}$). Interestingly, the contents of β-ZEL and α-ZEL in oolong and black tea were higher than those in the other teas. Pakshir et al. employed HPLC and observed similar incidences of OTA in green tea and black tea to those obtained in the present study: although most of the samples were contaminated with OTA, the contents were lower than the MRL (5 $\mu g \cdot kg^{-1}$) for OTA in coffee [17]. Moreover, in the present study, the average content of OTA was highest in dark tea (666.45 $\mu g \cdot kg^{-1}$), with three of the small green orange Pu-erh samples ("Xiao Qing Gan" or "Gan Pu") containing high concentrations of OTA. Li et al. observed that the content of CIT ranged from 7.8 to 206 $\mu g \cdot kg^{-1}$ among 107 tea samples, and two samples had high CIT contents (>200 $\mu g \cdot kg^{-1}$) [33,34]. The range of CIT contents in dark tea is similar to that in our study: one ripened Pu-erh tea sample out of the 80 samples had a CIT content above 200 $\mu g \cdot kg^{-1}$ (203.76 $\mu g \cdot kg^{-1}$).

The contents of the 16 mycotoxins during the production steps of ripened Pu-erh were also determined. The results illustrated that 15-Ac DON,3-Ac DON, α-ZAL, DON, NEO, β-ZAL, OTA, T-2 and CIT were not detected in the finished Pu-erh. In addition, seven other mycotoxins were detected at (0.10~72.61 $\mu g \cdot kg^{-1}$), which were lower than the corresponding MRLs listed above. Specifically, 15-Ac DON, DON, T-2, and NEO reached their highest concentrations in the samples from the first repiling step; the contents of OTA in the raw material, samples from the humidifying stage, and samples from the fifth repiling step were higher than those in the samples from other production stages. In addition, the content of CIT was highest in the samples from the third repiling step. Regarding the total amount of AFs and that of ZEN and its modified forms (i.e., α-ZAL, α-ZEL, and β-ZEL) [35], the contents were lowest in the semifinished Pu-erh and increased in the finished tea, but the total content of AFs was lower than 20 $\mu g \cdot kg^{-1}$ (Table S2). It is necessary to monitor mycotoxins in ripened Pu-erh during tea manufacturing, especially during later processing or storage. Such monitoring is conducive to eliminating high-risk tea samples as much as possible.

Table 4. Concentration range and mean of mycotoxins in all tea samples ($\mu g \cdot kg^{-1}$).

Mycotoxin	Green Tea			Oolong Tea			Black Tea			Dark Tea			TSP	rMRL$_S$
	Min	Max	Mean	Min	Max	Mean	Min	Max	Mean	Min	Max	Mean		
AFB1	<LOD	<LOD	ND	<LOD	<LOD	ND	<LOD	<LOD	ND	0.97	1.11	1.04	(0,0,0,0)/80	5 [a]
AFB2	<LOD	<LOD	ND	0.16	0.61	0.39	<LOD	0.15	0.15	0.13	0.32	0.22	(0,0,0,0)/80	5 [b]
AFG1	0.06	0.07	0.07	0.19	0.91	0.47	0.43	0.87	0.62	0.38	0.84	0.61	(0,0,0,0)/80	5 [b]
AFG2	0.90	6.75	2.97	2.98	13.36	8.71	0.79	7.51	3.46	3.27	23.49	8.98	(1,15,6,15)/80	5 [b]
AFs	0.90	6.75	2.98	2.98	14.27	9.03	1.22	7.51	3.71	0.20	24.07	8.83	(0,0,0,1)/80	20 [a]
ZEN	<LOD	0.26	0.26	<LOD	<LOD	ND	<LOD	<LOD	ND	<LOD	1.44	1.44	(0,0,0,0)/80	75 [c]
α-ZEL	10.30	24.71	19.72	29.27	48.03	39.77	19.19	116.22	51.28	25.31	71.09	36.55	(0,0.5,0)/80	75 [d]
β-ZEL	20.49	34.95	25.98	40.94	96.08	57.54	26.62	226.24	90.21	25.21	62.78	45.38	(0,2,6,0)/80	75 [d]
α-ZAL	<LOD	<LOD	ND	<LOD	12.87	12.87	<LOD	<LOD	ND	<LOD	<LOD	ND	(0,0,0,0)/80	75 [d]
β-ZAL	<LOD	<LOD	ND	<LOD	<LOD	ND	<LOD	<LOD	ND	<LOD	<LOD	ND	(0,0,0,0)/80	75 [d]
DON	11.42	89.89	42.72	667.35	1181.48	958.58	247.40	508.99	423.59	176.02	328.94	253.56	(0,5,0,0)/80	750 [c]
15AcDON	<LOD	<LOD	ND	<LOD	<LOD	ND	77.40	707.57	294.25	<LOD	425.70	192.65	(0,0,0,0)/80	750 [e]
3Ac DON	<LOD	<LOD	ND	<LOD	8.03	6.71	<LOD	<LOD	ND	<LOD	<LOD	ND	(0,0,0,0)/80	750 [e]
OTA	<LOD	<LOD	ND	<LOD	1.07	0.34	0.53	2.81	1.52	0.30	11354.64	666.45	(0,0,0,4)/80	5 [c]
NEO	<LOD	<LOD	ND	<LOD	1.30	0.91	1.28	4.61	2.50	3.49	13.56	7.79	(0,0,0,0)/80	200 [f]
T-2	<LOD	<LOD	ND	<LOD	<LOD	ND	<LOD	<LOD	ND	<LOD	<LOD	ND	(0,0,0,0)/80	200 [c]
CIT	4.16	9.41	6.73	4.84	91.90	25.91	8.90	93.27	54.59	12.20	203.76	62.76	(0,0,0,1)/80	200 [g]

ND: not detectable; LOD: limit of detection; TSP: Number of positive samples in green tea, oolong tea, black tea, or dark tea samples (n_i)/total samples (N). r-MRLs: relevant-Maximum Regulation Limits. [a] Limits for AFB1 and AFs in materials used for herbal tea infusions set by Argentina; [b] MRLs with respect to AFB1 MRL: 5 $\mu g \cdot kg^{-1}$; [c] MRLs of contaminants of foods in EU (EC Regulation No 1881/2006); [d] MRLs with respect to ZEN MRL: 75 $\mu g \cdot kg^{-1}$; [e] MRLs with respect to DON MRL: 750 $\mu g \cdot kg^{-1}$; [f] MRLs with respect to T-2 MRL: 200 $\mu g \cdot kg^{-1}$. [g] MRLs with respect to the evaluation of citrinin occurrence in Chinese Liupao tea by Li et al. [34].

3. Conclusions

Risk monitoring on multiple mycotoxins in tea matrices remains a challenge due to the high cost and complex operation. Compared to most commercial and reported columns, our homemade column is effective for more mycotoxins, which was wider in application. Based on a simple and low-cost sample pretreatment process with the help of our homemade purification column, a universal method for simultaneously detecting 16 mycotoxins in tea matrices was successfully established and validated. It is simple and universal, time-saving, and low-cost for the determination of a large number of tea samples. The proposed method was applied to 80 commercial samples and 25 process samples of ripened Pu-erh to analyze mycotoxin levels. Some MRLs were exceeded (for DON, OTA, α-ZEL, β-ZEL, AFG$_2$, or AFs) by the detected contents in some samples. Meanwhile, green tea had the lowest mycotoxin contamination among the four types of tea, which may be related to the fermentation process. This contamination investigation laid a good foundation for subsequent risk assessment consistent with drinking tea. The proposed strategy provides a practical and universal solution for the quantitative determination of multiple mycotoxins in complex tea matrices, which would help to monitor the potential risk of mycotoxins in the fields of tea production and consumption.

4. Materials and Methods

4.1. Chemicals and Reagents

ZEN (Art. No. Z 2125), α-ZEL (Art. No. Z 0166), β-ZEL (Art. No. Z 2000), α-ZAL (Art. No. Z 0292), β-ZAL (Art. No. Z 0417), DON (Art. No. D 0156), 3-Ac DON (Art. No. A 6166), 15-Ac DON (Art. No. A 1556), OTA (Art. No. O 1877), T-2 (Art. No. T 4887), and CIT (Art. No. C 1017) standards were obtained from Sigma-Aldrich (St. Louis, MO, USA). AFs (Part. No. 10000344) and NEO (Part. No. BRM S 92001) analytical standards were obtained from Romer Lab Biopure™. Cleanert MC (Art. No. LC-MYT10-B) was acquired from Agela Technologies (Shanghai, China). MFC260 (Art. No. M2600-25T) was obtained from Pribolab (Singapore). Romerlab MycoSpin 400 (Art. No. COCMY2400) was purchased from Shanghai Pu Yu Science and Trade, Inc. CNW 301 MFC (Art. No. SBEQ-CD301S) and an analytical-grade hydrophilic-lipophilic balance (HLB; Art. No. SBEQ-CA3100) were purchased from Anpel Laboratory Technologies (Shanghai, China). Analytical-grade carboxyl multiwalled carbon nanotubes (MWCNTs-COOH; Art. No. 190710134732) were acquired from Klamar Reagent, Inc. (Shanghai, China). Acetonitrile comes from Merck (Darmstadt, Germany) is HPLC-grade. Milli-Q quality water (Millipore, Billerica, MA, USA) was used throughout the experiments. Ammonium acetate, formic acid ($\geq$95%) and analytical-grade silica gel (SG; Art. No. 236799) were Sigma-Aldrich products (St. Louis, MO, USA).

4.2. Preparation of Stock Solutions

AFB$_1$ (2.03 μg·mL^{-1}), AFB$_2$ (0.503 μg·mL^{-1}), AFG$_1$ (2.02 μg·mL^{-1}), AFG$_2$ (0.502 μg·mL^{-1}), and NEO (100.5 μg·mL^{-1}) stock solutions were obtained as certified solutions in acetonitrile. Stock solutions of α-ZAL, β-ZAL, ZEN, α-ZEL, β-ZEL, DON, 15-Ac DON, 3-Ac DON, T-2, CIT (5 mg·mL^{-1}), and OTA (1 mg·mL^{-1}) were manufactured by diluting commercially available mycotoxins with pure acetonitrile. The mixture was freshly made and stored in the refrigerator ($-40\ ^\circ$C) before use. Tea extract from the blank sample was applied to prepare matrix-matched standards.

4.3. Samples Collection

Eighty commercial tea samples were randomly collected from Shanghai Difute International Tea Co., Ltd. (Shanghai, China). The 80 samples comprised 20 samples of green tea, 20 samples of oolong tea, 20 samples of black tea, and 20 samples of dark tea. Additionally, 25 samples of tea materials during processing were gathered from Yunnan Fengqing Longrun Co., Ltd. (Lincang, China), between May 2019 and August 2019, including samples of raw materials; samples of tea in the humidifying stage; samples from the first, second, third, fourth, and fifth repiling steps; and samples of piling-up tea, semifinished and finished Pu-erh. The number of samples in each stage of repiling was three, which are taken from the upper, middle, and lower layers. All samples were ground, divided, and stored before use. Each commercial sample was analyzed twice, and the results are expressed as the mean values.

4.4. Sample Preparation

A total of 1.0 g sample was placed in the 10 mL homogenizing centrifuge tube. To optimize the extraction process of multiple mycotoxins, we experimented with acetonitrile with different formic acid contents (0%, 1%, 2%, 5%, 10% and 15%). After ultrasound-assisted extraction (30 min), it was centrifuged (3220 g, 10 min). Optimization of the purification process was achieved by comparing the purification effects of different clean-up strategies. Finally, supernatant (1 mL) was selected to add the MWCNTs-COOH: HLB: SG. The centrifuge tube was vortexed (1 min) and centrifuged (13,800 g, 5 min). Finally, the sample was filtered out with the organic filter membrane (0.22 μm) for subsequent analysis on the newly proposed UPLC-MS/MS system. The recoveries of mycotoxins purification and extraction from tea matrices were compared and evaluated. To prepare spiked samples, each mixture was vortexed (30 s) and incubated (25 °C) for solvent evaporation after mycotoxins were added to the ground tea sample. Purification with other multifunctional purification columns was also performed according to the existing literature [15] and relevant product instructions (Supplementary Materials).

4.5. UPLC-MS/MS Analysis

Mycotoxins were monitored on an Ultimate 3000 UPLC system (Thermo Fisher Scientific, San Jose, CA, USA) coupled to a TSQ VantageTM triple stage quadrupole mass spectrometer (Thermo Fisher Scientific, San Jose, CA, USA). Analytes were separated on an Agilent Extend C18 column (150 × 3.0 mm, 3.5 μm, Art. No. 763954-302) for gradient elution with water containing 5 mM ammonium acetate and methanol (0.35 mL·min^{-1}, 30 °C, 10 μL) through the following elution program: 15% B (initial), 15% B (0–1 min), 15–25% B (1–3 min), 25–50% B (3–6.5 min), 50–100% B (6.5–12 min), 100% B (12–15 min), 100–15% B (15–17 min), and 15% B (17–20 min).

We monitored mycotoxin standards at middle concentrations in full-scan mode (m/z: 50–1000). Different optimal parent ions were selected, and different collision energies with argon were used to conduct collision-induced dissociation (CID) to obtain the product ion spectrum. The product ions and the collision energy were automatically optimized on the mass spectrometer. The optimized parameters for monitoring selected reactions (SRM), positive (ESI$^+$ 3.0 kV) and negative (ESI$^-$ 2.5 kV) ionization modes using mass spectrometry are: capillary temperature 300 °C; vaporizer temperature 250 °C; ion sweep and aux gas pressure are 5 psi; and sheath gas pressure 40 psi.

4.6. Method Evaluation and Application

According to the documented guidelines [23], the parameters of the UPLC-MS/MS were verified in four tea matrices containing 16 mycotoxins. The ME was evaluated by analyzing the slopes of two sets of mycotoxin standards prepared with organic solvent and the relevant tea extracts. MEs were calculated according to Equation (1) below, where K_a (matrices) and K_b (organic solvent) are the slopes of curves, respectively.

$$\text{ME (\%)} = 100 \times (K_a/K_b - 1) \tag{1}$$

The limits of detection (LODs) and quantification (LOQs) of this method are the lowest detectable concentrations with the signal-to-noise ratio (S/N) values of 3 and 10, respectively. The proposed strategy was validated at three levels in mycotoxin-spiked different types of tea [24]. Recoveries were obtained through Equation (2) below, where C_{ai} is the actual detection concentration for mycotoxin i-spiked samples, C_{bi} is the actual detection concentration for the nonspiked sample, and C_{Ai} is the theoretical concentration of analytes that was added to the sample. The range of added standard concentrations to mycotoxin-free tea with six replicates was 0.13~1200 $\mu g \cdot kg^{-1}$ (Table S1 and Figure S1). The precision and accuracy of the proposed strategy were checked through intra- and interday analysis. Intraday precision was assessed via the analysis of six replicate mycotoxin-fortified tea samples at three different concentration levels, and interday precision was evaluated consecutively on six different days. The collected commercial and processed tea samples were analyzed for multiple mycotoxins by the established method.

$$\text{Recovery (\%)} = 100 \times (C_{ai} - C_{bi})/C_{Ai} \tag{2}$$

4.7. Data Analysis

Use Thermo Xcalibur Qual Browser 4.0 to recognize UPLC-MS raw data. Mycotoxin recoveries for each solvent mixture and purification method were evaluated using two-way ANOVA at a significance level of 0.05; Tukey's multiple comparisons testing was used to evaluate the significance of difference within each group. GraphPad Prism 5.0 software is used for all statistical analysis and graphic production.

Supplementary Materials: The following supporting information can be downloaded at: https://www.mdpi.com/article/10.3390/toxins14030169/s1, Figure S1: The chromatography of blank matrix, bank matrix with mycotoxin standard and mycotoxin in real sample application (R): (a) the chromatography of blank matrix; (b) the chromatography of bank matrix with 16 mycotoxin standards; (c) the chromatography of OTA in real sample (D7); (d) the chromatography AFG2 real sample (D3); Table S1: Overview of the methodological characteristics including linear range, limits of detection, limits of quantification, spiked concentrations for various mycotoxins ($\mu g \cdot kg-1$); Table S2: Occurrence and concentration levels of 16 mycotoxins in pu-erh during different production steps.

Author Contributions: Conceptualization, H.Z., N.L. and A.W.; Methodology, H.Z. and Z.Y.; Software, S.Y. and H.Z.; Validation, Z.Y. and H.Z.; Formal Analysis, H.Z., S.Y. and Z.Y.; Investigation Z.Y. and H.Z.; Resources, A.W.; Data Curation, H.Z. and Z.Y.; Writing—Original Draft Preparation, H.Z.; Writing—Review and Editing, N.L. and A.W.; Visualization, Z.Y.; Supervision, N.L. and A.W.; Project Administration, H.Z., N.L. and A.W.; Funding Acquisition, A.W. and N.L. All authors have read and agreed to the published version of the manuscript.

Funding: This work was supported by the National Key Research and Development Program of China (2018YFC1604403). Shanghai Science and Technology Commission, China (17391901300).

Institutional Review Board Statement: Not applicable.

Informed Consent Statement: Not applicable.

Data Availability Statement: The data that support the findings of this study are available from the corresponding author upon reasonable request.

Conflicts of Interest: The authors declare that there is no known competing financial interests or personal relationships that could have appeared to influence the work reported in this paper.

References

1. Haas, D.; Pfeifer, B.; Reiterich, C.; Partenheimer, R.; Reck, B.; Buzina, W. Identification and quantification of fungi and mycotoxins from Pu-erh tea. *Int. J. Food Microbiol.* **2013**, *166*, 316–322. [CrossRef] [PubMed]
2. Zhong, J.; Chen, N.; Huang, S.; Fan, X.; Zhang, Y.; Ren, D.; Yi, L. Chemical profiling and discrimination of green tea and Pu-erh raw tea based on UPLC-Q-Orbitrap-MS/MS and chemometrics. *Food Chem.* **2020**, *326*, 126760. [CrossRef] [PubMed]
3. Jiang, C.; Zeng, Z.; Huang, Y.; Zhang, X. Chemical compositions of Pu'er tea fermented by Eurotium Cristatum and their lipid-lowering activity. *LWT* **2018**, *98*, 204–211. [CrossRef]
4. Asghar, M.A.; Zahir, E.; Shahid, S.M.; Khan, M.N.; Asghar, M.A.; Iqbal, J.; Walker, G. Iron, copper and silver nanoparticles: Green synthesis using green and black tea leaves extracts and evaluation of antibacterial, antifungal and aflatoxin B1 adsorption activity. *LWT* **2018**, *90*, 98–107. [CrossRef]
5. Rameshrad, M.; Razavi, B.M.; Hosseinzadeh, H. Protective effects of green tea and its main constituents against natural and chemical toxins: A comprehensive review. *Food Chem. Toxicol.* **2017**, *100*, 115–137. [CrossRef]
6. Ashiq, S.; Hussain, M.; Ahmad, B. Natural occurrence of mycotoxins in medicinal plants: A review. *Fungal Genet. Biol.* **2014**, *66*, 1–10. [CrossRef]

7.   Cladiere, M.; Delaporte, G.; Le Roux, E.; Camel, V. Multi-class analysis for simultaneous determination of pesticides, mycotoxins, process-induced toxicants and packaging contaminants in tea. *Food Chem.* **2018**, *242*, 113–121. [CrossRef]

8.   Sedova, I.; Kiseleva, M.; Tutelyan, V. Mycotoxins in Tea: Occurrence, Methods of Determination and Risk Evaluation. *Toxins* **2018**, *10*, 444. [CrossRef]

9.   Wang, L.-M.; Huang, D.-F.; Fang, Y.; Wang, F.; Li, F.-L.; Liao, M. Soil fungal communities in tea plantation after 10 years of chemical vs. integrated fertilization. *Chil. J. Agric. Res.* **2017**, *77*, 355–364. [CrossRef]

10. Redan, B.W. Processing Aids in Food and Beverage Manufacturing: Potential Source of Elemental and Trace Metal Contaminants. *J. Agric. Food Chem.* **2020**, *68*, 13001–13007. [CrossRef]

11. Ma, Y.; Ling, T.J.; Su, X.Q.; Jiang, B.; Nian, B.; Chen, L.J.; Liu, M.L.; Zhang, Z.Y.; Wang, D.P.; Mu, Y.Y.; et al. Integrated proteomics and metabolomics analysis of tea leaves fermented by *Aspergillus niger*, *Aspergillus tamarii* and *Aspergillus fumigatus*. *Food Chem.* **2021**, *334*, 127560. [CrossRef]

12. Zhao, M.; Su, X.Q.; Nian, B.; Chen, L.J.; Zhang, D.L.; Duan, S.M.; Wang, L.Y.; Shi, X.Y.; Jiang, B.; Jiang, W.W.; et al. Integrated Meta-omics Approaches to Understand the Microbiome of Spontaneous Fermentation of Traditional Chinese Pu-erh Tea. *mSystems* **2019**, *4*, 1–17. [CrossRef]

13. Frisvad, J.C.; Moller, L.L.H.; Larsen, T.O.; Kumar, R.; Arnau, J. Safety of the fungal workhorses of industrial biotechnology: Update on the mycotoxin and secondary metabolite potential of *Aspergillus niger*, *Aspergillus oryzae*, and *Trichoderma reesei*. *Appl. Microbiol. Biotechnol.* **2018**, *102*, 9481–9515. [CrossRef]

14. Ye, Z.; Wang, X.; Fu, R.; Yan, H.; Han, S.; Gerelt, K.; Cui, P.; Chen, J.; Qi, K.; Zhou, Y. Determination of six groups of mycotoxins in Chinese dark tea and the associated risk assessment. *Environ. Pollut.* **2020**, *261*, 114180. [CrossRef]

15. Ye, Z.; Cui, P.; Wang, Y.; Yan, H.; Wang, X.; Han, S.; Zhou, Y. Simultaneous Determination of Four Aflatoxins in Dark Tea by Multifunctional Purification Column and Immunoaffinity Column Coupled to Liquid Chromatography Tandem Mass Spectrometry. *J. Agric. Food Chem.* **2019**, *67*, 11481–11488. [CrossRef]

16. Reinholds, I.; Bogdanova, E.; Pugajeva, I.; Alksne, L.; Stalberga, D.; Valcina, O.; Bartkevics, V. Determination of Fungi and Multi-Class Mycotoxins in Camelia Sinensis and Herbal Teas and Dietary Exposure Assessment. *Toxins* **2020**, *12*, 555. [CrossRef]

17. Pakshir, K.; Mirshekari, Z.; Nouraei, H.; Zareshahrabadi, Z.; Zomorodian, K.; Khodadadi, H.; Hadaegh, A. Mycotoxins Detection and Fungal Contamination in Black and Green Tea by HPLC-Based Method. *J. Toxicol.* **2020**, *2020*, 2456210. [CrossRef]

18. Pallarés, N.; Font, G.; Mañes, J.; Ferrer, E. Multimycotoxin LC–MS/MS Analysis in Tea Beverages after Dispersive Liquid–Liquid Microextraction (DLLME). *J. Agric. Food Chem.* **2017**, *65*, 10282–10289. [CrossRef]

19. Monbaliu, S.; Wu, A.; Zhang, D.; Van Peteghem, C.; De Saeger, S. Multimycotoxin UPLC-MS/MS for tea, herbal infusions and the derived drinkable products. *J. Agric. Food Chem.* **2010**, *58*, 12664–12671. [CrossRef]

20. Watanabe, E. Review of sample preparation methods for chromatographic analysis of neonicotinoids in agricultural and environmental matrices: From classical to state-of-the-art methods. *J. Chromatogr. A* **2021**, *1643*, 462042. [CrossRef]

21. Guo, J.; Tong, M.; Tang, J.; Bian, H.; Wan, X.; He, L.; Hou, R. Analysis of multiple pesticide residues in polyphenol-rich agricultural products by UPLC-MS/MS using a modified QuEChERS extraction and dilution method. *Food Chem.* **2019**, *274*, 452–459. [CrossRef]

22. Zhan, J.; Zhang, R.R.; Shi, X.Z.; Huang, Z.; Cao, G.Z.; Chen, X.F.; Hu, L. A novel sample-preparation method for the generic and rapid determination of pesticides and mycotoxins in tea by ultra-performance liquid chromatography-tandem mass spectrometry. *J. Chromatogr. A* **2021**, *1636*, 461794. [CrossRef] [PubMed]

23. European Commission. Guidance Document on Method Validation and Quality Control Procedures for Pesticide Residues Analysis in Food and Feed (Document no. SANTE/12682/2019). Available online: https://www.eurl-pesticides.eu/userfiles/file/EurlALL/AqcGuidance_SANTE_2019_12682.pdf (accessed on 8 October 2021).

24. Commission Regulation (EC). Setting Maximum Levels for Certain Contaminants in Foodstuffs (Document no. EC/1881/2006). Available online: https://www.legislation.gov.uk/eur/2006/1881 (accessed on 10 August 2021).

25. Bogdanova, E.; Pugajeva, I.; Reinholds, I.; Bartkevics, V. Two-dimensional liquid chromatography—High resolution mass spectrometry method for simultaneous monitoring of 70 regulated and emerging mycotoxins in Pu-erh tea. *J. Chromatogr. A* **2020**, *1622*, 461145. [CrossRef] [PubMed]

26. Zhou, H.; Liu, N.; Yan, Z.; Yu, D.; Wang, L.; Wang, K.; Wei, X.; Wu, A. Development and Validation of the One-step Purification Method Coupled to LC-MS/MS for Simultaneous Determination of Four Aflatoxins in Fermented Tea. *Food Chem.* **2021**, *354*, 129497. [CrossRef] [PubMed]

27. Technical Regulations of the Customs Union. On Food Safety (Document No. TRCU/021/2011). Available online: http://www.eurexcert.com/TRCUpdf/TRCU-0021-On-food-safety.pdf (accessed on 16 November 2021).

28. Zhang, L.; Dou, X.W.; Zhang, C.; Logrieco, A.F.; Yang, M.H. A Review of Current Methods for Analysis of Mycotoxins in Herbal Medicines. *Toxins* **2018**, *10*, 65. [CrossRef]

29. Reinholds, I.; Bogdanova, E.; Pugajeva, I.; Bartkevics, V. Mycotoxins in herbal teas marketed in Latvia and dietary exposure assessment. *Food Addit. Contam. Part B Surveill.* **2019**, *12*, 199–208. [CrossRef]

30. Kiseleva, M.G.; Chalyy, Z.A.; Sedova, I.B.; Minaeva, L.P.; Sheveleva, S.A. Studying the contamination of tea and herbal infu-sions with mycotoxins (Message 2). *Health Risk Anal.* **2020**, *2020*, 38–51. [CrossRef]

31. Mannani, N.; Tabarani, A.; Abdennebi, E.H.; Zinedine, A. Assessment of aflatoxin levels in herbal green tea available on the Moroccan market. *Food Control* **2020**, *108*, 106882. [CrossRef]

32. Li, W.; Xu, K.; Xiao, R.; Yin, G.; Liu, W. Development of an HPLC-Based Method for the Detection of Aflatoxins in Pu-erh Tea. *Int. J. Food Prop.* **2015**, *18*, 842–848. [CrossRef]
33. Guo, W.; Zhao, M.; Chen, Q.; Huang, L.; Mao, Y.; Xia, N.; Teng, J.; Wei, B. Citrinin produced using strains of *Penicillium citrinum* from Liupao tea. *Food Biosci.* **2019**, *28*, 183–191. [CrossRef]
34. Li, Z.; Mao, Y.; Teng, J.; Xia, N.; Huang, L.; Wei, B.; Chen, Q. Evaluation of Mycoflora and Citrinin Occurrence in Chinese Liupao Tea. *J. Agric. Food Chem.* **2020**, *68*, 12116–12123. [CrossRef]
35. Lorenz, N.; Danicke, S.; Edler, L.; Gottschalk, C.; Lassek, E.; Marko, D.; Rychlik, M.; Mally, A. A critical evaluation of health risk assessment of modified mycotoxins with a special focus on zearalenone. *Mycotoxin Res.* **2019**, *35*, 27–46. [CrossRef]

Article

Toolbox for the Extraction and Quantification of Ochratoxin A and Ochratoxin Alpha Applicable for Different Pig and Poultry Matrices

Barbara Streit [1], Tibor Czabany [1], Georg Weingart [1], Martina Marchetti-Deschmann [2] and Shreenath Prasad [1,*]

[1]   BIOMIN Research Center, 3430 Tulln an der Donau, Austria; barbara.streit@dsm.com (B.S.);
      tibor.czabany@dsm.com (T.C.); georg.weingart@dsm.com (G.W.)
[2]   Institute of Chemical Technologies and Analytics, TU Wien, 1060 Vienna, Austria;
      martina.marchetti-deschmann@tuwien.ac.at
*   Correspondence: shreenath.prasad@dsm.com

Abstract: AbstractOchratoxin A (OTA) is one of the major mycotoxins causing severe effects on the health of humans and animals. Ochratoxin alpha (OTα) is a metabolite of OTA, which is produced through microbial or enzymatic hydrolysis, and one of the preferred routes of OTA detoxification. The methods described here are applicable for the extraction and quantification of OTA and OTα in several pig and poultry matrices such as feed, feces/excreta, urine, plasma, dried blood spots, and tissue samples such as liver, kidney, muscle, skin, and fat. The samples are homogenized and extracted. Extraction is either based on a stepwise extraction using ethyl acetate/sodium hydrogencarbonate/ethyl acetate or an acetonitrile/water mixture. Quantitative analysis is based on reversed-phase liquid chromatography coupled to tandem mass spectrometry (LC-MS/MS). Method validation was successfully performed and the linearity, limit of quantification, accuracy, precision as well as the stability of the samples, were evaluated. The analyte recovery of the spiked samples was between 80 and 120% (80–150% for spiked concentrations $\leq$ 1 ng/g or ng/mL) and the relative standard deviation was $\leq$15%. Therefore, we provide a toolbox for the extraction and quantification of OTA and OTα in all relevant pig and poultry matrices.

Keywords: mycotoxin; swine matrices; chicken matrices; liquid chromatography; tandem mass spectrometry; method validation

Key Contribution: A method for the quantitative determination of OTA and OTα in biological matrices from pig and chicken was developed and validated. The methods are applicable for matrices such as feed, feces/excreta, urine, plasma, dried blood spots, and tissue samples such as liver, kidney, muscle, skin, and fat. The analyte recovery of spiked samples was between 80 and 120% (80–150% for spiked concentrations $\leq$ 1 ng/g or ng/mL) and RSD was $\leq$15%.

Citation: Streit, B.; Czabany, T.; Weingart, G.; Marchetti-Deschmann, M.; Prasad, S. Toolbox for the Extraction and Quantification of Ochratoxin A and Ochratoxin Alpha Applicable for Different Pig and Poultry Matrices. *Toxins* **2022**, *14*, 432. https://doi.org/10.3390/toxins14070432

Received: 24 May 2022
Accepted: 17 June 2022
Published: 24 June 2022

Publisher's Note: MDPI stays neutral with regard to jurisdictional claims in published maps and institutional affiliations.

1. Introduction

Mycotoxins are toxic secondary metabolites produced by filamentous fungi contaminating food and animal feed [1]. One of the major mycotoxins is ochratoxin A (OTA), which was first reported in *Aspergillus ochraceus* [2], but later also found in other *Aspergillus* and *Penicillum* species [3]. OTA is considered to be the most toxic member of the ochratoxin group consisting also of ochratoxin B and ochratoxin C [3]. There are various effects of OTA reported on the health of human and animals. It is a nephrotoxin and has also been shown to be hepatoxic, teratogenic, immunotoxic, and carcinogenic in various species [4–10]. In addition, it is classified by the International Agency of Research on Cancer (IARC) as a possible human carcinogen group 2B [11,12].

There is a broad range of exposure routes to OTA. Different types of food and feed can be contaminated, including cereals like wheat, rice, rye, maize, and barley [3]. Therefore, cereal-based products, raisins and wine, grapes, nuts, spices, legumes, but also coffee and beer are affected [3]. OTA has a high affinity to proteins such as serum albumin, which promotes accumulation in animal products [5]. The highest levels of OTA were found in porcine blood-based sausages or liver products, but also in other animal products such as milk or meat OTA can also be detected [3,5,13,14]. Considering the harmful effects of OTA on human and animal health and welfare, the European Commission has set maximum levels for OTA in several food products in the range of 0.5–80 µg/kg (EC, 1881/2006) [15], and recommendations for OTA in animal feed between 10–250 µg/kg (EC, 2016/1319) [16], to minimize the risk of exposure and assure product safety.

Ingested OTA can easily be absorbed and further distributed to different organs. Nevertheless, natural detoxification into less-/nontoxic metabolites, such as ochratoxin alpha (OTα) and phenylalanine, is possible by enzymes and microbes in the gut [17–19]. The concentration of OTA metabolites in biological samples is dependent on the OTA dose and exposure time, and generally lower compared to measured OTA concentrations in feed. Even if there are lower concentrations, estimating OTα along with OTA in biological samples would be useful to monitor OTA exposure.

In human and animal matrices many methods for the detection of OTA are either based on liquid extraction from tissue or feed [20–26] or OTA levels are estimated by dried blood spot (DBS) analysis [27–30]. We provide here a comprehensive toolbox, which is easily established, for the extraction and reliable quantification of OTA and OTα in many different matrices. The methods are either based on the green solvent ethyl acetate or rather small volumes (<1 mL) of acetonitrile. Quantitative analysis is done by liquid chromatography coupled to tandem mass spectrometry (LC-MS/MS). The focus of this work is on animal matrices. Methods were developed and validated for the quantification of OTA and OTα in pig as well as chicken samples. The methods are applicable to many matrices of interest related to animals such as feed, feces/excreta, liver, kidney, muscle, skin, and fat. Furthermore, the extraction mechanism was adapted for liquid matrices such as urine or plasma samples. With the rising interest in DBS samples, this sampling method was included as well. Validation parameters cover the accuracy and precision of analysis as well as the linear quantification ranges, limits of quantification (LOQ), and limits of detection (LOD) for each matrix. The stability of extracted samples ready for measurement was evaluated at different storage temperatures and durations for each matrix. The presented methods are suitable for routine measurements to monitor OTA exposure of animals and the effects of OTA-mitigating agents supplemented in feed.

2. Results

The aim of this work was to develop simple methods for the extraction and quantification of OTA and OTα in different pig and chicken matrices. The matrices of interest were feces or excreta, urine (only available from pig), plasma, DBS, skin and fat (only from chicken), liver, kidney, muscle, and feed for pigs and poultry. For the method development and validation, blank matrices were homogenized and spiked with OTA and OTα at different levels, covering the full linear range of quantification for each different matrix. Afterwards, the samples were extracted using ethyl acetate followed by re-extraction in a reduced amount of sodium hydrogencarbonate solution, and another re-extraction in a reduced amount of ethyl acetate. A similar stepwise liquid/liquid extraction was already described by Monaci et al. [21] or Giacomo et al. [20]. Our simplified procedure allows the concentration of the analytes and clean-up of the samples to minimize matrix effects during this stepwise extraction. For DBS, a direct extraction using acetonitrile (ACN)/water (70/30 *v/v*) mixture worked best. Regardless of the matrix extracted, the analyte concentration was determined by reversed phase LC-MS/MS using external calibration in neat solvent. Possible extraction losses and matrix effects during ionization were compensated using ^{13}C-labelled OTA and OTα as internal standards, which were added to the samples prior to extraction. Figure 1A shows exemplarily the analyte recoveries with and without internal standard correction for pig feed. Without internal standard correction, the analyte recovery is approximately 55% and 75% for OTA and OTα, respectively. Applying internal standard correction for each analyte increases the analyte recovery to 80–120% of the initially spiked concentration. As the blood volume dried on protein saver cards can influence the results, blood was spiked with analyte, and different blood volumes (50 µL, 75 µL, and 100 µL) were spotted and extracted (Figure 1B). No influence of the spotted blood volume on the analyte recovery was visible. Especially in the case of chicken excreta, the available sample quantity can be very limited. Therefore, the amount of sample to be extracted can be reduced to 100 mg in combination with scale-down of all following steps. The same analyte recoveries for OTA and OTα are achieved for both sample amounts, at 1 g and 0.1 g, respectively (Figure 1C).

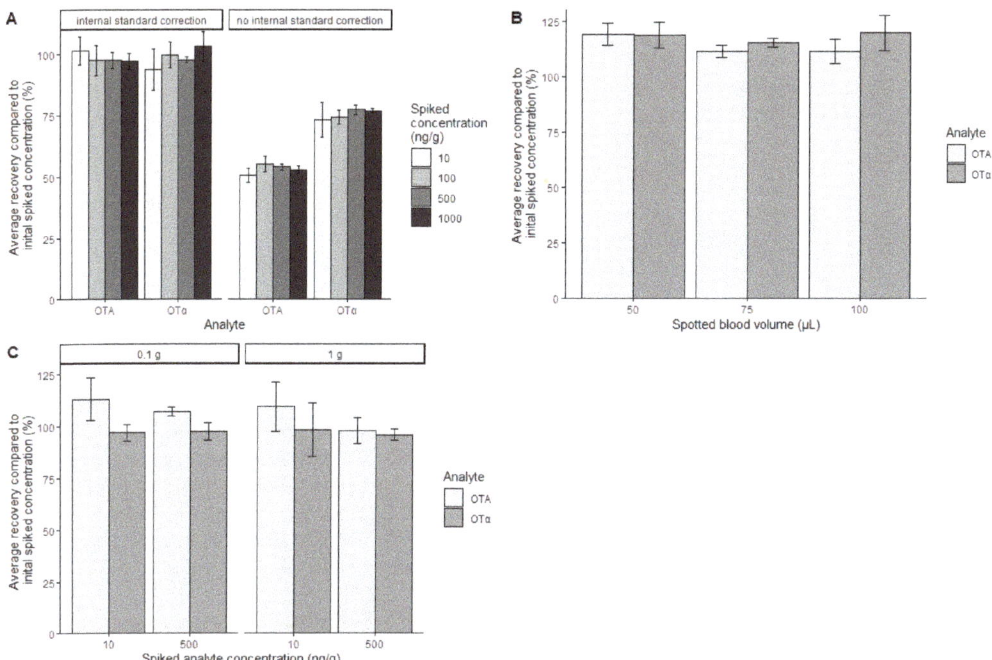

Figure 1. (**A**) Comparison between analyte recovery of OTA and OTα in pig feed with and without internal standard correction using ^{13}C-labelled analytes. (**B**) Influence of the blood volume spotted on protein saver cards on the analyte recovery for OTA and OTα. (**C**) Comparison of analyte recovery of different sample amounts for chicken excreta. For extraction, 1 g or 0.1 g excreta were used.

To determine if the method is fit for its intended purpose, a single-laboratory method validation was carried out, and the linearity of the calibration function, intraday and interday precision, accuracy, LOQ, LOD, specificity, and analyte stability in processed samples were evaluated.

The linearity of the external calibration for OTA and OTα was assessed by preparing four individual calibration series from seven individual concentration levels between 0.25 ng/mL and 250 ng/mL in neat solvent (Figure 2). For the calibration, internal standard correction (^{13}C-labelled OTA and OTAα) was applied for the linear regression with a weighting factor of 1/analyte concentration, as recommended by Gu et al. [31]. The calculated squared correlation coefficient (R^2) was >0.99.

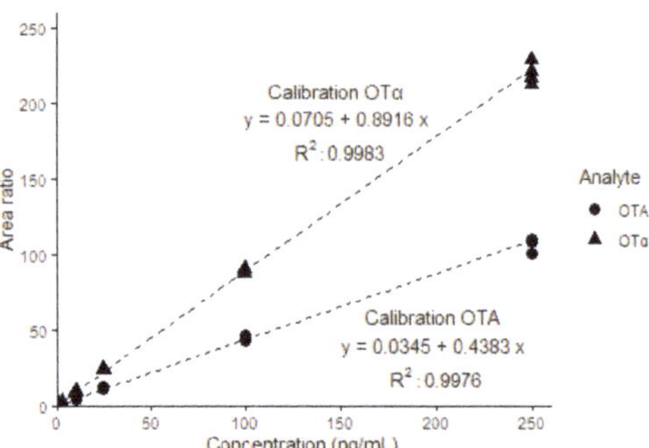

Figure 2. Calibration of OTA and OTα using four independent dilutions (seven individual calibration points) in neat solvent from 0.25 ng/mL to 250 ng/mL. ^{13}C-labelled analytes were used as internal standard for correction.

Accuracy and precision were determined in all matrices using blank samples spiked with OTA and OTα. Samples with different spiking levels were extracted on two independent days to calculate the interday precision. The intraday precision was calculated from an independent triplicate prepared on the same day. The intraday precision (calculated from a triplicate on the same day) and interday precision (calculated from independent extractions on two days, as it will be done in following routine measurements) showed a relative standard deviation (RSD) $\leq$ 15% for all matrices. For all different matrices evaluated, the accuracy of the method was between 81% and 147%. Analyte recovery of over 120% was determined only for low-level concentrations at the LOQ for poultry plasma samples, and liver, kidney, and muscle samples from pigs. Therefore, the set acceptance criteria for the analyte recovery of 80–120% (80–150% for spiked concentrations $\leq$ 1 ng/g or ng/mL) of the spiked concentration were met. To fulfill all validation criteria according to accuracy and precision, ^{13}C-labelled internal standard correction for OTA and OTα was used for all matrices except for DBS of pigs. DBS of pigs showed already without internal standard correction analyte recoveries between 88–116%, and therefore internal standard correction was not necessary. Tables 1 and 2 summarize all evaluated validation parameters for pig and poultry matrices.

The LOQ was estimated by spiking experiments. It is the smallest measured content of the analyte above which the determination can be made with a specified accuracy and precision, as recommended by the Guidance for Industry 208 from the Food and Drug Administration [32]. The LOD was estimated by calculating the analyte concentrations, which delivered an S/N ratio of 3/1 based on the S/N ratio of the analytes at the LOQ.

The specificity of the method is given by the retention time and the multiple reaction monitoring (MRM) transitions. Blank samples were extracted and measured to evaluate the endogenous response. No interfering peaks were detected at the specific retention time of OTA at 1.6 min and OTα at 1.3 min, except for pig plasma and pig DBS samples. In these samples, a small peak at the specific retention time of OTA was found, with an approximate S/N of LOD. As this peak was visible in all measured MRM transitions for OTA, we assume that this is not an interfering peak, but rather the analyte itself. We considered that the used plasma and blood samples were not completely OTA-free.

As analyte stability in processed samples is crucial for reliable results, the analyte stability of OTA and OTα in ready-to-inject samples was evaluated at different temperatures and storage periods. The analytes were stable in all evaluated matrices. At room temperature after 7 days of storage, slightly higher recoveries (up to 125% compared to the initial measured concentration) were observed in poultry matrices. All other tested storage conditions fulfilled the acceptance criteria of an analyte recovery of 80–120% compared to the initial measured value. Nevertheless, for long-term storage of processed samples, storage at −20 °C would be recommended.

Table 1. Summary of validation parameters for pig matrices.

Method Parameters		Feces OTA	Feces OTα	Urine OTA	Urine OTα	Plasma OTA	Plasma OTα	DBS OTA	DBS OTα	Liver OTA	Liver OTα	Kidney OTA	Kidney OTα	Muscle OTA	Muscle OTα	Feed OTA	Feed OTα
Range of analysis (ng/g or ng/mL)		10–500	10–500	5.0–500	5.0–500	5.0–500	5.0–500	5.0–500	5.0–500	0.50–100	0.50–100	0.50–100	0.50–100	0.50–100	0.50–100	10–1000	10–1000
LOQ (ng/g or ng/mL)		10	10	5.0	5.0	5.0	5.0	5.0	5.0	0.50	0.50	0.50	0.50	0.50	0.50	10	10
LOD (ng/g or ng/mL)		3.0	3.0	1.5	1.5	1.5	1.5	1.5	1.5	0.15	0.15	0.15	0.15	0.15	0.15	3.0	3.0
^{13}C internal standard correction		yes	yes	yes	yes	yes	yes	no	no	yes	yes	yes	yes	yes	yes	yes	yes
Analyte recovery (%)	Low level [1]	106–113	110–120	88–91	83–97	98–120	90–103	88–106	93–110	101–130	90–104	108–115	130–147	95–114	123–142	96–107	85–101
Analyte recovery (%)	High level [2]	90–98	108–117	93–102	96–97	94–104	93–114	97–114	110–116	89–93	87–91	95–106	93–97	84–113	88–100	94–101	92–110
Max. intraday precision (n = 3) RSD (%)		3.7	5.1	4.1	7.0	6.8	7.7	4.5	7.6	13	6.0	6.5	9.6	14	8.4	5.8	8.4
Max. interday precision (n = 2) RSD (%)		5.3	7.4	12	2.6	7.2	11	12	8.7	11	2.3	1.3	6.3	13	6.6	2.4	4.8
Processed sample stability (%)	RT (2 days)	100–101	95–100	94–100	102–103	90–97	93–95	116–120	112–116	93–98	106–110	94–99	104–106	94–101	99–108	93–98	98–109
Processed sample stability (%)	4 °C (7 days)	102–112	104–112	97–120	81–89	90–99	90–96	97–106	94–102	104–114	99–103	100–102	108–110	97–114	101–108	99–113	110–118
Processed sample stability (%)	−20 °C (30 days) [3]	95–98	100–108	85–93	104–106	96–100	96–98	109–110	109–113	103–108	103–114	100–117	96–105	95–107	100–104	89–99	106–114

[1] spiking at the stated lower range of analysis, which is equal to the LOQ for this matrix; [2] spiking at the stated upper range of analysis; [3] in liver, kidney, and muscle samples, the stability was tested for 7 days

Table 2. Summary of validation parameters for poultry matrices.

Method Parameters		Excreta		Plasma		DBS		Liver		Kidney		Muscle		Skin and Fat		Feed	
		OTA	OTα	OTA	OTα	OTA	OTα	OTA	OTα	OTA	OTα	OTA	OTα	OTA	OTα	OTA	OTα
Range of analysis (ng/g or ng/mL)		10–1000	10–1000	1.0–1000	1.0–1000	1.0–500	1.0–500	0.50–200	0.50–200	0.50–200	0.50–200	0.50–200	0.50–200	0.50–200	0.50–200	10–1000	10–1000
LOQ (ng/g or ng/mL)		10	10	1.0	1.0	1.0	1.0	0.5	0.5	0.5	0.5	0.5	0.5	0.5	0.5	10	10
LOD (ng/g or ng/mL)		3.0	3.0	0.30	0.30	0.30	0.30	0.15	0.15	0.15	0.15	0.15	0.15	0.15	0.15	3.0	3.0
^{13}C internal standard correction		yes	yes	yes	yes	yes	yes	yes	yes	yes	yes	yes	yes	yes	yes	yes	yes
Accuracy range (%)	Low level [1]	96–116	90–113	102–122	81–99	97–103	91–104	89–95	89–116	89–101	85–115	81–101	89–114	99–114	88–98	92–107	94–100
	High level [2]	93–105	93–98	103–110	102–108	87–97	88–101	96–105	94–102	91–105	98–104	94–104	94–105	85–99	85–90	98–107	94–105
Max. intraday precision ($n = 3$) RSD (%)		12	13	10	9.8	4.4	5.4	5.1	3.2	6.8	13	10	13	6.2	2.6	5.0	3.3
Max. interday precision ($n = 2$) RSD (%)		5.4	6.9	1.7	3.7	7.5	5.5	3.0	15	4.5	2.1	2.0	7.5	10	6.7	10	6.0
processed sample stability (%)	RT (7 days)	101–104	102–106	89–99	10–114	94–99	97–101	98–121	82–100	102–109	85–97	99–125	93–99	93–98	96–98	94–107	97–101
	4 °C (7 days)	101–105	90–106	86–100	100–103	103–106	103	98–99	88–97	105–109	90–98	103–111	92–106	92–94	96–99	98–99	100–103
	−20 °C (30 days)	93–97	98–101	85–95	95–117	99–101	99–101	97–108	97–100	99–103	100–103	95–103	97–100	93	98–106	99–105	99–101

[1] spiking at the stated lower range of analysis, which is equal to the LOQ for this matrix; [2] spiking at the stated upper range of analysis.

3. Discussion

The intention of this work was to develop and validate methods for extraction and quantification of OTA and OTα in different pig and poultry matrices. The European Commission (EC/2016/1319) has published recommendations for OTA in feed, with maximum thresholds of 50 μg/kg for pig feed and 100 μg/kg for poultry feed [16]. Considering these contamination levels of OTA in the feed, we expected OTA levels for example in plasma to be in the low ng/mL range [33,34]. We present methods applicable for pig and poultry with similar extraction procedures for matrices such as feed, feces/excreta, urine, plasma, DBS, liver, kidney, muscle, skin, and fat. The primary goal of the method development was the quantitative determination of OTA and OTα at biologically relevant concentration levels.

There are already many methods described for the detection and quantification of OTA, and sometimes also OTα, in different matrices. The methods presented here are similar to those described by Monaci et al. [21] or Giacomo et al. [20], as the same step-wise extraction and clean-up procedure using solid/liquid and liquid/liquid extraction mechanisms were used. However, in contrast to the already described protocols, our method is applicable to a larger variety of matrices and also simpler, as some extraction steps have been modified. Monaci et al. as well as Giacomo et al. also use the three-phase solid/liquid/liquid extraction system, but the tissue samples are extracted twice. So removal of the organic phase is required. We simplified the extraction by only applying a single extraction of the matrix. Then sample preparation is continued with a defined aliquot of the organic phase, which can be easily pipetted. Furthermore, other methods rely on the reduction of the sample volume to a specific amount [20]. In our methods, clean-up and concentration of analytes are achieved by re-extraction in reduced volumes of sodium hydrogencarbonate solution and ethyl acetate. Moreover, these steps also provide fat-removal from the samples. Therefore, the use of expensive SPE columns for sample clean-up, as reported for liver samples by Monaci et al. [21], is not necessary. Other methods for the extraction of OTA and OTα are often based on halogenated solvents such as dichloromethane [35] or an enzymatic digestion [36]. The methods described here used the green solvent ethyl acetate for the main extraction of the analytes from the matrix. For the extraction of DBS using ethyl acetate was not successful. Instead of using this green solvent, rather small volumes (<1 mL) of ACN/water mixtures were used for the extraction of DBS and showed good results.

As sometimes the available quantity of samples is very limited, the robustness of the method regarding the sample amount was evaluated. Starting from 1 g or 0.1 g of excreta, the extraction parameters were scaled down for the smaller sample size. However, no influence on the results was observed, confirming the ruggedness of our method. Moreover, different blood volumes dried on protein saver cards were evaluated regarding analyte recovery. The same results were obtained, regardless of the spotted blood volume. This is in accordance with results of Osteresch et al., where different blood volumes had almost no impact on the results for OTA [29]. The same principle of extraction is applied to solid samples such as tissue or feces as well as liquid samples such as urine and plasma. Hence, we provide here a widely applicable, robust toolbox for the extraction and quantification of OTA and OTα in all relevant pig and poultry matrices.

4. Conclusions and Outlook

The detection of OTA and its metabolite OTα in important pig and poultry matrices was established and validated. The developed methods allow the extraction and quantification of OTA and OTα in feed, plasma, DBS, liver, kidney, and muscle samples from pigs and chickens. Furthermore, extraction protocols for feces and urine samples from pigs as well as excreta, skin, and fat samples from chickens are available. Extraction is similar for all matrices, and either based on the green solvent ethyl acetate or on rather small volumes of ACN/water. Quantitative analysis of the analytes is done using LC-MS/MS. Method development was focused on the detection of biologically relevant concentrations. Therefore, the LOQ is between 0.5 ng/g in tissue matrices and 10 ng/g in feed or feces/excreta. The accuracy and precision of the methods were evaluated, and the analyte recovery was between 80 and 120% (80–150% for spiked concentrations ≤ 1 ng/g or ng/mL) of the initially spiked concentration and the RSD was $\leq 15\%$. Analyte stability in processed samples was tested at room temperature, at 4 °C and -20 °C for different storage periods. For long term storage -20 °C is recommended. The presented methods are easily established and feasible to monitor OTA exposure or for the use in feeding trials of pig and poultry to evaluate OTA-mitigating feed additives.

5. Materials and Methods

5.1. Chemicals and Reagents

The analytical standards for OTA, OTα, and ^{13}C-labelled OTA as internal standard were obtained from Romer Labs (Tulln an der Donau, Austria). The ^{13}C-labelled OTα was produced in-house by enzymatic degradation of ^{13}C-labelled OTA into ^{13}C-labelled OTα and phenylalanine. An analytical standard for creatinine was purchased from Merck (Darmstadt, Germany). All standards were stored as recommended by the supplier. Ortho-phosphoric acid 85%, formic acid, ACN HPLC grade, and acetic acid were obtained from VWR (Vienna, Austria). Ethyl acetate HPLC grade and ACN HPLC-MS/MS grade were purchased from Chem-Lab (Zedelgem, Belgium). Methanol (MeOH) LC-MS grade was bought from Honeywell (Charlotte, NC, USA). Sodium hydrogencarbonate and Whatman 903 protein saver cards were obtained from Merck (Darmstadt, Germany). Ultrapure water was produced in-house using a Milli-Q IQ 7000 water purification system from Merck (Darmstadt, Germany).

5.2. Preparation of Spiking Solutions and Spiking of the Samples

Standard stock solutions for OTA, OTα, and ^{13}C-labelled OTA were purchased with a concentration of 10 µg/mL already dissolved in ACN. The ^{13}C-labelled OTα was prepared by proprietary enzymatic degradation of ^{13}C-labelled OTA. Progress of conversion of ^{13}C-OTA to ^{13}C-OTα was checked by LC-MS/MS. After complete conversion ^{13}C-OTα was extracted five times with ethyl acetate. The combined ethyl acetate phases were dried completely under reduced pressure and reconstituted in ACN to a final concentration of 6.34 µg/mL. The same enzymatic reaction can be performed with any other OTA hydrolyzing enzymes, such as Carboxypeptidase A, under suitable reaction conditions.

Stock solutions of OTA and OTα were mixed and used directly or optionally further diluted in ACN/water/formic acid (50/49/1 *v/v/v*) to the respective target concentrations between 1 µg/mL and 10 ng/mL. The prepared solutions were used for spiking the matrices at different concentration levels. Internal standard was prepared by dilution and mixing of ^{13}C-OTA and ^{13}C-OTα in ACN/water/formic acid (50/49/1 *v/v/v*) to 1 µg/mL, 0.25 µg/mL, or 0.1 µg/mL. Samples were spiked with internal standard during sample preparation, as described below. Further dilutions of internal standard were prepared and used for the external calibration in neat solvent.

5.3. Biological Samples

For method development and validation, urine, plasma, blood, feces, excreta, feed (based on wheat–barley–corn–soy diet for pig and corn–soy diets for poultry), liver, kidney, muscle, skin, and fat samples were obtained in-house from the center of applied animal nutrition (BIOMIN, Tulln an der Donau, Austria) or from the local slaughterhouse. Blood was analyzed using dried blood spots. All samples except feed, lyophilized feces or excreta, and DBS samples were stored at -20 °C until sample preparation. Feed, lyophilized feces or excreta samples were stored at room temperature in the dark until sample preparation. DBS were stored at 4 °C in the dark after drying overnight at room temperature. For all matrices quantitative determination was based on external calibration in neat solvent.

5.3.1. Urine (Pig) and Plasma Samples (Pig and Poultry)

The concentration of analytes in urine depends, among others, on the amount of water the animal was drinking. To reduce variation in the results, urine samples were diluted to a specified concentration of creatinine. Determination of the creatinine concentration of each urine sample, and afterwards the individual dilution factor, urine samples were brought to room temperature, mixed shortly on a vortex shaker, and diluted 1:10,000 in ultrapure water. Creatinine concentration was determined by LC-MS/MS in positive mode. An Agilent 1290 Infinity II system was coupled to a Sciex Triple Quad 5500 mass spectrometer. Chromatographic separation was achieved on a Gemini 5 µm C18 column 150 × 4.6 mm with a suitable precolumn from Phenomenex (Aschaffenburg, Germany). The column compartment was heated to 30 °C and the injection volume was set to 2 µL. Solvent A consisted of MeOH/water/acetic acid (40/59.8/0.2 *v/v/v*) and MeOH/acetic acid (99.8/0.2 *v/v*) was used as solvent B. Flow rate was set to 0.8 mL/min and a gradient was used during the chromatographic run, containing 0 min–1.6 min 0% B constant, 1.6 min–1.65 min linear gradient to 100% B, 1.65 min–2 min 100% B constant, and 2 min–2.05 min linear gradient to 0% B. The total run time was 3 min. The measured MRM transitions were 114 > 86 (collision energy 15 volts, declustering potential 20 volts) as quantifier and 114 > 44 (collision energy 15 volts, declustering potential 20 volts) as qualifier. The retention time of creatinine was 1.36 min. Quantitative determination of creatinine was based on a serial dilution of standards in MeOH/water (10/90 *v/v*). After determination of the creatinine values, urine samples were diluted individually with ultrapure water to 10 µM creatinine. If the creatinine value was below 10 µM, the samples were used directly.

The samples (diluted urine samples to 10 µM creatinine or plasma samples) were thawed and mixed shortly on a vortex shaker. An aliquot of 200 µL was transferred to a fresh Eppendorf tube and spiked with 8 µL of internal standard solution containing 1 µg/mL ^{13}C-OTA and ^{13}C-OTα. Then 800 µL extraction solution consisting of ethyl acetate/phosphoric acid 85% (99/1 *v/v*) were added and mixed by manual shaking to make sure that the samples did not stick to the bottom of the vial, before mixing vigorously for 10 min on a vortex shaker with adapter for 2 mL reaction tubes. The samples were centrifuged for 5 min at 19,000 rcf. Finally, 50 µL of the supernatant were transferred to an HPLC vial with 0.2 mL silanized insert and mixed with 50 µL ACN.

5.3.2. Feed, Feces, and Excreta

Fresh feces and excreta were freeze dried. For extraction, 1 g of the sample was weighed in a 50 mL falcon tube and spiked with 30 µL of the mixed internal standard solution containing 1 µg/mL ^{13}C-OTA and ^{13}C-OTα. Then 6 mL 1 M phosphoric acid were added and mixed shortly, followed by the addition of 30 mL ethyl acetate. The samples were extracted on an end-over-end shaker for 60 min at 80 rpm. The samples were centrifuged for 10 min at 3200 rcf. Afterwards, 8 mL of the supernatant were transferred to a fresh 50 mL falcon tube and mixed with 4 mL 0.1 M sodium hydrogencarbonate solution (pH 8.2). The samples were mixed for 1 min on the end-over-end shaker at 80 rpm and centrifuged for 10 min at 3200 rcf. Then 3 mL of the water phase were transferred to a fresh 15 mL falcon tube and mixed with 70 µL ortho-phosphoric acid 85%. Finally, 1.5 mL ethyl acetate were added to the sample and the samples were mixed for 1 min on the end-over-end shaker at 80 rpm. The samples were centrifuged for 10 min at 3200 rcf and 50 µL of the supernatant were transferred to an HPLC vial with a 0.2 mL silanized insert and mixed with 50 µL ACN.

To test the robustness of the method, the used sample amount for extraction was reduced to 100 mg and all following steps were adapted to this smaller amount.

5.3.3. Liver, Kidney, Muscle, Skin, and Fat Samples

Fresh tissue samples were frozen and stored at −20 °C in sealed bags. For homogenization, the samples were cooled in liquid nitrogen for approximately 1 min and pulverized using a ball mill (MM 400 from Retsch, Haan, Germany) for 30 s with 30 Hz in two cycles. For extraction, 1 g of the sample was weighed in a 50 mL falcon tube and spiked with 30 µL of the mixed internal standard solution containing 0.1 µg/mL ^{13}C-OTA and ^{13}C-OTα. The same procedure as already described in 5.3.2 for feed, feces, and excreta samples was used for further sample preparation. As the expected analyte concentrations are lower in tissue compared to feed or feces/excreta, an additional step for analyte concentration was added. Therefore, 1 mL of the supernatant after the second ethyl acetate extraction step was transferred to a fresh Eppendorf tube and dried completely under reduced pressure. The samples were reconstituted in 100 µL ACN/water/formic acid (50/49/1 *v/v/v*) in the ultrasonic bath for 10 min. Afterwards the samples were mixed vigorously for 10 min on a vortex shaker. Subsequently, the samples were centrifuged for 5 min at 19,000 rcf before 50 µL of the supernatant were transferred to an HPLC vial with a 0.2 mL silanized insert.

5.3.4. Dried Blood Spots

For the preparation of DBS, 50 µL whole blood were spotted on the protein saver cards and dried overnight at room temperature. To test the influence of the spotted volume, in addition 75 µL and 100 µL blood were spotted.

For DBS of pigs, the whole blood spot (50 µL, 75 µL or 100 µL) was cut out from the protein saver card and placed in a 1.5 mL Eppendorf tube. Then 800 µL extraction solution containing ACN/water (70/30 v/v) were added and mixed. The samples were placed on an end-over-end shaker for 60 min at 80 rpm. Afterwards, 100 µL of the liquid were transferred to an HPLC vial with a 0.2 mL silanized insert.

A similar approach was used for DBS from chicken. Here, 50 µL of whole blood were spotted on protein saver cards and dried overnight. The entire spot was cut out and placed in a 1.5 mL Eppendorf tube. Additionally, 10 µL mixed internal standard solution containing 250 ng/mL of ^{13}C-OTA and ^{13}C-OTα were spiked in the tubes. Then 800 µL extraction solution of ACN/water (70/30 v/v) were added and mixed. The samples were placed for 60 min on an end-over-end shaker at 80 rpm. An aliquot of 500 µL of the supernatant was dried completely under reduced pressure at 60 °C and reconstituted afterwards in 62.5 µL ACN/water/formic acid (50/49/1 *v/v/v*). Samples were mixed vigorously for 10 min on a vortex shaker and afterwards centrifuged for 10 min at 19,000 rcf. Finally, 50 µL of the supernatant were transferred to an HPLC vial with a 0.2 mL silanized insert.

5.4. Chromatography and LC-MS/MS Parameters

Quantification was based on reversed-phase LC-MS/MS in negative MRM mode. Therefore, an Agilent 1290 Infinity II system was coupled to a Sciex QTRAP 6500+ mass spectrometer. Chromatographic separation was achieved on a Kinetex 2.6 µm EVO C18 column 150 × 2.1 mm with a suitable precolumn from Phenomenex (Aschaffenburg, Germany). The temperature of the column oven was set to 40 °C and 4 µL injection volume were used. The optimum chromatographic conditions were achieved with eluent A water/ACN (95/5 *v/v*) containing 0.1% formic acid and eluent B ACN/water (95/5 *v/v*) containing 0.1% formic acid. A gradient was used for separation: 0 min–0.25 min 20% B constant, 0.25 min–2 min linear gradient to 100% B, 2 min–2.50 min 100% B constant, and 2.50 min–2.51 min linear gradient to 20% B. The total run time was 3 min. Over the whole gradient the flow rate was set to 1 mL/min. The measured MRM transitions were optimized using direct infusion of the analytes. The optimized parameters for each analyte are listed in Table 3.

Table 3. Measured MRM transitions for each analyte.

Analyte	Measured Form	Precursor Ion (*m/z*)	Quantifier Ion [1] Qualifier Ions (*m/z*)	Declustering Potential (V)	Collision Energy (V)	Collision Cell Exit Potential (V)	Entrance Potential (V)	Retention Time (min)
OTA	[M−H]⁻	402	<u>167</u> 358 211	−85	−46 −26 −36	−17 −21 −23	−10	1.60
OTα	[M−H]⁻	255	<u>167</u> 211 123	−25	−34 −22 −40	−19 −13 −13	−10	1.30
¹³C-OTA	[M−H]⁻	422	<u>175</u> 377 221	−85	−48 −28 −38	−11 −23 −13	−10	1.60
¹³C-OTα	[M−H]⁻	266	<u>175</u> 221 130	−55	−32 −22 −40	−19 −13 −13	−10	1.30

[1] underline indicates the ion used quantification, all other ions are used as qualifier.

5.5. Method Validation

For method validation, blank matrix was spiked at two levels, extracted, and measured. The following parameters were evaluated: linearity, intraday and interday precision, accuracy, LOQ, LOD, specificity, and analyte stability in processed samples.

5.5.1. Linearity

Linearity was assessed by preparing four individual calibration series through diluting defined concentrations of OTA, OTα, ¹³C-OTA, and ¹³C-OTα in ACN/water/formic acid (50/49/1 *v/v/v*) in the range from 0.25 ng/mL to 250 ng/mL. Linear regression was performed with a weighting factor of 1/analyte concentration, as recommended by Gu et al. [31]. The R^2 was calculated and the acceptance criteria was set at $R^2 >0.99$.

5.5.2. Accuracy and Precision

Accuracy and precision were evaluated by analyzing blank samples, which were spiked with defined concentrations of OTA and OTα at two days at two different levels. The accuracy was calculated using the following equation:

$$Accuracy\ (\%) = \frac{measured\ concentration\ \left(\frac{ng}{g}\ or\ \frac{ng}{mL}\right)}{spiked\ concentration\ \left(\frac{ng}{g}\ or\ \frac{ng}{mL}\right)} \times 100 \tag{1}$$

The acceptance criteria for accuracy were for spiking levels above 1 ng/g or ng/mL an analyte recovery of 80–120% of the initial spiked concentration. For spiking levels less or equal to 1 ng/g or ng/mL 80–150% analyte recovery was accepted. For intraday precision, the samples were spiked and analyzed in triplicate on the same day. For the interday precision two independent extractions were analyzed, as planned for further routine analysis. For the intraday and interday precision the acceptance criteria were set to RSD $\leq$ 15%. To calculate the RSD, the following equation was used:

$$RSD\ (\%) = \frac{\sqrt{\frac{\sum\left(measured\ concentration\ \left(\frac{ng}{g}\ or\ \frac{ng}{mL}\right) - average\ concentration\ \left(\frac{ng}{g}\ or\ \frac{ng}{mL}\right)\right)^2}{number\ of\ replicates - 1}}}{spiked\ concentration\ \left(\frac{ng}{g}\ or\ \frac{ng}{mL}\right)} \times 100 \tag{2}$$

5.5.3. Limit of Quantification and Limit of Detection

For the determination of the LOQ, spiking experiments were used. As defined in the Guidance for Industry 208, LOQ is the smallest measured content of analyte which can be determined with specified accuracy and precision [32]. LOD was estimated by calculating the analyte concentrations which delivered a S/N ratio of 3/1 based on the S/N ratio of the analytes at LOQ.

5.5.4. Selectivity

Selectivity is given by the retention time and the specific MRM transitions of the analytes. Furthermore, the endogenous response of blank samples was evaluated. A signal of interferences at the retention time of the analyte below a $S/N \leq 3{:}1$ was acceptable.

5.5.5. Analyte Stability in Processed Samples

Samples were extracted and extracts ready to inject were stored at different storage conditions in triplicate. The stability in process samples was evaluated at room temperature up to 7 days, in the fridge at 4 °C up to 7 days, and in the freezer at −20 °C up to 7 days for liver, kidney, and muscle samples, as low analyte concentrations are expected in combination with a high matrix content after drying and reconstitution, or at −20 °C up to 30 days for all other matrices. The acceptance criteria were 80–120% analyte recovery compared to the initial measured concentration.

Author Contributions: T.C., S.P. and B.S. conceived and designed the experiments; B.S. performed the experiments, analyzed the data, and drafted the paper; T.C., S.P., G.W. and M.M.-D. carefully revised it. All authors have read and agreed to the published version of the manuscript.

Funding: This research was funded by the Austrian Research Promotion Agency (Österreichische Forschungsförderungsgesellschaft FFG, grant numbers 872270, 879467, and 887040).

Institutional Review Board Statement: Not applicable.

Informed Consent Statement: Not applicable.

Data Availability Statement: Not applicable.

Conflicts of Interest: The authors declare no conflict of interest.

References

1. Edite Bezerra da Rocha, M.; Freire, F.D.C.O.; Erlan Feitosa Maia, F.; Izabel Florindo Guedes, M.; Rondina, D. Mycotoxins and their effects on human and animal health. *Food Control* **2014**, *36*, 159–165. [CrossRef]
2. Van Der Merwe, K.J.; Steyn, P.S.; Fourie, L.; Scott, D.B.; Theron, J.J. Ochratoxin A, a Toxic Metabolite produced by Aspergillus ochraceus Wilh. *Nature* **1965**, *205*, 1112–1113. [CrossRef] [PubMed]
3. Bayman, P.; Baker, J.L. Ochratoxins: A global perspective. *Mycopathologia* **2006**, *162*, 215–223. [CrossRef] [PubMed]
4. Marroquín-Cardona, A.G.; Johnson, N.M.; Phillips, T.D.; Hayes, A.W. Mycotoxins in a changing global environment—A review. *Food Chem. Toxicol.* **2014**, *69*, 220–230. [CrossRef]
5. Heussner, A.H.; Bingle, L.E.H. Comparative Ochratoxin Toxicity: A Review of the Available Data. *Toxins* **2015**, *7*, 4253–4282. [CrossRef]
6. Damiano, S.; Andretta, E.; Longobardi, C.; Prisco, F.; Paciello, O.; Squillacioti, C.; Mirabella, N.; Florio, S.; Ciarcia, R. Effects of Curcumin on the Renal Toxicity Induced by Ochratoxin A in Rats. *Antioxidants* **2020**, *9*, 332. [CrossRef]
7. Damiano, S.; Longobardi, C.; Andretta, E.; Prisco, F.; Piegari, G.; Squillacioti, C.; Montagnaro, S.; Pagnini, F.; Badino, P.; Florio, S.; et al. Antioxidative Effects of Curcumin on the Hepatotoxicity Induced by Ochratoxin A in Rats. *Antioxidants* **2021**, *10*, 125. [CrossRef]
8. Bondy, G.S.; Curran, I.H.C.; Coady, L.C.; Armstrong, C.; Bourque, C.; Bugiel, S.; Caldwell, D.; Kwong, K.; Lefebvre, D.E.; Maurice, C.; et al. A one-generation reproductive toxicity study of the mycotoxin ochratoxin A in Fischer rats. *Food Chem. Toxicol.* **2021**, *153*, 112247. [CrossRef]
9. Sun, Y.; Huang, K.; Long, M.; Yang, S.; Zhang, Y. An update on immunotoxicity and mechanisms of action of six environmental mycotoxins. *Food Chem. Toxicol.* **2022**, *163*, 112895. [CrossRef]
10. Stoev, S.D. New Evidences about the Carcinogenic Effects of Ochratoxin A and Possible Prevention by Target Feed Additives. *Toxins* **2022**, *14*, 380. [CrossRef]

11. IARC Monographs on the Identification of Carcinogenic Hazards to Humands. Available online: https://monographs.iarc.who.int/list-of-classifications (accessed on 23 May 2022).
12. Claeys, L.; Romano, C.; De Ruyck, K.; Wilson, H.; Fervers, B.; Korenjak, M.; Zavadil, J.; Gunter, M.J.; De Saeger, S.; De Boevre, M.; et al. Mycotoxin exposure and human cancer risk: A systematic review of epidemiological studies. *Compr. Rev. Food Sci. Food Saf.* **2020**, *19*, 1449–1464. [CrossRef] [PubMed]
13. Denli, M.; Perez, J.F. Ochratoxins in feed, a risk for animal and human health: Control strategies. *Toxins* **2010**, *2*, 1065–1077. [CrossRef] [PubMed]
14. Becker-Algeri, T.A.; Castagnaro, D.; de Bortoli, K.; de Souza, C.; Drunkler, D.A.; Badiale-Furlong, E. Mycotoxins in bovine milk and dairy products: A review. *J. Food Sci.* **2016**, *81*, R544–R552. [CrossRef] [PubMed]
15. EUR-Lex Access to European Union law. Available online: https://eur-lex.europa.eu/eli/reg/2006/1881 (accessed on 23 May 2022).
16. EUR-Lex Access to European Union law. Available online: https://eur-lex.europa.eu/eli/reco/2016/1319 (accessed on 23 May 2022).
17. Schatzmayr, G.; Zehner, F.; Täubel, M.; Schatzmayr, D.; Klimitsch, A.; Loibner, A.P.; Binder, E.M. Microbiologicals for deactivating mycotoxins. *Mol. Nutr. Food Res.* **2006**, *50*, 543–551. [CrossRef]
18. Dellafiora, L.; Gonaus, C.; Streit, B.; Galaverna, G.; Moll, W.-D.; Vogtentanz, G.; Schatzmayr, G.; Dall'Asta, C.; Prasad, S. An In Silico Target Fishing Approach to Identify Novel Ochratoxin A Hydrolyzing Enzyme. *Toxins* **2020**, *12*, 258. [CrossRef]
19. Joo, Y.D.; Kang, C.W.; An, B.K.; Ahn, J.S.; Borutova, R. Effects of ochratoxin a and preventive action of a mycotoxin-deactivation product in broiler chickens. *Vet. Ir Zootech.* **2013**, *61*, 22–29.
20. Giacomo, L.; Michele, V.; Guido, F.; Danilo, M.; Luigi, I.; Valentina, M. Determination of ochratoxin A in pig tissues using enzymatic digestion coupled with high-performance liquid chromatography with a fluorescence detector. *MethodsX* **2016**, *3*, 171–177. [CrossRef]
21. Monaci, L.; Tantillo, G.; Palmisano, F. Determination of ochratoxin A in pig tissues by liquid–liquid extraction and clean-up and high-performance liquid chromatography. *Anal. Bioanal. Chem.* **2004**, *378*, 1777–1782. [CrossRef]
22. Iqbal, S.Z.; Nisar, S.; Asi, M.R.; Jinap, S. Natural incidence of aflatoxins, ochratoxin A and zearalenone in chicken meat and eggs. *Food Control.* **2014**, *43*, 98–103. [CrossRef]
23. Meerpoel, C.; Vidal, A.; di Mavungu, J.D.; Huybrechts, B.; Tangni, E.K.; Devreese, M.; Croubels, S.; De Saeger, S. Development and validation of an LC–MS/MS method for the simultaneous determination of citrinin and ochratoxin a in a variety of feed and foodstuffs. *J. Chromatogr. A* **2018**, *1580*, 100–109. [CrossRef]
24. Nakhjavan, B.; Ahmed, N.S.; Khosravifard, M. Development of an Improved Method of Sample Extraction and Quantitation of Multi-Mycotoxin in Feed by LC-MS/MS. *Toxins* **2020**, *12*, 462. [CrossRef] [PubMed]
25. Paoloni, A.; Solfrizzo, M.; Bibi, R.; Pecorelli, I. Development and validation of LC-MS/MS method for the determination of Ochratoxin A and its metabolite Ochratoxin α in poultry tissues and eggs. *J. Environ. Sci. Health Part B* **2018**, *53*, 327–333. [CrossRef] [PubMed]
26. Mackay, N.; Marley, E.; Leeman, D.; Poplawski, C.; Donnelly, C. Analysis of Aflatoxins, Fumonisins, Deoxynivalenol, Ochratoxin A, Zearalenone, HT-2, and T-2 Toxins in Animal Feed by LC–MS/MS Using Cleanup with a Multi-Antibody Immunoaffinity Column. *J. AOAC Int.* **2022**, qsac035. [CrossRef] [PubMed]
27. Lauwers, M.; Croubels, S.; De Baere, S.; Sevastiyanova, M.; Romera Sierra, E.M.; Letor, B.; Gougoulias, C.; Devreese, M. Assessment of Dried Blood Spots for Multi-Mycotoxin Biomarker Analysis in Pigs and Broiler Chickens. *Toxins* **2019**, *11*, 541. [CrossRef] [PubMed]
28. Cramer, B.; Osteresch, B.; Muñoz, K.A.; Hillmann, H.; Sibrowski, W.; Humpf, H.-U. Biomonitoring using dried blood spots: Detection of ochratoxin A and its degradation product 2′R-ochratoxin A in blood from coffee drinkers. *Mol. Nutr. Food Res.* **2015**, *59*, 1837–1843. [CrossRef]
29. Osteresch, B.; Cramer, B.; Humpf, H.-U. Analysis of ochratoxin A in dried blood spots—Correlation between venous and finger-prick blood, the influence of hematocrit and spotted volume. *J. Chromatogr. B* **2016**, *1020*, 158–164. [CrossRef]
30. Osteresch, B.; Viegas, S.; Cramer, B.; Humpf, H.U. Multi-mycotoxin analysis using dried blood spots and dried serum spots. *Anal. Bioanal. Chem.* **2017**, *409*, 3369–3382. [CrossRef]
31. Gu, H.; Liu, G.; Wang, J.; Aubry, A.-F.; Arnold, M.E. Selecting the Correct Weighting Factors for Linear and Quadratic Calibration Curves with Least-Squares Regression Algorithm in Bioanalytical LC-MS/MS Assays and Impacts of Using Incorrect Weighting Factors on Curve Stability, Data Quality, and Assay Performance. *Anal. Chem.* **2014**, *86*, 8959–8966. [CrossRef]
32. U.S. Food & Drug Administration. Available online: https://www.fda.gov/regulatory-information/search-fda-guidance-documents/cvm-gfi-208-vich-gl49-studies-evaluate-metabolism-and-residue-kinetics-veterinary-drugs-food (accessed on 23 May 2022).
33. Hult, K.; Hökby, E.; Hägglund, U.; Gatenbeck, S.; Rutqvist, L.; Sellyey, G. Ochratoxin A in pig blood: Method of analysis and use as a tool for feed studies. *Appl. Environ. Microbiol.* **1979**, *38*, 772–776. [CrossRef]
34. Devreese, M.; Croubels, S.; De Baere, S.; Gehring, R.; Antonissen, G. Comparative Toxicokinetics and Plasma Protein Binding of Ochratoxin A in Four Avian Species. *J. Agric. Food Chem.* **2018**, *66*, 2129–2135. [CrossRef]

35. Milićević, D.; Jurić, V.; Stefanović, S.; Baltić, T.; Janković, S. Evaluation and Validation of Two Chromatographic Methods (HPLC-Fluorescence and LC–MS/MS) for the Determination and Confirmation of Ochratoxin A in Pig Tissues. *Arch. Environ. Contam. Toxicol.* **2010**, *58*, 1074–1081. [CrossRef] [PubMed]
36. Luci, G.; Intorre, L.; Ferruzzi, G.; Mani, D.; Giuliotti, L.; Pretti, C.; Tognetti, R.; Bertini, S.; Meucci, V. Determination of ochratoxin A in tissues of wild boar (*Sus scrofa* L.) by enzymatic digestion (ED) coupled to high-performance liquid chromatography with a fluorescence detector (HPLC-FLD). *Mycotoxin Res.* **2018**, *34*, 1–8. [CrossRef] [PubMed]

Article

Mycotoxins in Tea ((*Camellia sinensis* (L.) Kuntze)): Contamination and Dietary Exposure Profiling in the Chinese Population

Haiyan Zhou, Zheng Yan, Aibo Wu and Na Liu *

SIBS-UGENT-SJTU Joint Laboratory of Mycotoxin Research, CAS Key Laboratory of Nutrition, Metabolism and Food Safety, Shanghai Institute of Nutrition and Health, University of Chinese Academy of Sciences, Shanghai 200030, China; zhouhaiyan2018@sibs.ac.cn (H.Z.); zyan@sibs.ac.cn (Z.Y.); abwu@sibs.ac.cn (A.W.)
* Correspondence: liuna@sibs.ac.cn; Tel.: +86-21-54920716

Abstract: Tea is popular worldwide with multiple health benefits. It may be contaminated by the accidental introduction of toxigenic fungi during production and storage. The present study focuses on potential mycotoxin contamination in tea and the probable dietary exposure assessments associated with consumption. The contamination levels for 16 mycotoxins in 352 Chinese tea samples were determined by ultra-performance liquid chromatography–tandem mass spectrometry. Average concentrations of almost all mycotoxins in tea samples were below the established regulations, except for ochratoxin A in the dark tea samples. A risk assessment was performed for the worst-case scenarios by point evaluation and Monte Carlo assessment model using the obtained mycotoxin levels and the available green, oolong, black, and dark tea consumption data from cities in China. Additionally, we discuss dietary risk through tea consumption as beverages and dietary supplements. In conclusion, there is no dietary risk of exposure to mycotoxins through tea consumption in the Chinese population.

Keywords: mycotoxins; tea; dietary exposure; risk assessment

Key Contribution: We investigated and comparatively analyzed the contamination levels for 16 mycotoxins from 7 groups in 352 Chinese tea samples (81 green tea, 64 oolong tea, 104 black tea, and 103 dark tea) by the UPLC–MS/MS method, obtaining contamination data from a large number of tea samples. Considering the different human preferences and needs regarding the aroma, flavor, taste, and active constituents, we assessed the risk of dietary exposure to mycotoxins through various types of tea consumption as beverages (green, oolong, black, and dark tea) or dietary supplements (green tea powder).

Citation: Zhou, H.; Yan, Z.; Wu, A.; Liu, N. Mycotoxins in Tea ((*Camellia sinensis* (L.) Kuntze)): Contamination and Dietary Exposure Profiling in the Chinese Population. *Toxins* **2022**, *14*, 452. https://doi.org/10.3390/toxins14070452

Received: 26 May 2022
Accepted: 20 June 2022
Published: 1 July 2022

Publisher's Note: MDPI stays neutral with regard to jurisdictional claims in published maps and institutional affiliations.

1. Introduction

Tea (*Camellia sinensis* L.) has long been consumed, primarily as an aromatic beverage, in many countries worldwide, with the properties of multiple health benefits and good taste [1–3]. The application and consumption of commercial dried tea in many fields is also gradually expanding due to the continuous innovation of tea comprehensive processing technology, forming diversified tea products [4]. Botanical dietary supplements based on traditional raw tea materials have entered our market in different forms [5,6]. The global average consumption of dried tea is approximately 2.5 million metric tons per year [7]. China plays a vital role among the major tea-producing countries. According to the differences in processing technologies and fermentation degrees, Chinese tea can mainly be divided into green tea (undergoing steaming and without fermentation), dark tea (undergoing steaming and postfermentation), oolong tea (semifermentation, not undergoing steaming) and black tea (complete fermentation, not undergoing steaming) [8].

Mycotoxins in tea, as potential chemical pollutants, are of great concern because of the incidental introduction of toxigenic fungi during the long manufacturing and supply of tea from cultivation to final consumption [9–12]. Filamentous fungi have been found in the postfermentation and storage processes, and some are essential contributors to the quality of dark tea [13–16]. However, there are concerns that some strains of *Fusarium* spp., *Aspergillus* spp., and *Penicillium* spp. from other environments might be OTA (ochratoxin A), AFs (aflatoxins), ZEN (zearalenone), DON (deoxynivalenol), or CIT (citrinin) producers [17–19]. The associated health risks from the potential multiple mycotoxin contamination of raw tea materials used for beverages or dietary supplements have also received increasing research attention [6,20,21]. Many countries have made their own regulations on AFB_1 and AFs in raw tea materials [22].

In recent years, there has been progress in the quantitative detection of mycotoxins in tea [5,6,12,22]. Many scholars have also focused on risk assessments of AFs and OTA in dark tea [23,24], AFs in black tea [25], and AFs and ZEN in green tea, black tea, and even herbal teas [26]. However, there are limited studies on the risk assessment of multiple mycotoxins involved in tea (*Camelia Sinensis* L.) [27–29] compared to herbal teas [30–35]. The total risk of exposure to ZEN is also rarely assessed, underestimating the potential toxicity of biotransformation through biochemical components or fungi [36].

Thus, this work aims to investigate and comparatively analyze the contamination levels for 16 mycotoxins from 7 groups in 352 Chinese tea samples (81 green tea, 64 oolong tea, 104 black tea, and 103 dark tea) using UPLC–MS/MS (ultra-performance liquid chromatography–tandem mass spectrometry). In this study, considering the different human preferences and needs regarding the aroma, flavor, taste, and active constituents, risks of dietary exposure to mycotoxins through various types of tea consumption are assessed by point evaluation and the Monte Carlo assessment model, using mycotoxin levels and available tea consumption data obtained from most cities in China. A risk ranking of both tea categories and mycotoxins is also performed. Moreover, the nonneoplastic and neoplastic toxicities of OTA and the carcinogenic risks of individual lifetime AF exposure through drinking tea are also considered. In particular, the risk of green tea powder as a dietary supplement is evaluated preliminarily for Chinese adults.

2. Results and Discussion

2.1. Occurrence of Mycotoxins

A total of 352 Chinese tea samples were detected and analyzed for 16 mycotoxins in the present study. Descriptive statistics of the concentrations of 16 mycotoxins in different tea categories are summarized in Table S2. In general, the obtained results demonstrate that the occurrence of mycotoxins was different in various types of tea samples, while DON and OTA had significant differences. This may be influenced by various processing methods or biochemical components. Green tea, without a fermentation process and containing abundant polyphenols, had the lowest levels of mycotoxin contamination, while dark tea had a higher possibility of contamination during long-term processing and storage. We did not exclude the influence of differences in the detailed tea varieties and other complicated factors (brand, source, or date of production). The United States Pharmacopoeia has set the limit for AFB_1 and AFs in raw medicinal herb materials at 5 and 20 $\mu g \cdot kg^{-1}$, which are the same as the Canadian and Argentine regulations [6]. Custom union countries (Kazakhstan, Armenia, Kyrgyzstan, Belarus, and Russia) have also established a regulation for the AFB_1 value in raw tea material at 5 $\mu g \cdot kg^{-1}$ [37].

In this paper, the content of AFB_1 in all tea samples did not exceed 5 $\mu g \cdot kg^{-1}$. The positive detection rates of AFs in dark, black, oolong, and green tea were 6.80%, 1.92%, 0%, and 0%, respectively. The co-occurrence rate of DON and its derivatives (7.39%) in all tea samples was also lower than the result obtained by Reinholds et al. [32]. Their derivatives in all tea samples were far below the MRLs (maximum regulation limits: 750 $\mu g \cdot kg^{-1}$) for DON in cereals for direct consumption [38]. However, the positive detection rates of DON in dark, oolong, black, and green tea were 7.41%, 9.38%, 2.88%, and 0%, respectively.

Bonferroni posttests showed that the concentration of DON in green tea was significantly lower than that in oolong or dark tea ($p \leq 0.01$). Furthermore, the concentration of DON in oolong tea was also significantly higher than that in black tea ($p \leq 0.01$). The contents of OTA and CIT in dark tea exceeded 5.0 and 200 µg·kg^{-1}. Additionally, the mean concentration of OTA in dark tea (129.76 µg·kg^{-1}) was significantly ($p \leq 0.01$) higher than that in other tea (0.09–1.33 µg·kg^{-1}) (Table S2). The distribution of contamination levels for OTA in different types of tea samples is shown in Figure 1. However, we did not detect high concentrations of ZEN (0.17–2.10 µg·kg^{-1}), which is different from Ye et al. [27]. The contents of NEO and T-2 in all tea samples were far below 200 µg·kg^{-1} (Table S2).

Figure 1. The distribution of contamination levels for OTA in different tea categories. The solid and dotted lines in the violin graphics represent the median and quartile distributions.

2.2. Risk Assessments

2.2.1. Deterministic Estimation

A deterministic assessment of exposure to six groups of noncarcinogenic mycotoxins through four types of tea consumption was completed (Equations (1) and (2)), and is shown in Table 1 and Figure 2. HQ values (lower, middle, and upper bound) for single mycotoxins from various types of tea consumption were far below 1.0, and the HQ values at the upper bound for green tea decreased in the order of ZEN > DON > T-2 > NEO > OTA > CIT. Furthermore, the upper bound HQ values for oolong, black, and dark tea decreased in the order of: ZEN > DON > CIT > OTA > T-2 > NEO; ZEN > DON > OTA > CIT > T-2 > NEO; and OTA > ZEN > DON > CIT > NEO > T-2, respectively. The HI values at the upper bound of the six groups of mycotoxins obtained from the deterministic estimation of green, oolong, black, and dark tea consumption were 1.90×10^{-2}, 4.41×10^{-2}, 4.99×10^{-2}, and 5.46×10^{-1}, respectively, without significant differences. Moreover, HI values were all <1.0 in the following order: dark tea > black tea > oolong tea > green tea. These values indicated no significant noncarcinogenic health risk from mycotoxins in tea consumption (Figure 2). Statistical analyses suggested significant differences between the upper bound HQ values of OTA in dark tea and green, oolong, and black tea in the deterministic estimation. Meanwhile, concerning the differences in neoplastic and nonneoplastic effects for OTA, the MOE approach for the exposure assessment was performed by Equations (6) and (7). MOE$_1$ exceeded 200 in all exposure scenarios, indicating no health risk. Meanwhile, MOE$_2$ exceeded 10,000 for almost all exposure scenarios, revealing a low risk apart from dark tea exposure, which indicated a potential health problem when genotoxicity was direct (Table 2). The carcinogenic risk of individual lifetime AF exposure from tea consumption was estimated by Equation (8). The results showed that the carcinogenic risks of individual lifetime AF exposure for four types of tea consumption were far below the acceptable level of 10^{-5} (Table 3).

Table 1. The deterministic estimation of six groups of mycotoxin exposure for four types of tea consumption ($\mu g \cdot kg^{-1} \cdot day^{-1}$).

Mycotoxins	DONs	ZENs	NEO	OTA	T-2	CIT
PMTDI	1	0.25	0.1	0.0143	0.1	0.2
Lower Bound						
Green Tea	9.21×10^{-4}	3.36×10^{-3}	3.81×10^{-5}	2.99×10^{-6}	5.71×10^{-6}	6.01×10^{-5}
Oolong Tea	1.13×10^{-2}	6.21×10^{-3}	3.55×10^{-5}	1.86×10^{-5}	1.49×10^{-6}	5.60×10^{-4}
Black Tea	7.75×10^{-3}	7.59×10^{-3}	3.65×10^{-5}	8.30×10^{-5}	9.66×10^{-6}	4.34×10^{-4}
Dark Tea	1.07×10^{-2}	4.42×10^{-3}	2.50×10^{-4}	7.23×10^{-3}	6.35×10^{-5}	1.16×10^{-3}
Middle Bound						
Green Tea	1.81×10^{-3}	3.41×10^{-3}	5.31×10^{-5}	4.53×10^{-6}	5.29×10^{-5}	6.20×10^{-5}
Oolong Tea	1.21×10^{-2}	6.26×10^{-3}	5.50×10^{-5}	1.97×10^{-5}	4.78×10^{-5}	5.61×10^{-4}
Black Tea	8.55×10^{-3}	7.64×10^{-3}	5.84×10^{-5}	8.31×10^{-5}	5.62×10^{-5}	4.35×10^{-4}
Dark Tea	1.13×10^{-2}	4.46×10^{-3}	2.64×10^{-4}	7.24×10^{-3}	1.03×10^{-4}	1.16×10^{-3}
Upper Bound						
Green Tea	2.71×10^{-3}	3.47×10^{-3}	6.80×10^{-5}	6.07×10^{-6}	1.00×10^{-4}	6.38×10^{-5}
Oolong Tea	1.29×10^{-2}	6.31×10^{-3}	7.45×10^{-5}	2.08×10^{-5}	9.41×10^{-5}	5.62×10^{-4}
Black Tea	9.35×10^{-3}	7.68×10^{-3}	8.03×10^{-5}	8.33×10^{-5}	1.03×10^{-4}	4.37×10^{-4}

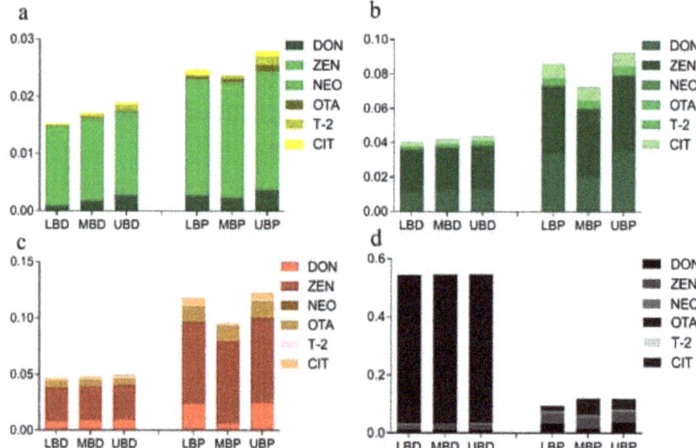

Figure 2. The (HI) values of six noncarcinogenic mycotoxins from tea consumption are the sum of the HQ values and MOS values (P95) from deterministic and probabilistic estimation with three bounds (lower, middle, and upper): (**a**) green tea, (**b**) oolong tea, (**c**) black tea, (**d**) dark tea. LBD/P: HQ/MOS values of deterministic/probabilistic estimation; MBD/P: HQ/MOS values of deterministic/probabilistic estimation; UBD/P: HQ/MOS values of deterministic/probabilistic estimation.

Table 2. The estimation of OTA exposure on four types of tea consumption with different methods based on exposure data from the upper bound deterministic and probabilistic estimation.

Tea	HQ/MOS	MOE$_1$	MOE$_2$
	The upper bound deterministic estimation		
Green Tea	4.24×10^{-4}	779,689.09	2,390,167.40
Oolong Tea	1.45×10^{-3}	227,572.33	697,631.89
Black Tea	5.83×10^{-3}	56,762.80	174,008.60
Dark Tea	0.51	653.59	2003.59
	The upper bound probabilistic estimation (Mean)		
Green Tea	4.24×10^{-4}	779,653.11	2,390,057.09
Oolong Tea	1.45×10^{-3}	227,583.84	697,667.16
Black Tea	5.90×10^{-3}	56,082.27	171,922.38
Dark Tea	8.12×10^{-3}	40,746.62	124,910.34
	The upper bound probabilistic estimation (P50)		
Green Tea	3.21×10^{-4}	1,030,801.16	3,159,961.29
Oolong Tea	1.01×10^{-3}	327,078.29	1,002,671.28
Black Tea	4.74×10^{-3}	69,726.95	213,750.68
Dark Tea	1.21×10^{-3}	273,674.19	838,958.92
	The upper bound probabilistic estimation (P95)		
Green Tea	1.14×10^{-3}	290,790.79	891,430.54
Oolong Tea	4.11×10^{-3}	80,440.80	246,594.41
Black Tea	1.47×10^{-2}	22,503.27	68,984.65
Dark Tea	1.90×10^{-2}	17,411.17	53,374.64

Table 3. The upper bound deterministic and probabilistic estimation of carcinogenic risk (R) of aflatoxins from four types of tea consumption.

Mycotoxins	Deterministic Estimation	Probabilistic Estimation		
		Mean	P50	P95
	AFB$_1$			
Green Tea	2.05×10^{-8}	1.86×10^{-8}	1.86×10^{-8}	2.29×10^{-8}
Oolong Tea	1.69×10^{-8}	1.69×10^{-8}	1.69×10^{-8}	1.69×10^{-8}
Black Tea	1.75×10^{-8}	1.75×10^{-8}	1.73×10^{-8}	1.86×10^{-8}
Dark Tea	3.06×10^{-8}	3.05×10^{-8}	2.57×10^{-8}	6.16×10^{-8}
	AFB$_2$			
Green Tea	1.37×10^{-8}	1.36×10^{-8}	1.23×10^{-8}	2.24×10^{-8}
Oolong Tea	1.67×10^{-8}	1.66×10^{-8}	1.40×10^{-8}	3.30×10^{-8}
Black Tea	1.18×10^{-8}	1.17×10^{-8}	1.07×10^{-8}	1.83×10^{-8}
Dark Tea	2.77×10^{-8}	2.75×10^{-8}	2.13×10^{-8}	6.78×10^{-8}
	AFG$_1$			
Green Tea	1.80×10^{-8}	1.80×10^{-8}	1.75×10^{-8}	3.01×10^{-8}
Oolong Tea	1.55×10^{-7}	1.56×10^{-7}	3.59×10^{-8}	6.75×10^{-7}
Black Tea	3.29×10^{-7}	3.26×10^{-7}	2.30×10^{-7}	9.49×10^{-7}
Dark Tea	2.68×10^{-7}	2.65×10^{-7}	1.88×10^{-7}	7.70×10^{-7}
	AFG$_2$			
Green Tea	7.50×10^{-7}	7.41×10^{-7}	5.16×10^{-7}	2.20×10^{-6}
Oolong Tea	3.28×10^{-6}	3.27×10^{-6}	2.98×10^{-6}	7.15×10^{-6}
Black Tea	3.01×10^{-6}	2.99×10^{-6}	2.50×10^{-6}	6.59×10^{-6}
Dark Tea	4.99×10^{-6}	5.07×10^{-6}	4.64×10^{-6}	$\mathbf{1.11 \times 10^{-5}}$

The bold font represents R > 1.0×10^{-5}.

2.2.2. Probabilistic Estimation

Concerning the variability and sampling uncertainty of the deterministic assessment, a probabilistic evaluation for mycotoxins was necessary to exclude the influence of individual differences, especially for OTA in dark tea. Table 4 shows the exposure distribution statistics of the six groups of mycotoxins for the four types of tea in the mean, 50th, and

95th percentiles performed with Equation (1) (Tables S3 and S4). At the same time, the noncarcinogenic risks of different tea types from three bounds of probabilistic estimation (P95) performed by Equation (3) are also shown in Figures 2 and S1.Statistical analyses suggested that there was no significant difference in the MOS values of mycotoxins in green, oolong, black, and dark tea from the probabilistic estimation, which was similar to the above ranking order of deterministic assessments, except for NEO and T-2 in black tea and OTA in dark tea. Interestingly, the upper bound MOS (P95) value of OTA from the probability assessment was significantly lower than the upper bound HQ value from the deterministic assessment, which resulted in a change in tea risk order. In summary, the MOS values at three bounds for the six groups of mycotoxins in the four types of tea were all far below 1.0. Even the HI values calculated by Equation (5) in the four types of tea were still <1.0 (black tea > dark tea > oolong tea > green tea); thus, no significant noncarcinogenic risk was observed from the probabilistic estimation in this study. According to the probabilistic estimation data for OTA, MOE_1 and MOE_2 exceeded 200 and 10,000 in all exposure scenarios, also indicating no nonneoplastic and neoplastic concern from OTA exposure through tea consumption (Table 2). The carcinogenic risk of individual lifetime AF exposure through drinking tea was almost below the acceptable level (Table 3). More detailed exposure assessments for AFG_2 were also performed by a Monte Carlo assessment model using dark tea consumption and body weight data in different age groups from the Kunming, Pu'er, and Ulan Bator of Mongolia questionnaire results (Table S5). The 95th percentile R values of AFG_2 in Pu'er for the 30–39 age group (1.13×10^{-5}) and Ulan Bator and Pu'er for the male group (1.35×10^{-5}, 1.34×10^{-5}) and >50 age group (1.52×10^{-5}, 1.82×10^{-5}) were close to the acceptable carcinogenic risk level. Interestingly, some of the groups were consistent with the high-exposure groups in previous studies, such as the age group (>50 years old) in Ulan Bator and Pu'er, and the male group in Pu'er. Although contaminant content is the most essential determining factor for mycotoxin exposure in dark tea [27], mycotoxin exposure may be correlated with differences in tea drinking habits and preferences (gender- or age-related).

Table 4. The upper bound probabilistic estimation of six groups of mycotoxin exposure on four types of tea consumption ($\mu g \cdot kg^{-1} \cdot day^{-1}$).

Mycotoxins	DONs	ZENs	NEO	OTA	T-2	CIT
PMTDI	1	0.25	0.1	0.0143	0.1	0.2
			Mean			
Green Tea	2.31×10^{-3}	3.46×10^{-3}	5.81×10^{-5}	6.07×10^{-6}	1.05×10^{-4}	6.31×10^{-5}
Oolong Tea	1.28×10^{-2}	6.49×10^{-3}	5.06×10^{-5}	2.08×10^{-5}	9.41×10^{-5}	5.53×10^{-4}
Black Tea	9.26×10^{-3}	7.68×10^{-3}	8.00×10^{-5}	8.43×10^{-5}	9.40×10^{-5}	4.33×10^{-4}
Dark Tea	1.17×10^{-2}	4.75×10^{-3}	3.08×10^{-4}	1.16×10^{-4}	1.43×10^{-4}	1.15×10^{-3}
			P50			
Green Tea	2.09×10^{-3}	3.34×10^{-3}	5.05×10^{-5}	4.59×10^{-6}	1.00×10^{-4}	4.48×10^{-5}
Oolong Tea	9.38×10^{-3}	5.91×10^{-3}	4.81×10^{-5}	1.45×10^{-5}	9.41×10^{-5}	3.82×10^{-4}
Black Tea	6.97×10^{-3}	5.91×10^{-3}	6.98×10^{-5}	6.78×10^{-5}	9.40×10^{-5}	2.99×10^{-4}
Dark Tea	6.60×10^{-3}	4.02×10^{-3}	6.79×10^{-5}	1.73×10^{-5}	1.24×10^{-4}	7.96×10^{-4}
			P95			
Green Tea	3.68×10^{-3}	4.88×10^{-3}	1.16×10^{-4}	1.63×10^{-5}	1.59×10^{-4}	1.82×10^{-4}
Oolong Tea	3.49×10^{-2}	1.10×10^{-2}	7.29×10^{-5}	5.89×10^{-5}	9.43×10^{-5}	1.67×10^{-3}
Black Tea	2.42×10^{-2}	1.88×10^{-2}	1.46×10^{-4}	2.10×10^{-4}	1.14×10^{-4}	1.30×10^{-3}
Dark Tea	3.55×10^{-2}	9.33×10^{-3}	7.56×10^{-4}	2.72×10^{-4}	2.61×10^{-4}	3.47×10^{-3}

2.2.3. Dietary Risk Profiling

We started by monitoring raw tea materials and observed no risk of exposure to mycotoxins through drinking tea in the worst-case scenarios, including nonneoplastic or neoplastic risks for OTA and risks of noncarcinogenic and carcinogenic mycotoxins. Information about the change in mycotoxin content after brewing with hot water showed that the transfer rates of these mycotoxins in spiked loose green tea were below 90%, especially AF_S (<45%) [39]; the mean transfer rates of AFs in black tea were 30.6% or 53.02% [25,26]. Clearly, there is also no risk of exposure to mycotoxins through drinking tea infused in hot or cold water. In addition to drinking tea, ready-to-eat green tea powder as an alternative strategy for weight loss is becoming increasingly available and is more acceptable to the Chinese population than other forms of dietary supplements (tablets or capsules). The concentrations of mycotoxins in botanical supplements are almost consistent with those in raw materials [6]. Therefore, we also performed a preliminary evaluation of green tea powder using the average body weight of 64.3 kg for adults and the recommended dosage of green tea powder (1000 mg/day). The probable values of EDI (from 1.02×10^{-6} to 7.83×10^{-4}) obtained in any exposure scenario were far below the values of PMTDI established (Table S6) and the values of EDI for drinking tea in the worst-case scenarios, indicating no risk through the consumption of green tea powder at the recommended dosage for Chinese adults.

3. Conclusions

The safety of tea and its consumption has been widely investigated. After the investigation and comparison of mycotoxin contamination levels in major tea categories from Chinese tea samples, differences between green tea and black or dark tea were found due to the existence of high contamination levels in individual varieties from the same category, highlighting the need for early monitoring in the processing process in order to keep the contamination levels as low as possible, especially for fermented tea with complex ingredients or long production chains. Taking into account differences in human preferences and needs, preliminary evaluations found that there is no dietary risk of exposure to mycotoxins through tea consumption as a beverage or dietary supplement. All in all, more up-to-date detailed data on different tea consumption should be encouraged to be recorded in future to further assess the health risks associated with tea consumption. In addition, tea is only one dietary source of possible exposure to mycotoxins, and more comprehensive dietary risk assessments should be carried out in the future based on the contributions of different diets.

4. Materials and Methods

4.1. Materials and Reagents

ZEN (Art. No. Z 2125), α-zearalenol (α-ZEL, Art. No. Z 0166), β-zearalenol (β-ZEL, Art. No. Z 2000), α-zearalanol (α-ZAL, Art. No. Z 0292), β-zearalanol (β-ZAL, Art. No. Z 0417), DON (Art. No. D 0156), 3-acetyl deoxynivalenol (3-Ac DON, Art. No. A 6166), 15-acetyl deoxynivalenol (15-Ac DON, Art. No. A 1556), OTA (Art. No. O 1877), T-2 toxin (T-2, Art. No. T 4887), and CIT (Art. No. C 1017) were obtained from Sigma-Aldrich (St. Louis, MO, USA). Ammonium acetate, silica gel (SG, Art. No. 236799), and formic acid (≥95%) were also from Sigma-Aldrich. AFs (Part. No. 10000344) and neosolaniol (NEO, Part. No. BRM S 92001) were from Romer Lab Biopure™. MWCNTs-COOH (carboxyl multiwalled carbon nanotubes, Art. No. 190710134732) was acquired from Klamar Reagent, Inc (Shanghai, China). HLB (hydrophilic–lipophilic balance; Art. No. SBEQ-CA3100) was purchased from Anpel Laboratory Technologies. Acetonitrile came from Merck (Darmstadt, Germany) and was HPLC grade.

4.2. Sampling and Preparation

A total of 352 tea samples were randomly collected from two representative enterprises: Shanghai Difute International Tea Co., Ltd. (Shanghai, China), and the Yunnan Fengqing Longrun Co., Ltd. (Guiyang, China) Each tea category (green, oolong, black, or dark tea) contained more than 60 tea samples, including at least 10 major tea consumption varieties. Specifically, green tea included Longjing, Maojian, Biluochun, Yunwu, Maofeng and other green tea varieties from different origins; oolong tea included Tieguanyin, Dahongpao and other oolong tea varieties from different origins; black tea included wild black tea,

Lapsang souchong, Jinjunmei, Wuyi bohea and other black tea varieties from different origins; dark tea included Puerh tea, Kang brick tea, Fu brick tea, Liubao tea, Tibetan tea and other varieties. The simultaneous determination of mycotoxins was performed using the UPLC–MS/MS method reported by our group [40]. Briefly, 1.0 g sample was ultrasound-assisted extracted using 10 mL acetonitrile with 1% formic acid and 1 mL of the co-extract was purified by a simple clean-up tube (MWCNTS-COOH, HLB, SG; weight ratio: 1:7.5:7.5). Then, the sample solution was filtered through the organic membrane (0.22 µm) for further determination on an Ultimate 3000 UPLC system (Thermo Fisher Scientific, San Jose, CA, USA) coupled to a TSQ Vantage TM triple-stage quadrupole mass spectrometer (Thermo Fisher Scientific, San Jose, CA, USA). Analytes were separated on an Agilent Extend C_{18} column (150 × 3.0 mm, 3.5 µm, Art. No. 763954-302) for gradient elution with methanol and water containing 5 mM ammonium acetate (0.35 mL·min^{-1}, 30 °C, 10 µL) in terms of the following elution program: 15% B (initial), 15% B (0–1 min), 15–25% B (1–3 min), 25–50% B (3–6.5 min), 50–100% B (6.5–12 min), 100% B (12–15 min), 100–15% B (15–17 min), and 15% B (17–20 min).

4.3. Tea Consumption Data

The differences in tea samples, drinking habits, and preferences for various tea categories could influence the assessment of population exposure to mycotoxins through drinking tea. We divided the tea samples into four tea categories, i.e., green tea (23.01%), oolong tea (18.18%), black tea (29.55%), and dark tea (29.26%). To evaluate the risk of mycotoxin exposure when drinking different tea categories, we used tea consumption and body weight data from questionnaires obtained from previous studies (Table S1) [27,41–43]. Specifically, we referred to the China Kadoorie Biobank (CKB) study which included a cohort of 512,891 adults from 5 urban areas and 5 rural areas in China, thus providing a complete database to support further research. Guan et al. focused on tea drinking behavior, and 512,824 samples were analyzed after the elimination of invalid data based on CKB. These collected data showed that tea drinkers in most of the project areas mainly drank green tea. Another questionnaire survey concerned the tea consumption of 219 residents from Kunming and Pu'er. Meanwhile, the relevant tea consumption data from Ulan Bator in Mongolia were also gathered by Ye et al. Participants in Kunming, Pu'er, and Ulan Bator in Mongolia mainly drank dark tea. In the latest national statistics for 2015–2019, more than 50% of total adults were overweight (BMI: 24.0–27.9) or obese (BMI > 28), with an average body weight of 64.3 kg [44]. The recommended dosage of green tea powder (1000 mg/day) was obtained from the USDA Dietary Supplement Ingredient Database (DSID; https://dsid.usda.nih.gov (accessed on 18 May 2022)).

4.4. Assessment of Health Risk

The risk of exposure to ingested mycotoxins by drinking green, oolong, black, and dark tea was assessed by point evaluation combined with the Monte Carlo assessment model.

4.4.1. Health Risk: Deterministic Estimation

Tea consumption data and contamination levels in tea were used to assess the dietary risk of exposure to mycotoxins through tea drinking. The hazard quotient (HQ) was used to assess the noncarcinogenic risk of exposure to each mycotoxin (e.g., DON, ZEN, T-2, OTA, NEO, and CIT). The estimated daily intake (EDI, µg·kg^{-1} bw·day^{-1}) of each mycotoxin and HQ were calculated using the following formulas:

$$EDI = (C \times CA)/BW \tag{1}$$

$$HQ = EDI/PMTDI \tag{2}$$

where C (µg·kg^{-1}) is the average level of mycotoxins. In our study, if the pollutant level was not detected or lower than LOD, alternative values of 0, 1/2 LOD and LOD were applied to assess exposure. CA and BW represent daily tea consumption amount (g·person^{-1}·day^{-1})

and the body weight of participants (kg). The CA and BW values used in the model refer to previous studies. For Equation (2), the HQ is evaluated by the ratio of PMTDI (provisional maximal tolerable daily intake), which is proposed by the WHO (World Health Organization) and recommended by the USEPA (United States Environmental Protection Agency), and comes from the PTWI (provisional tolerable weekly intake). The PMTDI values of DON and its acetylated derivatives, ZEN and its modified forms, T-2, and OTA were 1.0, 0.25, 0.1, and 0.0143 $\mu g \cdot kg^{-1}$ bw·day^{-1}, respectively [45–47]. The PTDI value of CIT was 0.2 $\mu g \cdot kg^{-1}$ bw·day^{-1} [48,49], and the PMTDI value of NEO referenced T-2. If HQ > 1, it means that mycotoxin is of concern. Otherwise, no significant health risk is found, meaning humans are not threatened by that exposure.

4.4.2. Health Risk: Probabilistic Estimation

The Monte Carlo model is one of the applicable models to quantify and decrease the uncertainty in health risk assessments. In our study, we performed Monte Carlo simulations for probabilistic estimation with @-Risk Industrial 7.5 (Palisade Corporation, New York, NY, USA) software and Microsoft Excel 2016. The data on the amount of daily tea consumption, the mycotoxin content of the tea samples, and body weight were input into the @RISK software. The best distribution was chosen based on Kolmogorov–Smirnov and Anderson–Darling tests in the @Risk software. Monte Carlo simulations were performed using 10,000 iterations (confidence interval > 90%). The MOS (margin of safety) values at the mean and the 50th and 95th percentile (mean, P50, P95) were calculated using the following formula. P50 and P95, more statistically significant representatives in the distribution of exposure levels, were selected as middle and universal estimates, respectively:

$$MOSi = Pi/PMTDI \tag{3}$$

4.4.3. Risk Ranking

Weight ratios of exposure to six groups of mycotoxins in four types of tea for the total population were regarded as the same in this work. The HQ values of mycotoxins from deterministic estimation were ranked. Meanwhile, the risk ranking could also be obtained by the MOS values of mycotoxins from the probabilistic evaluation. The HI (hazard index) value represents the combined noncarcinogenic risks of multiple mycotoxins. However, the interaction mechanisms and relevant coefficients among the six groups of mycotoxins were unknown and unavailable. In our study, the HI was calculated by adding the HQ or MOS for each group of mycotoxins from the deterministic and probabilistic estimation:

$$HI = HQ_{OTA} + HQ_{ZEN} + HQ_{DON} + HQ_{T-2} + HQ_{NEO} + HQ_{CIT} \tag{4}$$

$$HI = MOS_{OTA} + MOS_{ZEN} + MOS_{DON} + MOS_{T-2} + MOS_{NEO} + MOS_{CIT} \tag{5}$$

Conventionally, an HI less than 1.0 indicates that the total exposure does not exceed the level considered "acceptable", and people are unlikely to be exposed at a toxic level with possible consequences for health. In contrast, if it exceeds one, there is a possibility of suffering adverse effects.

4.4.4. Risk Assessment for OTA and AFs

Considering the differences in potential nonneoplastic and neoplastic risk for OTA, the margin of exposure (MOE) using the benchmark dose (BMD) approach of the European Food Safety Authority was also applied. MOE was calculated using the following equations:

$$MOE_1 = nonneoplastic\_BMDL10/ADD \tag{6}$$

$$MOE_2 = neoplastic\_BMDL10/ADD \tag{7}$$

BMDL10 values of 4.73 and 14.5 $\mu g \cdot kg^{-1}$ bw·day^{-1} were obtained from kidney lesions (nonneoplastic) observed in pigs and kidney tumors (neoplastic) seen in rats [50]. The ADD (average daily intake, mg·kg^{-1}·day^{-1}) data were from deterministic and probabilistic evaluations on OTA. In addition, MOE$_1$ and MOE$_2$ values exceeding 200 and 10,000 indicated no health concern.

AFs are considered to be carcinogens without associated PMTDI values to date. Thus, deterministic and probabilistic estimations could not be implemented for their risk assessment. In our study, the carcinogenic risk of individual lifetime AF exposure through tea consumption was estimated using the following formula [27,41]:

$$R = ADD \times SF \tag{8}$$

where R is cancer rate for the individual lifetime, SF represents the carcinogenic slope factor (HBsAg^{-}, 9 mg·kg^{-1}·day^{-1}), and ADD is the daily aflatoxin intake from deterministic (EDI) and probabilistic evaluation exposure (Pi) [27]. The acceptable level of carcinogenic risk for environmental risk management may change with the environmental policy. Considering the carcinogenic grade of AFs to humans, the lifetime cancer risk of AFs in this study was 10^{-5}, which means one additional cancer in 100,000 humans ingesting tea containing aflatoxins over 70 years [41].

4.5. Data Analysis

Raw data from UPLC–MS/MS were recognized using Thermo Xcalibur Qual Browser 4.0., and GraphPad Prism 8.4 software was used for all statistical analysis and graphic production. A probability level of <0.01 was considered statistically significant.

Supplementary Materials: The following supporting information can be downloaded at: https://www.mdpi.com/article/10.3390/toxins14070452/s1, Figure S1: The (HI) values of six noncarcinogenic mycotoxins from tea consumption are the sum of MOS values (P50) from probabilistic estimation with three bounds: (a) green tea, (b) oolong tea, (c) black tea, (d) dark tea; Table S1: Average dietary consumption of tea by persons in different areas and their average body weight; Table S2: Concentration range and mean of mycotoxins calculated with upper bound in all tea samples (µg·kg^{-1}); Table S3: The lower bound probabilistic estimation of six groups of mycotoxin exposure for four types of tea consumption (µg·kg^{-1}·day^{-1}); Table S4: The middle bound probabilistic estimation of six groups of mycotoxin exposure for four types of tea consumption (µg·kg^{-1}·day^{-1}); Table S5: The upper bound probabilistic estimation of carcinogenic risk (R) of AFG$_2$ through dark tea consumption in different regions; Table S6: The deterministic estimation of mycotoxin exposure on green tea powder consumption (µg·kg^{-1}·day^{-1}).

Author Contributions: Conceptualization, H.Z., N.L. and A.W.; methodology, H.Z. and Z.Y.; software, Z.Y. and H.Z.; validation, Z.Y. and H.Z.; formal Analysis, H.Z. and Z.Y.; investigation Z.Y. and H.Z.; resources, A.W.; data curation, H.Z. and Z.Y.; writing—original draft preparation, H.Z.; writing—review and editing, N.L. and A.W.; visualization, Z.Y.; supervision, N.L. and A.W.; project administration, H.Z., N.L. and A.W.; funding acquisition, A.W. and N.L. All authors have read and agreed to the published version of the manuscript.

Funding: This work was supported by the National Key Research and Development Program of China (2018YFC1604403).

Institutional Review Board Statement: Not applicable.

Informed Consent Statement: Not applicable.

Data Availability Statement: The data that support the findings of this study are available from the corresponding author upon reasonable request.

Conflicts of Interest: The authors declare that there is no known competing financial interest or personal relationships that could have appeared to influence the work reported in this paper.

References

1. Rameshrad, M.; Razavi, B.M.; Hosseinzadeh, H. Protective effects of green tea and its main constituents against natural and chemical toxins: A comprehensive review. *Food Chem. Toxicol.* **2017**, *100*, 115–137. [CrossRef] [PubMed]
2. Jiang, C.; Zeng, Z.; Huang, Y.; Zhang, X. Chemical compositions of Pu'er tea fermented by Eurotium Cristatum and their lipid-lowering activity. *LWT* **2018**, *98*, 204–211. [CrossRef]
3. Bag, S.; Mondal, A.; Majumder, A.; Banik, A. Tea and its phytochemicals: Hidden health benefits & modulation of signaling cascade by phytochemicals. *Food Chem.* **2022**, *371*, 131098. [CrossRef] [PubMed]
4. Liang, S.; Granato, D.; Zou, C.; Gao, Y.; Zhu, Y.; Zhang, L.; Yin, J.-F.; Zhou, W.; Xu, Y.-Q. Processing technologies for manufacturing tea beverages: From traditional to advanced hybrid processes. *Trends Food Sci. Technol.* **2021**, *118*, 431–446. [CrossRef]
5. Yu, X.L.; Sun, D.W.; He, Y. Emerging techniques for determining the quality and safety of tea products: A review. *Compr. Rev. Food Sci. Food Saf.* **2020**, *19*, 2613–2638. [CrossRef]
6. Pallarés, N.; Tolosa, J.; Ferrer, E.; Berrada, H. Mycotoxins in raw materials, beverages and supplements of botanicals: A review of occurrence, risk assessment and analytical methodologies. *Food Chem. Toxicol. Int. J. Publ. Br. Ind. Biol. Res. Assoc.* **2022**, *165*, 113013. [CrossRef]
7. Zhong, J.; Chen, N.; Huang, S.; Fan, X.; Zhang, Y.; Ren, D.; Yi, L. Chemical profiling and discrimination of green tea and Pu-erh raw tea based on UPLC-Q-Orbitrap-MS/MS and chemometrics. *Food Chem.* **2020**, *326*, 126760. [CrossRef]
8. Monbaliu, S.; Wu, A.; Zhang, D.; Van Peteghem, C.; De Saeger, S. Multimycotoxin UPLC-MS/MS for tea, herbal infusions and the derived drinkable products. *J. Agric. Food Chem.* **2010**, *58*, 12664–12671. [CrossRef]
9. Sedova, I.; Kiseleva, M.; Tutelyan, V. Mycotoxins in Tea: Occurrence, Methods of Determination and Risk Evaluation. *Toxins* **2018**, *10*, 444. [CrossRef]
10. Wang, L.M.; Huang, D.F.; Fang, Y.; Wang, F.; Li, F.L.; Liao, M. Soil fungal communities in tea plantation after 10 years of chemical vs. integrated fertilization. *Chil. J. Agric. Res.* **2017**, *77*, 355–364. [CrossRef]
11. Gil-Serna, J.; Vázquez, C.; Patiño, B. Mycotoxins in Functional Beverages: A Review. *Beverages* **2020**, *6*, 52. [CrossRef]
12. Pandey, A.K.; Samota, M.K.; Silva, A.S. Mycotoxins along the tea supply chain: A dark side of an ancient and high valued aromatic beverage. *Crit. Rev. Food Sci. Nutr.* **2022**, 2061908. [CrossRef] [PubMed]
13. Ma, Y.; Ling, T.J.; Su, X.Q.; Jiang, B.; Nian, B.; Chen, L.J.; Liu, M.L.; Zhang, Z.Y.; Wang, D.P.; Mu, Y.Y.; et al. Integrated proteomics and metabolomics analysis of tea leaves fermented by Aspergillus niger, Aspergillus tamarii and Aspergillus fumigatus. *Food Chem.* **2021**, *334*, 127560. [CrossRef]
14. Zhao, M.; Su, X.Q.; Nian, B.; Chen, L.J.; Zhang, D.L.; Duan, S.M.; Wang, L.Y.; Shi, X.Y.; Jiang, B.; Jiang, W.W.; et al. Integrated Meta-omics Approaches To Understand the Microbiome of Spontaneous Fermentation of Traditional Chinese Pu-erh Tea. *mSystems* **2019**, *4*, e00680-19. [CrossRef] [PubMed]
15. Zhou, B.; Ma, C.; Ren, X.; Xia, T.; Zheng, C.; Liu, X. Correlation analysis between filamentous fungi and chemical compositions in a pu-erh type tea after a long-term storage. *Food Sci. Nutr.* **2020**, *8*, 2501–2511. [CrossRef] [PubMed]
16. Li, Q.; Chai, S.; Li, Y.; Huang, J.; Luo, Y.; Xiao, L.; Liu, Z. Biochemical Components Associated With Microbial Community Shift During the Pile-Fermentation of Primary Dark Tea. *Front. Microbiol.* **2018**, *9*, 1509. [CrossRef]
17. Frisvad, J.C.; Moller, L.L.H.; Larsen, T.O.; Kumar, R.; Arnau, J. Safety of the fungal workhorses of industrial biotechnology: Update on the mycotoxin and secondary metabolite potential of Aspergillus niger, Aspergillus oryzae, and Trichoderma reesei. *Appl. Microbiol. Biotechnol.* **2018**, *102*, 9481–9515. [CrossRef]
18. Guo, W.; Zhao, M.; Chen, Q.; Huang, L.; Mao, Y.; Xia, N.; Teng, J.; Wei, B. Citrinin produced using strains of Penicillium citrinum from Liupao tea. *Food Biosci.* **2019**, *28*, 183–191. [CrossRef]
19. Li, Z.; Mao, Y.; Teng, J.; Xia, N.; Huang, L.; Wei, B.; Chen, Q. Evaluation of Mycoflora and Citrinin Occurrence in Chinese Liupao Tea. *J. Agric. Food Chem.* **2020**, *68*, 12116–12123. [CrossRef]
20. Veprikova, Z.; Zachariasova, M.; Dzuman, Z.; Zachariasova, A.; Fenclova, M.; Slavikova, P.; Vaclavikova, M.; Mastovska, K.; Hengst, D.; Hajslova, J. Mycotoxins in Plant-Based Dietary Supplements: Hidden Health Risk for Consumers. *J. Agric. Food Chem.* **2015**, *63*, 6633–6643. [CrossRef]
21. Taroncher, M.; Vila-Donat, P.; Tolosa, J.; Ruiz, M.J.; Rodríguez-Carrasco, Y. Biological activity and toxicity of plant nutraceuticals: An overview. *Curr. Opin. Food Sci.* **2021**, *42*, 113–118. [CrossRef]
22. Zhang, L.; Dou, X.W.; Zhang, C.; Logrieco, A.F.; Yang, M.H. A Review of Current Methods for Analysis of Mycotoxins in Herbal Medicines. *Toxins* **2018**, *10*, 65. [CrossRef] [PubMed]
23. Yan, H.; Zhang, L.; Ye, Z.; Wu, A.; Yu, D.; Wu, Y.; Zhou, Y. Determination and Comprehensive Risk Assessment of Dietary Exposure to Ochratoxin A on Fermented Teas. *J. Agric. Food Chem.* **2021**, *69*, 12021–12029. [CrossRef] [PubMed]
24. Cui, P.; Yan, H.; Granato, D.; Ho, C.T.; Ye, Z.; Wang, Y.; Zhang, L.; Zhou, Y. Quantitative analysis and dietary risk assessment of aflatoxins in Chinese post-fermented dark tea. *Food Chem. Toxicol.* **2020**, *146*, 111830. [CrossRef] [PubMed]
25. Ismail, A.; Akhtar, S.; Riaz, M.; Gong, Y.Y.; Routledge, M.N.; Naeem, I. Prevalence and Exposure Assessment of Aflatoxins Through Black Tea Consumption in the Multan City of Pakistan and the Impact of Tea Making Process on Aflatoxins. *Front. Microbiol* **2020**, *11*, 446. [CrossRef] [PubMed]
26. Duarte, S.C.; Salvador, N.; Machado, F.; Costa, E.; Almeida, A.; Silva, L.J.G.; Pereira, A.M.P.T.; Lino, C.; Pena, A. Mycotoxins in teas and medicinal plants destined to prepare infusions in Portugal. *Food Control.* **2020**, *115*, 107290. [CrossRef]

27. Ye, Z.; Wang, X.; Fu, R.; Yan, H.; Han, S.; Gerelt, K.; Cui, P.; Chen, J.; Qi, K.; Zhou, Y. Determination of six groups of mycotoxins in Chinese dark tea and the associated risk assessment. *Environ. Pollut.* **2020**, *261*, 114180. [CrossRef]

28. El Jai, A.; Juan, C.; Juan-Garcia, A.; Manes, J.; Zinedine, A. Multi-mycotoxin contamination of green tea infusion and dietary exposure assessment in Moroccan population. *Food Res. Int.* **2021**, *140*, 109958. [CrossRef]

29. Reinholds, I.; Bogdanova, E.; Pugajeva, I.; Alksne, L.; Stalberga, D.; Valcina, O.; Bartkevics, V. Determination of Fungi and Multi-Class Mycotoxins in Camelia Sinensis and Herbal Teas and Dietary Exposure Assessment. *Toxins* **2020**, *12*, 555. [CrossRef]

30. Pallarés, N.; Berrada, H.; Font, G.; Ferrer, E. Mycotoxins occurrence in medicinal herbs dietary supplements and exposure assessment. *J. Food Sci. Technol.* **2022**, *59*, 2830–2841. [CrossRef]

31. Mannani, N.; Tabarani, A.; Abdennebi, E.H.; Zinedine, A. Assessment of aflatoxin levels in herbal green tea available on the Moroccan market. *Food Control.* **2020**, *108*, 106882. [CrossRef]

32. Reinholds, I.; Bogdanova, E.; Pugajeva, I.; Bartkevics, V. Mycotoxins in herbal teas marketed in Latvia and dietary exposure assessment. *Food Addit Contam Part. B Surveill* **2019**, *12*, 199–208. [CrossRef] [PubMed]

33. Pallares, N.; Berrada, H.; Fernandez-Franzon, M.; Ferrer, E. Risk Assessment and Mitigation of the Mycotoxin Content in Medicinal Plants by the Infusion Process. *Plant. Foods Hum. Nutr.* **2020**, *75*, 362–368. [CrossRef] [PubMed]

34. El Jai, A.; Zinedine, A.; Juan-Garcia, A.; Manes, J.; Etahiri, S.; Juan, C. Occurrence of Free and Conjugated Mycotoxins in Aromatic and Medicinal Plants and Dietary Exposure Assessment in the Moroccan Population. *Toxins* **2021**, *13*, 125. [CrossRef] [PubMed]

35. Pallares, N.; Tolosa, J.; Manes, J.; Ferrer, E. Occurrence of Mycotoxins in Botanical Dietary Supplement Infusion Beverages. *J. Nat. Prod.* **2019**, *82*, 403–406. [CrossRef]

36. Lorenz, N.; Danicke, S.; Edler, L.; Gottschalk, C.; Lassek, E.; Marko, D.; Rychlik, M.; Mally, A. A critical evaluation of health risk assessment of modified mycotoxins with a special focus on zearalenone. *Mycotoxin Res.* **2019**, *35*, 27–46. [CrossRef] [PubMed]

37. Technical Regulations of the Customs Union, on Food Safety (Document No. TRCU/021/2011). Available online: http://www.eurexcert.com/TRCUpdf/TRCU-0021-On-food-safety.pdf (accessed on 16 November 2021).

38. Commission Regulation (EC), Setting Maximum Levels for Certain Contaminants in Foodstuffs (Document No. EC/1881/2006). Available online: https://www.legislation.gov.uk/eur/2006/1881 (accessed on 10 August 2021).

39. Kiseleva, M.; Chalyy, Z.; Sedova, I. Tea: Transfer of Mycotoxins from the Spiked Matrix into an Infusion. *Toxins* **2021**, *13*, 404. [CrossRef]

40. Zhou, H.; Yan, Z.; Yu, S.; Wu, A.; Liu, N. Development of a Novel UPLC-MS/MS Method for the Simultaneous Determination of 16 Mycotoxins in Different Tea Categories. *Toxins* **2022**, *14*, 169. [CrossRef]

41. Cao, H.; Qiao, L.; Zhang, H.; Chen, J. Exposure and risk assessment for aluminium and heavy metals in Puerh tea. *Sci. Total Env.* **2010**, *408*, 2777–2784. [CrossRef]

42. Chen, Z.; Chen, J.; Collins, R.; Guo, Y.; Peto, R.; Wu, F.; Li, L. China Kadoorie Biobank of 0.5 million people: Survey methods, baseline characteristics and long-term follow-up. *Int. J. Epidemiol.* **2011**, *40*, 1652–1666. [CrossRef]

43. Guan, X.; Yang, J.-F.; Xie, X.-Y.; Lin, C.; Li, J.-Y. Research on the Behavior of Tea Consumption in China with the CKB Data. *J. Tea Sci.* **2018**, *38*, 287–295. [CrossRef]

44. Pan, X.F.; Wang, L.; Pan, A. Epidemiology and determinants of obesity in China. *Lancet Diabetes Endocrinol.* **2021**, *9*, 373–392. [CrossRef]

45. Joint FAO/WHO Expert Committee on Food Additives (JECFA). *56 St Report of JECFA-Safty Evaluation of Certain Mycotoxins in Food*; World Health Organization: Geneva, Switzerland, 2002.

46. Walker, R.; Meyland, I.; Tritscher, A. Joint FAO/WHO Expert Committee on Food Additives, Seventy-Second Meeting (JECFA/72/SC). 2010. Available online: https://www.researchgate.net/publication/242567110 (accessed on 16 November 2021).

47. EFSA Panel on Contaminants in the Food Chain (CONTAM). Appropriateness to set a group health-based guidance value for zearalenone and its modified forms. *EFSA J.* **2016**, *14*, e04425. [CrossRef]

48. Meerpoel, C.; Vidal, A.; Andjelkovic, M.; De Boevre, M.; Tangni, E.K.; Huybrechts, B.; Devreese, M.; Croubels, S.; De Saeger, S. Dietary exposure assessment and risk characterization of citrinin and ochratoxin A in Belgium. *Food Chem. Toxicol.* **2021**, *147*, 111914. [CrossRef] [PubMed]

49. European Commission. Scientific Opinion on the risks for public and animal health related to the presence of citrinin in food and feed. *EFSA J.* **2012**, *10*, 2605. [CrossRef]

50. Schrenk, D.; Bodin, L.; Chipman, J.K.; del Mazo, J.; Grasl Kraupp, B.; Hogstrand, C.; Hoogenboom, L.; Leblanc, J.C.; Nebbia, C.S.; Nielsen, E.; et al. Risk assessment of ochratoxin A in food. *EFSA J.* **2020**, *18*, e06113. [CrossRef]

Article

Development and Validation of an Automated Magneto-Controlled Pretreatment for Chromatography-Free Detection of Aflatoxin B_1 in Cereals and Oils through Atomic Absorption Spectroscopy

Jin Ye [1,2,3,4], Mengyao Zheng [4,5], Haihua Ma [1,2,3,*], Zhihong Xuan [4], Wei Tian [4], Hongmei Liu [4], Songxue Wang [4,5] and Yuan Zhang [1,2,3,*]

1   Key Laboratory of Grain Information Processing and Control, Henan University of Technology, Ministry of Education, Zhengzhou 450001, China; yj@ags.ac.cn
2   Henan Key Laboratory of Grain Photoelectric Detection and Control, Henan University of Technology, Zhengzhou 450001, China
3   College of Information Science and Engineering, Henan University of Technology, Zhengzhou 450001, China
4   Academy of National Food and Strategic Reserves Administration, Beijing 102600, China; zhengmengyao2021@126.com (M.Z.); xzh@ags.ac.cn (Z.X.); tw@ags.ac.cn (W.T.); lhm@ags.ac.cn (H.L.); wsx@ags.ac.cn (S.W.)
5   College of Food Science and Engineering, Henan University of Technology, Zhengzhou 450001, China
*   Correspondence: mahaihua@haut.edu.cn (H.M.); zhangyuan@haut.edu.cn (Y.Z.)

Citation: Ye, J.; Zheng, M.; Ma, H.; Xuan, Z.; Tian, W.; Liu, H.; Wang, S.; Zhang, Y. Development and Validation of an Automated Magneto-Controlled Pretreatment for Chromatography-Free Detection of Aflatoxin B_1 in Cereals and Oils through Atomic Absorption Spectroscopy. *Toxins* **2022**, *14*, 454. https://doi.org/10.3390/toxins14070454

Received: 12 June 2022
Accepted: 29 June 2022
Published: 1 July 2022

Publisher's Note: MDPI stays neutral with regard to jurisdictional claims in published maps and institutional affiliations.

Abstract: A chromatography-free detection of aflatoxin B_1 (AFB_1) in cereals and oils through atomic absorption spectroscopy (AAS) has been developed using quantum dots and immunomagnetic beads. A magneto-controlled pretreatment platform for automatic purification, labeling, and digestion was constructed, and AFB_1 detection through AAS was enabled. Under optimal conditions, this immunoassay exhibited high sensitivity for AFB_1 detection, with limits of detection as low as 0.04 µg/kg and a linear dynamic range of 2.5–240 µg/kg. The recoveries for four different food matrices ranged from 92.6% to 108.7%, with intra- and inter-day standard deviations of 0.7–6.3% and 0.6–6.9%, respectively. The method was successfully applied to the detection of AFB_1 in husked rice, maize, and polished rice samples, and the detection results were not significantly different from those of liquid chromatography-tandem mass spectrometry. The proposed method realized the detection of mycotoxins through AAS for the first time. It provides a new route for AFB_1 detection, expands the application scope of AAS, and provides a reference for the simultaneous determination of multiple poisonous compounds (such as mycotoxins and heavy metals).

Keywords: aflatoxin B_1; atomic absorption spectroscopy; automated pretreatment system; quantum dots; magnetic-based immunosensor

Key Contribution: Using a combination of quantum dots and immunomagnetic beads; a magnetron platform for automatic purification; labeling; and digestion was constructed; AFB_1 was detected through atomic absorption spectroscopy.

1. Introduction

Aflatoxins (AFTs) are the most common mycotoxins [1] produced by *Aspergillus flavus*, *A. parasiticus*, and A. *nomius* [2–4] in cereal grains in the entire food chain, from farm to factory, under favorable temperature and humidity conditions [5]. AFTs include aflatoxin B_1 (AFB_1), aflatoxin B_2, aflatoxin G_1, and aflatoxin G_2. Among these, AFB_1 is the most toxic and has the highest detection frequency [6–8]. The AFB_1 contamination of food products is the most common contamination problem in the food chain and is potentially hazardous to humans and animals because it causes high carcinogenicity, mutagenicity, suppression of immunity, and liver damage [8,9]. To minimize the human health risk, many countries and regions have set extremely low maximum limits (MLs) for AFB_1 in food. For example, the European Commission established 2–12 μg/kg as the ML of AFB_1 in food, while China established 5–20 μg/kg as the ML of AFB_1 [10,11]. Therefore, it is necessary to address the contamination problem by assessing the risk of AFB_1 contamination in the food chain for food safety and human health protection.

Currently, conventional analytical approaches used for AFB_1 monitoring in food include advanced instrumental analysis and fast detection techniques. Instrumental analysis methods mainly rely on chromatography-based techniques, such as high-performance liquid chromatography [11–13] and liquid chromatography–tandem mass spectrometry (LC-MS/MS) [14–16], because of their high selectivity, excellent accuracy, and reproducibility. However, these analytical procedures have several limitations. Chromatographic methods incur high equipment costs and involve complicated and time-consuming sample operation and analysis procedures, which limit their use to skilled operators [10]. In addition, chromatography also requires consuming some chromatographic consumables, such as chromatographic columns, etc. These limitations are overcome by fast detection methods, such as thin-layer chromatography [17–21], colloidal gold immunochromatography [22–25], and enzyme-linked immunosorbent assay [26–28]; these are commonly used in the routine monitoring of AFB_1, especially for on-site or field detection, because of their low cost and convenience of operation. However, because limited sample clean-up strategies are used, the fast detection methods are often affected by matrix effects, leading to false-positive/negative results. Hence, a rapid, accurate, sensitive, and robust methodology based on inexpensive and easy-to-operate techniques for high-frequency and precise monitoring of trace AFB_1 must be developed.

Atomic absorption spectroscopy (AAS) is one of the most commonly used techniques for tracing heavy-metal element determination in food, owing to its robustness, accuracy, speed, and simplicity. However, AAS-based methods have received considerably less attention in the analysis of organic contaminants such as mycotoxins primarily because they are typically regarded as "elemental analytical methods [29]." In recent years, with the continuous development of molecular labeling methods and the high specificity of antigen–antibody reactions, the target signal (organic contaminant signal) to be detected can be successfully converted to metal ion signals, which can be detected by using an elemental analysis method. Wang [30] and Hansen [31] et al. completed the conversion of DNA and protein signals to metal ion signals with the help of modified quantum dots, which allowed for the detection of target DNA and proteins by electrochemical means. This suggests that the gap between inorganic and organic analyses can be bridged. In addition, nanomaterials such as quantum dots (QDs) are aggregates of atoms and molecules on the nanometer scale, which can generate a large number of atoms after digestion, thereby improving signal amplification and detection sensitivity [32,33]. Furthermore, an AAS-based analysis system has a lower instrument cost than chromatography-based systems. As a necessary instrument for food analysis labs (many end users), the analytical potential of AAS for mycotoxins is worth exploring.

In conventional analysis, several manual steps such as extraction, filtering, and purification are required. Consequently, it is only applicable to high-end laboratories with skilled technicians. Previously, we reported a fully automated pretreatment platform for sample enrichment, purification, and elution based on immunoaffinity magnetic beads (IMBs) [10,34–36], which consumes less time, costs, and labor as well as introduces fewer errors compared to manual processing of mycotoxin assays.

In this study, we proposed an AAS-based analysis method for the specific and sensitive detection of AFB_1 in complex food samples by using QDs as labels. To overcome the limitations of cumbersome steps, in this work, we used the IMB-based sample pretreatment platform and the AAS system in combination to generate new automated magneto-controlled analytical approaches for achieving programmable immunoassay operations, including automatic enrichment, purification, QD labeling, release, and analysis. In-house magneto-controlled systems present significant advantages over other immunosensors, which minimize labor and eliminate operational errors. The proposed method overcomes the shortcomings of existing methods and has several advantages such as simple manipulation, high sensitivity, acceptable linear calibration range and reproducibility, low-cost detection, ease of automation, and high analyte throughput. To the best of our knowledge, this study is the first to use AAS to detect AFB_1 in different food matrices.

2. Results and Discussion

2.1. Method Principle

In this study, by exploiting the specificity of immunomagnetic beads and the signal amplification effect of QDs, an automatic platform for enrichment, purification, labeling, and digestion was constructed to detect AFB_1 through AAS. This principle is outlined in Figure 1. AFB_1 in the sample extract was first specifically captured by IMBs according to the antigen–antibody reaction. After the reaction, the IMBs were transferred to wash wells to prevent nonspecific adsorption of the matrix from affecting subsequent steps and to improve the stability of the method. The complex of AFB_1 and the IMBs was then transferred to a reaction well, where the bovine serum albumin (BSA)-assisted QD-labeled intact AFB_1 antigen (AFB_1-BSA-QD) is located, and AFB_1-BSA-QD occupied the remaining adsorption sites. After washing, the IMB complexes were transferred to digestion well, and the QDs captured on the IMB were digested by the acid solution, releasing the corresponding metal ions (Cd^{2+}). When the content of AFB_1 in solution is high, a low amount of the AFB_1-BSA-QD conjugate should be bound to the IMB, and low content of Cd^{2+} must be released via digestion. Therefore, the content of Cd^{2+} in the digestion solution is inversely proportional to the content of AFB_1. Finally, the released Cd^{2+} was measured using an atomic absorption spectrometer to detect AFB_1.

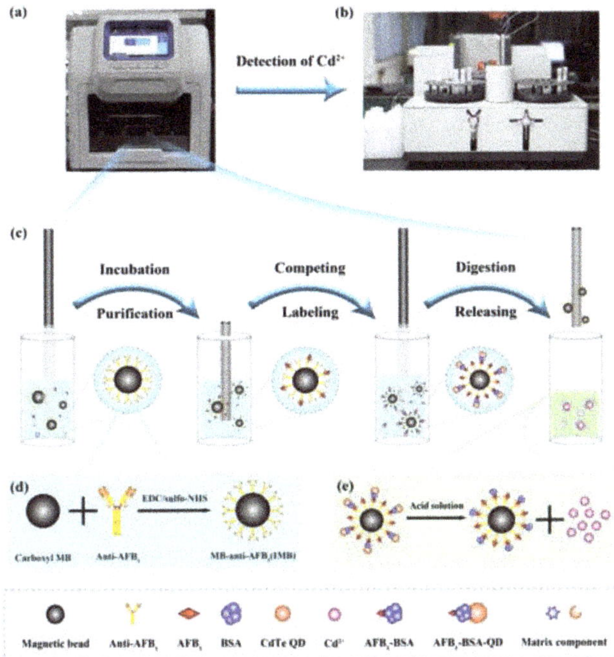

Figure 1. Schematic diagrams of the integrated detection platform: (**a**) automatic purifier; (**b**) atomic absorption spectrometer; (**c**) the purification process of the automatic purifier; (**d**) synthesis of immunomagnetic beads (IMB); (**e**) release of Cd^{2+} from quantum dot digestion.

2.2. Optimization of Experimental Parameters

2.2.1. Extraction Solution

Adequate extraction of the target from the sample matrix is the first step toward realizing accurate detection. In this experiment, blank rice was spiked with AFB_1 at a known concentration and different proportions of methanol and acetonitrile (30–80% methanol in water and 30–84% acetonitrile in water) were used to investigate the extraction efficiencies of AFB_1. The detection results were compared using the least significant difference method; the results are shown in Figure 2A. When 50% and 84% acetonitrile in water were used as the extract, the recovery rate of AFB_1 was close to 100%, and there was no significant difference between the two. To reduce the use of organic solvents, 50% acetonitrile in water was selected as the optimal extraction solution for AFB_1.

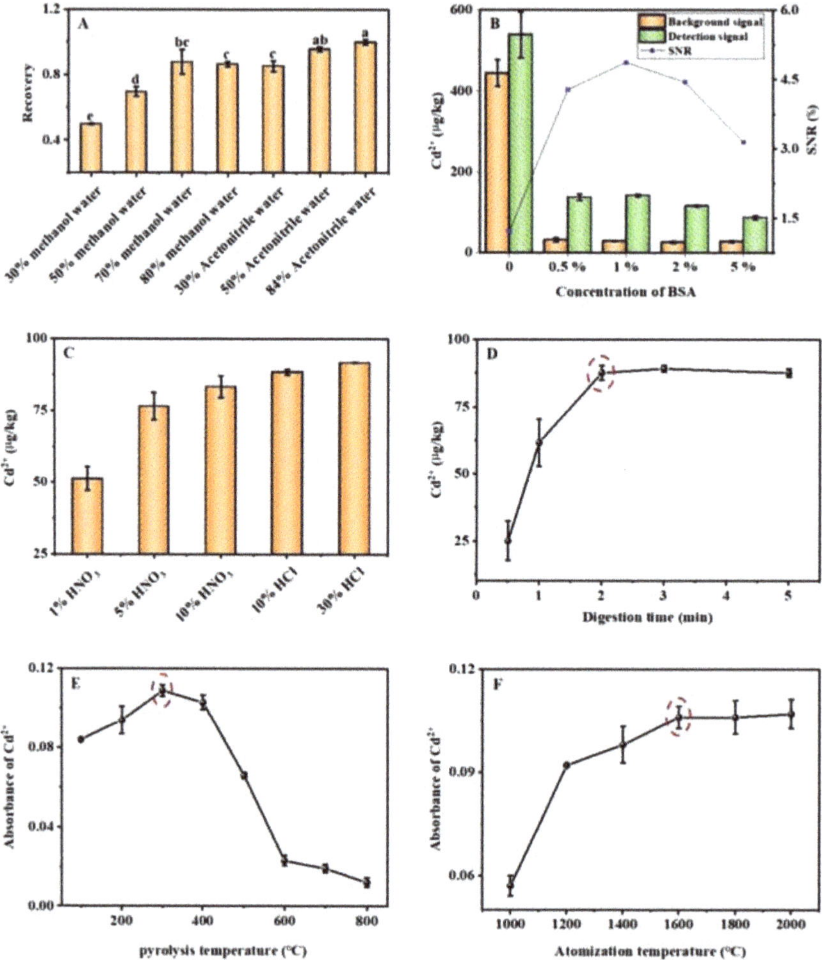

Figure 2. Optimization results of (**A**) extraction solution (the same small letters indicate no significant difference, $p > 0.05$), (**B**) BSA concentration, (**C**) digestion solution, (**D**) digestion time, (**E**) pyrolysis temperature, and (**F**) atomization temperature. The red circles mark the best conditions.

2.2.2. Nonspecific Adsorption

Nonspecific adsorption in the sample matrix leads to strong background interference. To eliminate background interference, BSA was used to block the inactive sites; however, the BSA concentration was too high, which negatively affected the strong binding. Therefore, the concentration of BSA in the reaction system was optimized; the results are shown in Figure 2B. As the BSA concentration increased from 0% to 5%, the signal-to-noise ratio of Cd^{2+} first increased and then decreased; when the BSA concentration was 1%, the signal-to-noise ratio of Cd^{2+} reached the maximum value. Therefore, 1% was determined to be the optimal BSA concentration.

2.2.3. Digestion Conditions

The digestion effect of acid on the QDs directly affects the generation of Cd^{2+}. To maximize the digestion of the QDs and generate Cd^{2+} in the shortest amount of time, the digestion solution and time were optimized. Figure 2C shows that as the concentration of HNO_3 solution increased from 1% to 10%, the concentration of Cd^{2+} trended upward, whereas as the concentration of the HCl solution increased from 10% to 30%, the concentration of Cd^{2+} exhibited only a slight upward trend. Therefore, 10% HCl was determined to be the optimal digestion solution. The digestion time was optimized by using an optimal digestion solution. As shown in Figure 2D, with the extension of the digestion time, the concentration of Cd^{2+} gradually increased, reaching a maximum at 2 min, and then gradually stabilized. Therefore, 2 min was determined as the optimal digestion time.

2.2.4. Atomic Absorption Conditions

Pyrolysis is a crucial stage in sample pretreatment. To ensure that the measured elements are not lost, an appropriate pyrolysis temperature at which matrix interference and pyrolysis time are reduced must be chosen. As shown in Figure 2E, with an increase in the pyrolysis temperature, the absorbance value of Cd^{2+} first increased and then decreased, reaching a maximum of 300 °C. This is because an appropriate increase in the pyrolysis temperature can allow the removal of the coexisting matrix and interfering components, the reduction or elimination of the background peak, and an increase in the absorbance value. When the temperature increased beyond a certain level, the pyrolysis-induced loss of Cd^{2+} increased, resulting in a decrease in the absorbance value. Therefore, 300 °C was determined to be the optimal pyrolysis temperature.

The atomization temperature is determined on the basis of the properties of the elements and their corresponding compounds. If the temperature is too high, the sensitivity is reduced, and the service life of the graphite tube is shortened. A low atomization temperature can prolong the service life of the graphite tube. However, if the atomization temperature is too low, the complete atomization of the measured element cannot be guaranteed, resulting in decreased sensitivity and low reproducibility. As shown in Figure 2F, with an increase in the atomization temperature, the absorbance of Cd^{2+} gradually increased, reached a maximum at 1600 °C, and then gradually became stable. Therefore, 1600 °C was determined to be the optimal atomization temperature.

2.3. Establishment and Specificity of Standard Curves in a Standard Solution and Four Types of Matrix Fluids

Under optimal conditions, taking the log value of the AFB_1 concentration as the abscissa and the Cd^{2+} concentration as the ordinate, the S-shaped curve and standard curve of the spiked standard solution and the four matrix solutions can be obtained. The results for the spiked standard solutions are shown in Figure 3A. The correlation coefficient of the fitted curve was 0.9986, and when the concentration of AFB_1 was 5–240 µg/kg, the log value of its concentration was linearly related to the concentration of Cd^{2+} with an R^2 of 0.9859. The limits of detection and quantification were 0.04 and 0.12 µg/kg, respectively. The standard addition results of the sample matrix (wheat, corn, peanut oil, and husked rice) are shown in Figure 3B–E, and the correlation coefficients of the curve fits were all greater than 0.99, approximately 1. In addition, the log value of the AFB_1 concentration was linearly related to the concentration of Cd^{2+} within a certain range. The relevant parameters are listed in Table 1.

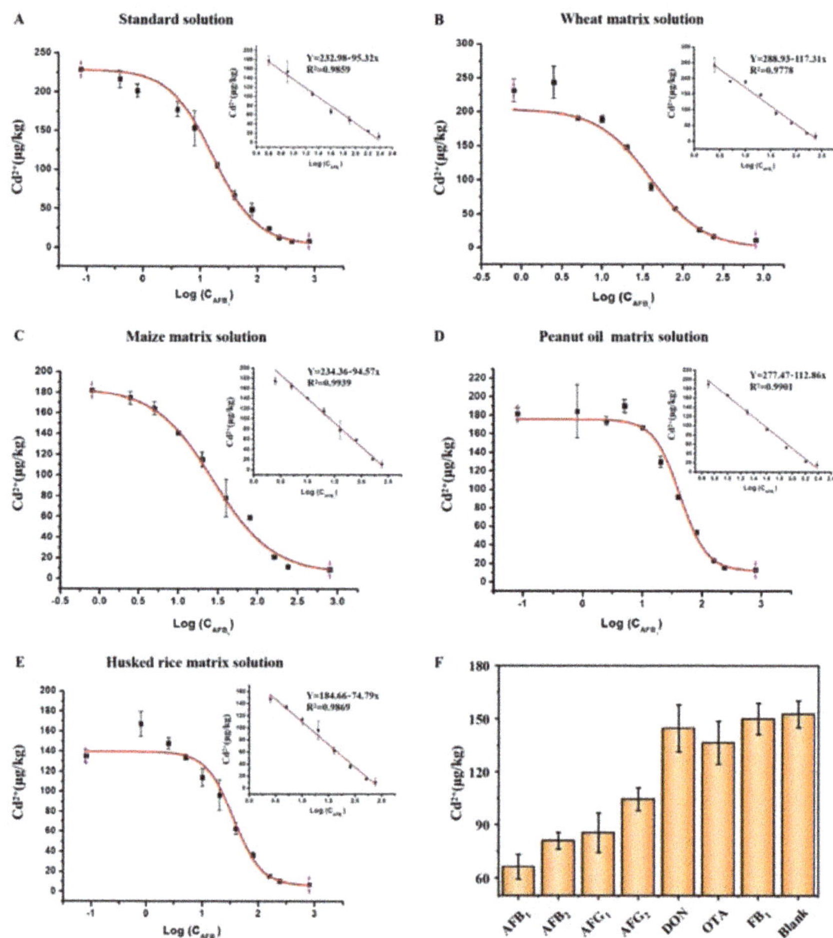

Figure 3. S-curves and standard curves of (**A**) a standard solution and (**B–E**) four matrix solutions spiked with aflatoxin B_1 (AFB_1). (**F**) Specificity detection results.

Table 1. Analysis parameters related to the sigmoid curves and standard curves of the standard solution and the four matrix solutions.

	Curve Range (μg/kg)	Curve Coefficient	Linear Range (μg/kg)	Linear Coefficient
Standard solution	0.08–800	0.9986	5–240	0.9859
Wheat	0.8–800	0.9944	2.5–240	0.9778
Maize	0.8–800	0.9951	2.5–240	0.9939
Peanut oil	0.08–800	0.9977	2.5–240	0.9901
Husked rice	0.08–800	0.9989	5–240	0.9869

To evaluate the selectivity of the detection platform for AFB_1, three analogs of AFB_1, including aflatoxin B_2 (AFB_2), aflatoxin G_1 (AFG_1), and aflatoxin G_2 (AFG_2), and common mycotoxins deoxynivalenol (DON) ochratoxin (OTA)and fumonisin B_1 (FB_1) were investigated. All toxin concentrations were 20 µg/kg. Figure 3F shows that AFB_1 produced a significantly lower metal ion signal than that of the blank control, while the signal variations induced by the DON, OTA, and FB_1 could be ignored. Although the metal ion signal values of the three analogs similar in structure to AFB_1 are slightly lower, in the future, the use of antibodies with better specificity can reduce the cross-reactivity to AFB_1 structural analogs and improve the specificity of the method.

2.4. Accuracy, Repeatability, and Reproducibility of the AAS Method

To evaluate the accuracy, repeatability, and reproducibility of the AAS method, four samples of corn, wheat, husked rice, and peanut oil were used, and they were spiked with three concentration gradients of low, medium, and high matrices (10, 20, and 40 µg/kg for maize and peanut oil; 2.5, 5, and 10 µg/kg for wheat; and 5, 10, and 20 µg/kg for husked rice). The recovery rates and intra- and inter-day relative standard deviations were measured and calculated; the results are shown in Table 2. The recoveries for the four matrices ranged from 92.6% to 108.7%, with intra- and inter-day standard deviations of 0.7–6.3% and 0.6–6.9%, respectively. These data demonstrate that the proposed method has high accuracy and precision.

Table 2. Recoveries and intra- and inter-day relative standard deviations of the sample matrices.

Matrix	Recovery ± RSD (%, $n = 3$)			Inter-Day RSD (%)
	Low	Medium	High	
Maize	99.5 ± 4.2	94.9 ± 3.2	101.7 ± 1.7	6.2
Wheat	92.9 ± 4.7	103.8 ± 2	100.3 ± 4.4	5.8
Husked rice	94.7 ± 5.2	95.8 ± 0.8	92.6 ± 4.8	0.6
Peanut oil	108.7 ± 6.3	104.4 ± 1.8	101.9 ± 0.7	6.9

The accuracy of this method was also verified using (certified) reference material. In this study, three AFB_1 reference materials (maize, peanut oil, and husked rice) provided by ASAG were selected. Among them, the uncertainty range of maize (GBW(E)100386) is 24–30 µg/kg; the uncertainty range of peanut oil(JTZK-002) is 13.9–17.7 µg/kg; the uncertainty range of husked rice(JTZK-007) is 22.1–29.9 µg/kg. It can be seen from Table 3 that all the detected values of the AFB_1 standard material are within the uncertainty range.

Table 3. The detection results for the certified reference material and reference materials ($n = 3$).

AFB_1 Reference Material	Lot Number	Detected Amount (µg/kg)	Certificate Value (µg/kg) ± SD
Maize	GBW(E)100386	26.5	27 ± 3
Peanut oil	JTZK-007	15.2	15.8 ± 1.9
Husked rice	JTZK-002	25.5	26 ± 3.9

2.5. Analysis of Real Samples

To verify the feasibility of the proposed method, naturally contaminated husked rice, maize, and rice samples were analyzed, and AFB_1 concentrations were determined by the proposed method and LC-MS/MS [37]. A paired sample T-test was used to compare whether there were significant differences in the detection results of the two methods. The measurement and analysis results are summarized in Table 4. The Sig. (two-tailed) values of the husked rice, maize, and rice samples are all greater than 0.05. There is no significant difference in the detection results of AFB_1 between this method and LC-MS/MS; thus, the proposed method can be used for the detection of AFB_1 in actual samples.

Table 4. Analyses of the results of the AAS and liquid chromatography–tandem mass spectrometry (LC-MS/MS) methods.

Sample	This Method (μg/kg)	LC-MS/MS (μg/kg)	t	df	Sig. (Two-Tailed)
Husked rice	31.70	31.88	0.305	2	0.789
	32.31	32.40			
	33.77	33.32			
Maize	38.32	31.00	1.855	2	0.205
	31.75	29.00			
	31.75	31.00			
Rice	6.90	8.60	1.095	2	0.388
	16.69	11.30			
	14.59	11.40			

3. Conclusions

In this study, an automatic platform for enrichment, purification, labeling, and digestion was successfully constructed, allowing for the detection of AFB_1 by AAS. The experimental data showed that the method has high sensitivity, specificity, accuracy, and precision, a wide range of linearity, and is applicable for the analysis of most samples. There was no significant difference in the detection results of the proposed method and LC-MS/MS during real sample analysis, indicating that this method can be used for the detection of AFB_1 in food. An automatic magnetron pretreatment system based on a quantum dot immunosensor was combined with an atomic absorption spectrometer to realize signal conversion between mycotoxin AFB_1 and metal ion Cd^{2+}. This method expands the application scope of AAS and provides a reference for the simultaneous determination of multiple poisonous compounds (such as mycotoxins and heavy metals).

4. Materials and Methods

4.1. Materials

AFB_1, dimethyl formamide (DMF), pyridine,O-(carboxymethyl) hydroxylamine hemihydrochloride (CMO), trichloromethane, 1-ethyl-3-(3-dimethylaminopropyl) carbodiimide hydrochloride (EDC), sodium hydroxide, N-hydroxysuccinimide (NHS), hydrochloric acid (HCl), nitric acid (HNO_3), bovine serum albumin (BSA), and phosphate buffer (PBS) were purchased from SigmaAldrich (St. Louis, MO, USA). HPLC grade methanol (MeOH) was obtained from Merck (Darmstadt, Germany). Deionized water (H_2O) was purchased from Watsons (Hong Kong, China). Blank maize, rice, wheat, husked rice, and peanut oil samples were obtained from a market in China. Certified reference material and reference materials are provided by the Academy of National Food and Strategic Reserves Administration(ASAG) (Beijing, China; GBW(E)100386, JTZK-007, JTZK-002).

4.2. Synthesis of Immunomagnetic Beads (IMB)

First, take 500 μL of NHS-activated carboxyl magnetic beads, discard the supernatant after magnetic separation, add 1 mL of dilute hydrochloric acid pre-cooled at 2–8 °C and mix well. After magnetic separation, the supernatant was discarded, and 500 μL of 2 mg/mL AFB_1 antibody was added. After homogeneous, it was placed on a mixer for 2 h at room temperature. Next, the reaction mixture was magnetically separated, the supernatant was discarded, 1 mL of blocking buffer (Tris-HCl) was added and mixed, and then placed on a mixer for 2 h at room temperature to form IMB. Finally, the mixture was magnetically separated, and the supernatant was discarded, washed with 1 mL PBST (2 times) and 1 mL PBS (1 time) IMB, and then added 500 μL PBS to resuspend and stored at 4°C for later use.

4.3. Synthesis of Aflatoxin Haptens (AFB$_1$-CMO)

Since there is no active group on the surface of the AFB$_1$ molecule that can be coupled with the protein, the AFB$_1$ molecule needs to be derivatized before coupling with BSA. We used O-(carboxymethyl)hydroxylamine hemihydrochloride (CMO) as a derivatizing agent to make AFB$_1$ carry a carboxyl group, which can then react with the amino group on BSA to achieve the coupling effect. The derivatization scheme was as follows: 1 mg of AFB$_1$ was dissolved in 0.6 mL of methanol–water–pyridine (4:1:1) solution, 2 mg of CMO was added, and the reaction was carried out under magnetic stirring in a water bath at 70 °C for 6 h and allowed to stand overnight at room temperature in the dark. The reaction solution was blown dry with nitrogen, and the precipitate was dissolved in 1 mL of chloroform solution and extracted three times with an equal volume of ultrapure water. The organic phase was collected and dried with nitrogen, and the precipitate was dissolved in 200 μL of Dimethyl Formamide (DMF) solution, which was the derivative product.

4.4. Synthesis of Aflatoxin Complete Antigen (AFB$_1$-BSA)

The complete antigen synthesis is based on the hapten synthesis by adding the coupling protein BSA. The coupled protein BSA was first dissolved in carbonate buffer. Then, 1 mg of AFB$_1$-CMO was dissolved in 200 μL of DMF solution, 2.4 mg of EDC and 1.3 mg of NHS were added, and the reaction was conducted under magnetic stirring at room temperature for 12 h. The reaction solution was added dropwise to the carrier protein solution (BSA: 7.54 mg), and the reaction was magnetically stirred at room temperature overnight. The coupled product was ultrafiltered with an ultrafiltration tube, and the retentate was resuspended in PBS and stored at −20 °C for later use. The successful synthesis of AFB$_1$-BSA was characterized by the UV-Vis absorption spectra of BSA and AFB$_1$-BSA. The characteristic absorption peak of BSA is at 278 nm, the characteristic absorption peak of AFB$_1$ is at 265 nm and 360 nm, and the characteristic absorption peak of AFB$_1$-BSA is between 265–278 nm, which is mainly due to the superposition of the characteristic absorption peaks of AFB$_1$ and BSA. The successful synthesis of the complete antigen can be well characterized by the UV-Vis absorption pattern.

4.5. Coupling of Pegylated Quantum Dots with Complete Antigen AFB$_1$-BSA

The cross-linking principle of polyethylene glycol-modified quantum dots and AFB$_1$-BSA is based on the fact that after heat treatment of BSA, a part of the internal hydrophobic structure will be exposed, which can be adsorbed with PEG on the surface of quantum dots. The cross-linking steps are as follows: 22.2 μL of 100 mM PBS was added to 200 μL of 1 μM PEG-modified quantum dot solution to prepare a 10 mM PBS quantum dot solution (pH 7.4). Subsequently, 67 μg of AFB$_1$-BSA was added and mixed, followed by a boiling water bath for 10 min, centrifugation at 25,000 rpm for 15 min, the supernatant was discarded, and the quantum dots were resuspended.

4.6. Sample Automatic Processing

Representative samples were thoroughly ground and homogenized according to the Codex General Guidelines on Sampling from the FAO and WHO with minor modifications [38].Cereal samples are treated as follows: 3 points are randomly selected for sampling, and the sampling amount of each point is 0.5 kg as the laboratory sample size. All laboratory samples were pulverized with a particle size of 0.5 mm. After fully mixing, 5 g of the sample was weighed for processing. Oil samples are treated as follows: after the peanut oil sample is fully mixed, 5 g of the sample is weighed for processing. Sequentially, 5 g [37] sample and 20 mL extraction solution were vortexed at 2500 rpm for 20 min in a centrifuge tube (50-mL). Finally, the centrifuge tube was centrifuged at 7000 rpm for 5 min, with the supernatants for further analysis. Proper mixing frequency, mixing amplitude, and sufficient reaction time can ensure sufficient reaction, washing, and elution of the sample. In order to prevent the solvent from splashing out during the mixing process, the mixing range was set to 80%, and other reaction conditions were shown in Table 5. After the pretreatment, the eluates in 5 wells were collected for detection by atomic absorption spectrometer or LC-MS/MS [37].

Table 5. The mixing frequency and sequence of the automatic clean-up procedure.

Step	Well	Mixing Time/min	Collection Time/min	Mixing Frequency /Hz
Transfer	2	1.0	0.5	6.5
Reaction	1	3.0	1.0	1.5
Wash 1	2	1.0	0.5	6.5
Competing	3	1.0	0.5	6.5
Wash 2	4	1.0	0.5	6.5
Digestion	5	1.0	0.5	7.5
Collection	2	1.0	0.5	6.5

4.7. Atomic Absorption Detection

The determination of cadmium (Cd^{2+}) was done by using the graphite furnace atomic absorption spectrometer (CPG2S, China). A Cd hollow cathode lamp operating at the 228.8 nm analytical line (4 mA current and a 0.8 nm spectral bandpass) was used for absorbance measurement, and the deuterium lamp was used to correct the background. Other experimental conditions and heating procedures are shown in Table 6.

Table 6. Instrumental operating conditions and heating program for the determination of Cd^{2+}.

Spectrometer Conditions	Cd	Step	Temperature (°C)	Ramp (s)	Hold (s)	Argon
Wavelength (nm)	228.8	Drying 1	75	5	2	ON
Bandpass (nm)	0.8	Drying 2	90	5	2	ON
sample volume (µL)	12	Drying 3	110	10	2	ON
Lamp current (mA)	4	Pyrolysis	300	5	5	ON
		Atomization	1600	2	1	OFF
		Cleaning	1650	1	1	ON

4.8. Method Verification

The linear range, limit of detection (LOD), limit of quantification (LOQ), recovery, intra-day relative standard deviation, and inter-day relative standard deviation were determined for the method, and real samples were analyzed together by LC-MS/MS. Where the limit of detection (3 σ/s) and the limit of quantification (10 σ/s) are calculated from the calibration, "σ" is the standard deviation of the 11 blank measurements, and "s" is the slope of the calibration curve.

Author Contributions: Conceptualization, J.Y., M.Z. and Z.X.; methodology, J.Y., Z.X. and W.T.; software, M.Z. and W.T.; validation, H.M., Y.Z. and S.W.; formal analysis, H.L.; investigation, J.Y.; writing—original draft preparation, Z.X. and M.Z.; writing—review and editing, J.Y. and M.Z.; supervision, H.M. and S.W.; project administration, J.Y. and Y.Z.; funding acquisition, J.Y. All authors have read and agreed to the published version of the manuscript.

Funding: This research was funded by Open Fund Project of Key Laboratory of Grain Information Processing & Control, Ministry of Education, Henan University of Technology (KFJJ-2020-102), the National Natural Science Foundation of China (31901806), Young Elite Scientists Sponsorship Program by CAST (2021QNRC001).

Institutional Review Board Statement: Not applicable.

Informed Consent Statement: Not applicable.

Data Availability Statement: Not applicable.

Acknowledgments: We thank PGENERAL instrument Co., LTD for donations in graphite furnace atomic absorption spectrophotometer.

Conflicts of Interest: The authors declare no conflict of interest.

References

1. Liu, H.; Zhao, Y.; Lu, A.; Ye, J.; Wang, J.; Wang, S.; Luan, Y. An aptamer affinity column for purification and enrichment of aflatoxin B1 and aflatoxin B2 in agro-products. *Anal. Bioanal. Chem.* **2020**, *412*, 895–904. [CrossRef] [PubMed]
2. Rahimi, F.; Roshanfekr, H.; Peyman, H. Ultra-sensitive electrochemical aptasensor for label-free detection of Aflatoxin B1 in wheat flour sample using factorial design experiments. *Food Chem.* **2021**, *343*, 128436. [CrossRef] [PubMed]
3. Xie, Y.; Wang, W.; Zhang, S. Purification and identification of an aflatoxin B1 degradation enzyme from Pantoea sp. T6. *Toxicon* **2019**, *157*, 35–42. [CrossRef] [PubMed]
4. Fouad, A.M.; Ruan, D.; El-Senousey, H.K.; Chen, W.; Jiang, S.; Zheng, C. Harmful effects and control strategies of aflatoxin b1 produced by Aspergillus flavus and Aspergillus parasiticus strains on poultry. *Toxins* **2019**, *11*, 176. [CrossRef]
5. Fan, T.; Xie, Y.; Ma, W. Research progress on the protection and detoxification of phytochemicals against aflatoxin B1-Induced liver toxicity. *Toxicon* **2021**, *195*, 58–68. [CrossRef]
6. Ma, J.; Liu, Y.; Guo, Y.; Ma, Q.; Ji, C.; Zhao, L. Transcriptional profiling of aflatoxin B1-induced oxidative stress and inflammatory response in macrophages. *Toxins* **2021**, *13*, 401. [CrossRef]
7. Hussain, Z.; Khan, M.Z.; Khan, A.; Javed, I.; Saleemi, M.K.; Mahmood, S.; Asi, M.R. Residues of aflatoxin B1 in broiler meat: Effect of age and dietary aflatoxin B1 levels. *Food Chem. Toxicol.* **2010**, *48*, 3304–3307. [CrossRef]
8. Yu, L.; Zhang, Y.; Hu, C.; Wu, H.; Yang, Y.; Huang, C.; Jia, N. Highly sensitive electrochemical impedance spectroscopy immunosensor for the detection of AFB1 in olive oil. *Food Chem.* **2015**, *176*, 22–26. [CrossRef]
9. Xia, L.; Rasheed, H.; Routledge, M.N.; Wu, H.; Gong, Y.Y. Super-Sensitive LC-MS Analyses of Exposure Biomarkers for Multiple Mycotoxins in a Rural Pakistan Population. *Toxins* **2022**, *14*, 193. [CrossRef]
10. Xuan, Z.; Liu, H.; Ye, J.; Li, L.; Tian, W.; Wang, S. Reliable and disposable quantum dot–based electrochemical immunosensor for aflatoxin B1 simplified analysis with automated magneto-controlled pretreatment system. *Anal. Bioanal. Chem.* **2020**, *412*, 7615–7625. [CrossRef]
11. Xuan, Z.; Ye, J.; Zhang, B.; Li, L.; Wu, Y.; Wang, S. An automated and high-throughput immunoaffinity magnetic bead-based sample clean-up platform for the determination of aflatoxins in grains and oils using UPLC-FLD. *Toxins* **2019**, *11*, 583. [CrossRef]
12. Wu, Y.; Ye, J.; Xuan, Z.; Li, L.; Wang, H.; Wang, S.; Liu, H.; Wang, S. Development and validation of a rapid and efficient method for simultaneous determination of mycotoxins in coix seed using one-step extraction and UHPLC-HRMS. *Food Addit. Contam. Part A* **2021**, *38*, 148–159. [CrossRef]
13. Egner, P.A.; Groopman, J.D.; Wang, J.-S.; Kensler, T.W.; Friesen, M.D. Quantification of aflatoxin-b1-n 7-guanine in human urine by high-performance liquid chromatography and isotope dilution tandem mass spectrometry1. *Chem. Res. Toxicol.* **2006**, *19*, 1191–1195. [CrossRef]
14. Er Demirhan, B.; Demirhan, B. Investigation of Twelve Significant Mycotoxin Contamination in Nut-Based Products by the LC–MS/MS Method. *Metabolites* **2022**, *12*, 120. [CrossRef]
15. Sarwat, A.; Rauf, W.; Majeed, S.; De Boevre, M.; De Saeger, S.; Iqbal, M. LC-MS/MS based appraisal of multi-mycotoxin co-occurrence in poultry feeds from different regions of Punjab, Pakistan. *Food Addit. Contam. Part B* **2022**, *15*, 106–122. [CrossRef]
16. Kudumija, N.; Vulić, A.; Lešić, T.; Vahčić, N.; Pleadin, J. Aflatoxins and ochratoxin A in dry-fermented sausages in Croatia, by LC-MS/MS. *Food Addit. Contam. Part B* **2020**, *13*, 225–232. [CrossRef]
17. Pradhan, S.; Ananthanarayan, L. Standardization and validation of a high-performance thin-layer chromatography method for the quantification of aflatoxin B1 and its application in surveillance of contamination level in marketed food commodities from the Mumbai region. *JPC–J. Planar Chromatogr.–Mod. TLC* **2020**, *33*, 617–630. [CrossRef]
18. Castro, L.; Vargas, E.A. Determining aflatoxins B1, B2, G1 and G2 in maize using florisil clean up with thin layer chromatography and visual and densitometric quantification. *Food Sci. Technol.* **2001**, *21*, 115–122. [CrossRef]
19. Qu, L.-L.; Jia, Q.; Liu, C.; Wang, W.; Duan, L.; Yang, G.; Han, C.-Q.; Li, H. Thin layer chromatography combined with surface-enhanced raman spectroscopy for rapid sensing aflatoxins. *J. Chromatogr. A* **2018**, *1579*, 115–120. [CrossRef]
20. Lin, L.; Zhang, J.; Wang, P.; Wang, Y.; Chen, J. Thin-layer chromatography of mycotoxins and comparison with other chromatographic methods. *J. Chromatogr. A* **1998**, *815*, 3–20. [CrossRef]
21. Stroka, J.; Anklam, E. Development of a simplified densitometer for the determination of aflatoxins by thin-layer chromatography. *J. Chromatogr. A* **2000**, *904*, 263–268. [CrossRef]
22. Sojinrin, T.; Liu, K.; Wang, K.; Cui, D.; J Byrne, H.; Curtin, J.F.; Tian, F. Developing gold nanoparticles-conjugated aflatoxin B1 antifungal strips. *Int. J. Mol. Sci.* **2019**, *20*, 6260. [CrossRef]
23. Sun, Q.; Zhu, Z.; Deng, Q.-m.; Liu, J.-m.; Shi, G.-Q. A "green" method to detect aflatoxin B 1 residue in plant oil based on a colloidal gold immunochromatographic assay. *Anal. Methods* **2016**, *8*, 564–569. [CrossRef]
24. Masinde, L.A.; Sheng, W.; Xu, X.; Zhang, Y.; Yuan, M.; Kennedy, I.R.; Wang, S. Colloidal gold based immunochromatographic strip for the simple and sensitive determination of aflatoxin B1 and B2 in corn and rice. *Microchim. Acta* **2013**, *180*, 921–928. [CrossRef]
25. Chen, X.Q.; Lu, S.S.; Sun, Q.; Yang, J.C.; Shi, G.Q. *Development of a Colloidal Gold-Based Immunochromatographic Assay for the Rapid Detection of Aflatoxin B1 in Food Samples*; Advanced Materials Research; Trans Tech Publications: Zurich, Switzerland, 2013; pp. 1279–1282.

26. Lee, N.A.; Wang, S.; Allan, R.D.; Kennedy, I.R. A rapid aflatoxin B1 ELISA: Development and validation with reduced matrix effects for peanuts, corn, pistachio, and soybeans. *J. Agric. Food Chem.* **2004**, *52*, 2746–2755. [CrossRef]
27. Oplatowska-Stachowiak, M.; Sajic, N.; Xu, Y.; Haughey, S.A.; Mooney, M.H.; Gong, Y.Y.; Verheijen, R.; Elliott, C.T. Fast and sensitive aflatoxin B1 and total aflatoxins ELISAs for analysis of peanuts, maize and feed ingredients. *Food Control* **2016**, *63*, 239–245. [CrossRef]
28. Eslami, M.; Mashak, Z.; Heshmati, A.; Shokrzadeh, M.; Mozaffari Nejad, A.S. Determination of aflatoxin B1 levels in Iranian rice by ELISA method. *Toxin Rev.* **2015**, *34*, 125–128.
29. Souza, J.P.; Cerveira, C.; Miceli, T.M.; Moraes, D.P.; Mesko, M.F.; Pereira, J.S. Evaluation of sample preparation methods for cereal digestion for subsequent As, Cd, Hg and Pb determination by AAS-based techniques. *Food Chem.* **2020**, *321*, 126715. [CrossRef]
30. Wang, J.; Liu, G.; Merkoçi, A. Electrochemical coding technology for simultaneous detection of multiple DNA targets. *J. Am. Chem. Soc.* **2003**, *125*, 3214–3215. [CrossRef]
31. Hansen, J.A.; Wang, J.; Kawde, A.-N.; Xiang, Y.; Gothelf, K.V.; Collins, G. Quantum-dot/aptamer-based ultrasensitive multi-analyte electrochemical biosensor. *J. Am. Chem. Soc.* **2006**, *128*, 2228–2229. [CrossRef]
32. Molaei, M.J. Principles, mechanisms, and application of carbon quantum dots in sensors: A review. *Anal. Methods* **2020**, *12*, 1266–1287. [CrossRef]
33. Pérez, E.; Marco, F.M.; Martínez-Peinado, P.; Mora, J.; Grindlay, G. Evaluation of different competitive immunoassays for aflatoxin M1 determination in milk samples by means of inductively coupled plasma mass spectrometry. *Anal. Chim. Acta* **2019**, *1049*, 10–19. [CrossRef] [PubMed]
34. Chen, J.; Ye, J.; Li, L.; Wu, Y.; Liu, H.; Xuan, Z.; Chen, M.; Wang, S. One-step automatic sample pretreatment for rapid, simple, sensitive, and efficient determination of aflatoxin M1 in milk by immunomagnetic beads coupled to liquid chromatography-tandem mass spectrometry. *Food Control* **2022**, *137*, 108927. [CrossRef]
35. Xuan, Z.; Wu, Y.; Liu, H.; Li, L.; Ye, J.; Wang, S. Copper Oxide Nanoparticle-Based Immunosensor for Zearalenone Analysis by Combining Automated Sample Pre-Processing and High-Throughput Terminal Detection. *Sensors* **2021**, *21*, 6538. [CrossRef]
36. Ye, J.; Xuan, Z.; Zhang, B.; Wu, Y.; Li, L.; Wang, S.; Xie, G.; Wang, S. Automated analysis of ochratoxin A in cereals and oil by immunoaffinity magnetic beads coupled to UPLC-FLD. *Food Control* **2019**, *104*, 57–62. [CrossRef]
37. Ye, J.; Wu, Y.; Guo, Q.; Lu, M.; Wang, S.; Xin, Y.; Xie, G.; Zhang, Y.; Mariappan, M.; Wang, S. Development and Interlaboratory Study of a Liquid Chromatography Tandem Mass Spectrometric Methodfor the Determination of Multiple Mycotoxins inCereals Using Stable Isotope Dilution. *J. AOAC Int.* **2018**, *101*, 667–676. [CrossRef]
38. Codex, S. *STAN 193-1995 General Standard for Contaminants and Toxins in Food and Feed*; FAO/WHO: Quebec City, QC, Canada, 1995.

 *toxins*

 MDPI

Article

Growth and Toxigenicity of *A. flavus* on Resistant and Susceptible Peanut Genotypes

Theophilus Kwabla Tengey [1,*], Frederick Kankam [2], Dominic Ngagmayan Ndela [2], Daniel Frempong [2] and William Ofori Appaw [3]

[1] Council for Scientific and Industrial Research-Savanna Agricultural Research Institute (CSIR-SARI), Nyankpala NL-1032-0471, Ghana
[2] Department of Crop Science, Faculty of Agriculture, Food and Consumer Sciences, University for Development Studies, Nyankpala NL-1029-6240, Ghana
[3] Department of Food Science and Technology, College of Science, Kwame Nkrumah University of Science and Technology, Kumasi AK-448-1125, Ghana
* Correspondence: tktengey@gmail.com; Tel.: +233-249-437-879

Abstract: Aflatoxin contamination poses serious health concerns to consumers of peanut and peanut products. This study aimed at investigating the response of peanuts to *Aspergillus flavus* infection and aflatoxin accumulation. Isolates of *A. flavus* were characterised either as aflatoxigenic or non-aflatoxigenic using multiple cultural techniques. The selected isolates were used in an in vitro seed colonisation (IVSC) experiment on two *A. flavus*-resistant and susceptible peanut genotypes. Disease incidence, severity, and aflatoxin accumulation were measured. Genotypes differed significantly ($p < 0.001$) in terms of the incidence and severity of aflatoxigenic and non-aflatoxigenic *A. flavus* infection with the non-aflatoxigenic isolate having significantly higher incidence and severity values. There was no accumulation of aflatoxins in peanut genotypes inoculated with non-aflatoxigenic isolate, indicating its potential as a biocontrol agent. Inoculations with the aflatoxigenic isolate resulted in the accumulation of aflatoxin B_1 and G_1 in all the peanut genotypes. Aflatoxin B_2 was not detected in ICGV–03401 (resistant genotype), while it was present and higher in Manipinta (susceptible genotype) than L027B (resistant genotype). ICGV–03401 can resist fungal infection and aflatoxin accumulation than L027B and Manipinta. Non-aflatoxigenic isolate detected in this study could further be investigated as a biocontrol agent.

Keywords: aflatoxigenic; in vitro seed colonisation; non-aflatoxigenic; host plant resistance and susceptible

Key Contribution: It was established that non-aflatoxigenic *A. flavus* isolates do not lead to aflatoxin production and these isolates grow faster than aflatoxigenic isolates, making them a good biocontrol against aflatoxin contamination under field conditions. Additionally, peanut genotypes with resistance to post-harvest aflatoxin accumulation will resist the growth of *A. flavus* and subsequent aflatoxin accumulation.

Citation: Tengey, T.K.; Kankam, F.; Ndela, D.N.; Frempong, D.; Appaw, W.O. Growth and Toxigenicity of *A. flavus* on Resistant and Susceptible Peanut Genotypes. *Toxins* **2022**, *14*, 536. https://doi.org/10.3390/toxins14080536

Received: 6 June 2022
Accepted: 18 July 2022
Published: 5 August 2022

Publisher's Note: MDPI stays neutral with regard to jurisdictional claims in published maps and institutional affiliations.

1. Introduction

Peanut (*Arachis hypogaea* L.) is an excellent source of plant nutrient mostly cultivated in tropical and sub-tropical regions [1–3]. Peanut originates from Latin America and was introduced in the 16th century into the African continent by the Portuguese [4,5]. Peanut production serves as a source of livelihood for farmers [6,7]. It is a source of vegetable oil (40%–60%), protein (20%–40%), carbohydrate content (10%–13%), numerous vitamins and minerals, haulm, and cake as food for humans and feed for livestock [1,5,8].

Peanut production is faced with challenges such as inadequate inputs, unreliable rains, the use of unimproved varieties, and disease and pest infestation [9]. In addition to the above, peanut is reported to be prone to *Aspergillus* species infection [10–12], which leads to a decline in its quality and quantity [13,14]. Among the *Aspergillus* species, *A. flavus* and *A. parasiticus* are the major contributors to aflatoxin production at the pre and postharvest stages, although *A. flavus* is reported to be the most prevalent in Ghana [14,15]. Aflatoxin contamination is a food safety concern due to its detrimental health impact and its resulting economic loss to peanut growers and dealers [16]. Aflatoxin contamination can also result in the reduction in the quality of grain and oilseed and depletion in their nutritional value [11].

Research discoveries have established that *A. flavus* is the most dominant and the most frequently isolated species from peanut seeds, products, and farmers' fields leading to aflatoxin contaminations [13,17–20]. However, due to gene mutations in the aflatoxin biosynthesis pathway, not all *A. flavus* isolates are aflatoxigenic [13]. The *Aspergillus section flavi* are grouped into aflatoxin-producing and non-aflatoxin producing species, which have similarities in their morphological characteristics [21,22].

In order to deploy and use host plant resistance, *A. flavus* and aflatoxin genetics must be well understood to develop effective methods of identifying genes of resistance and the availability of low-cost aflatoxin assays to correctly verify aflatoxin accumulation is needed [23].

Resistance to *A. flavus* infection and aflatoxin accumulation is quantitative in nature [24,25], and different mechanisms probably contribute to resistance in different peanut species, possibly under different environmental conditions [26]. Resistance may result from prevention of fungal infection on the peanut seed, prevention of growth of the fungus once the infection has occurred, inhibition of aflatoxin production following infection;, and degradation of aflatoxin by products or enzymes produced by the plant or by the fungus itself [27,28].

This study sought to culturally characterise *A. flavus* isolates based on their aflatoxigenicity and distinguish resistance mechanisms acting on resistant and susceptible peanut genotypes following inoculations with aflatoxigenic and non-aflatoxigenic *A. flavus* in vitro.

Measurements of fungal incidence, severity, and aflatoxin accumulations in the *A. flavus* resistant peanut genotypes (ICGV–03401 and L027B) and susceptible peanut genotype (Manipinta) will help to distinguish the different resistance mechanisms acting in them.

2. Results

2.1. Characterisation of A. Flavus into Aflatoxigenic and Non-Aflatoxigenic Isolates

A. flavus isolates cultured on PDA and YESA+β-cyclodextrin media were observed through UV light at 312 nm as shown in Figure 1.

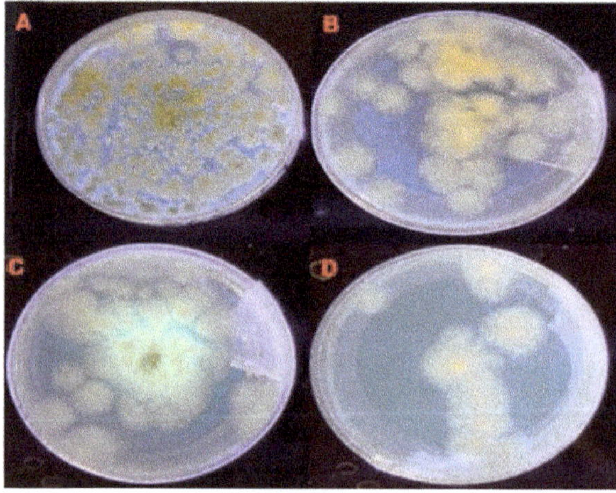

Figure 1. *A. flavus* colony observed under UV light at 312 nm. (**A**) *A. flavus* isolate grown on PDA fluoresces under UV light indicating aflatoxigenicity. (**B**) *A. flavus* isolate grown on PDA did not fluoresce under UV light indicating, non-aflatoxigenicity. (**C**) *A. flavus* isolate grown on YESA produced a blue fluorescence under UV light, indicating aflatoxigenicity and (**D**) *A. flavus* isolate grown on YESA did not produce a blue fluorescence under UV light, indicating non-aflatoxigenicity.

Additionally, the isolates of *A. flavus* on PDA in each Petri dish were exposed to 2 mL of concentrated Ammonia solution for 4 to 5 min (Figure 2).

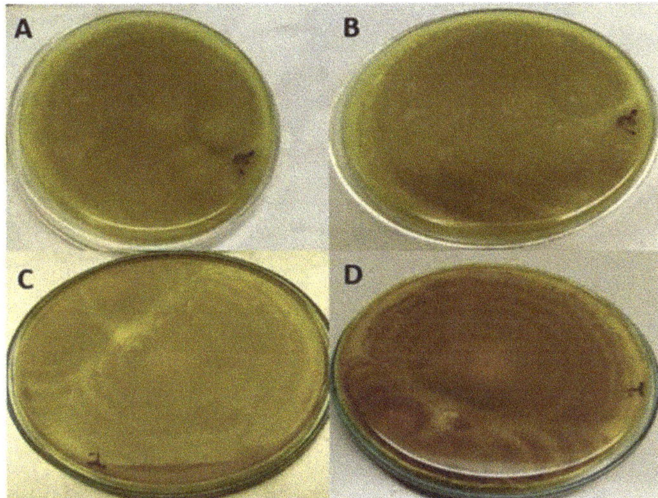

Figure 2. Colour of the underside of *A. flavus* colony on PDA when exposed to ammonia solution. (**A**) Non-aflatoxigenic *A. flavus* isolate before exposure and (**B**) non-aflatoxigenic isolate after exposure. (**C**) Aflatoxigenic isolate of *A. flavus* before exposure to Ammonia solution and (**D**) aflatoxigenic isolate after exposure to ammonia solution.

The results revealed that, on the PDA, 62.5% were characterised as aflatoxigenic whilst 37% were non-aflatoxigenic *A. flavus*. Similarly, on the YESA+β-cyclodextrin, 81.25% were aflatoxigenic while 18.75% were non-aflatoxigenic (Table 1). Additionally, it was observed that 13 out of 16 *A. flavus* isolates, which represents 81.25%, were aflatoxigenic while 3 isolates (18.75%) were non-aflatoxigenic when *A. flavus* culture plates were exposed to concentrated ammonia solution (Table 1).

Table 1. Screening aflatoxigenic and non-aflatoxigenic *A. flavus* using cultural methods.

Isolates Code	UV Fluorescence		Concentrated Ammonia Solution
	PDA	YESA+β-CD	PDA
AF01	+	+	+
AF02	+	+	+
AF03	+	−	+
AF04	+	+	+
AF05	+	+	+
AF06	+	+	+
AF07	+	+	+
AF08	+	+	+
AF09	−	+	+
AF10	−	+	+
AF11	−	+	−
AF12	−	+	−
AF13	−	−	−
AF14	−	−	+
AF15	+	+	+
AF16	+	+	+

AF—*Aspergillus flavus*.

2.2. Morphological Variation in the Growth of Aflatoxigenic and Non-Aflatoxigenic A. flavus

The results on the radial growth of aflatoxigenic isolate 8 (A8) and isolate 2 (A2), and non-aflatoxigenic *A. flavus* isolate 11 (NA11) and isolate 12 (NA12) revealed that *A. flavus* increased in diameter on the PDA as the days progressed, as shown in (Figure 3). The radial growth of the non-aflatoxigenic *A. flavus* isolates were higher than the aflatoxigenic *A. flavus* isolates. Additionally, the area under the disease progress curve (AUDPC) values of aflatoxigenic *A. flavus* isolates (A2 and A8) were significantly lower than AUDPC values for non-aflatoxigenic *A. flavus* isolates (NA11 and NA12) during the 8-day incubation period (Figure 4).

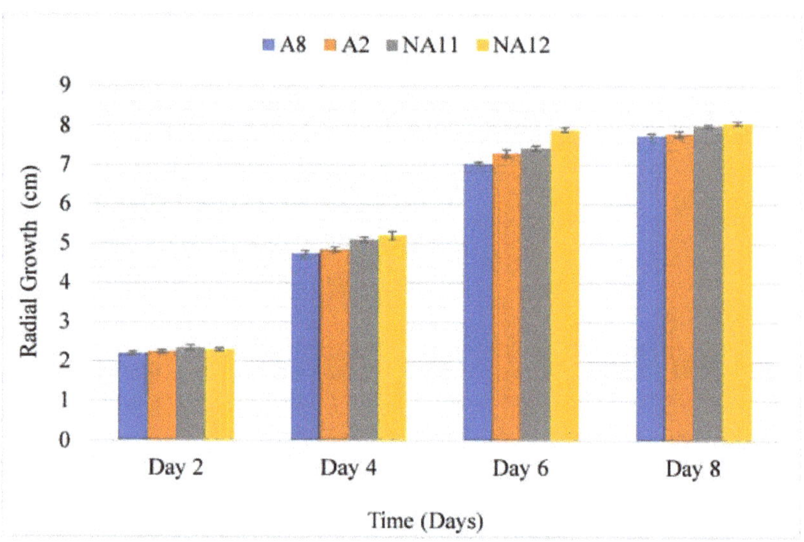

Figure 3. Radial growth of aflatoxigenic (A) and non-aflatoxigenic (NA) *A. flavus* on PDA. Error bars represent standard errors of means. A2—Aflatoxigenic isolate 2, A8—Aflatoxigenic isolate 8, NA11—Non-aflatoxigenic isolate 11, NA12—Non-aflatoxigenic isolate 12.

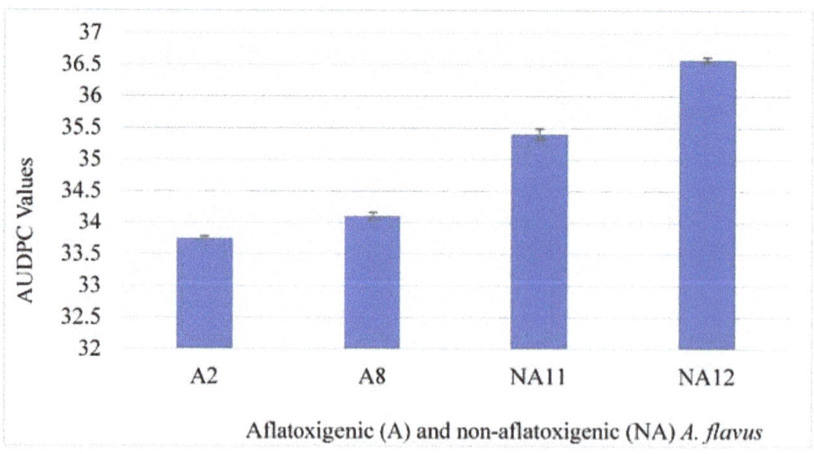

Figure 4. Area under disease progress curve (AUDPC) value for aflatoxigenic and non-aflatoxigenic *A. flavus* isolates on PDA. Error bars represent standard errors of means. A2—Aflatoxigenic isolate 2, A8—Aflatoxigenic isolate 8, NA11—Non-aflatoxigenic isolate 11, NA12—Non-aflatoxigenic isolate 12.

2.3. Incidence of Aflatoxigenic and Non-Aflatoxigenic A. flavus

There was highly significant ($p < 0.001$) variation in incidence of the three peanut genotypes (ICGV–03401, L027B, and Manipinta) inoculated with aflatoxigenic and non-aflatoxigenic *A. flavus*. The resistant genotypes (ICGV–03401 and L027B) had incidence values less than 40% for aflatoxigenic and non-aflatoxigenic *A. flavus* whereas the susceptible genotype (Manipinta) had the highest (78%–80%). However, the percentage incidence of non-aflatoxigenic *A. flavus* on all the genotypes was higher than the percentage incidence of aflatoxigenic *A. flavus* (Figure 5).

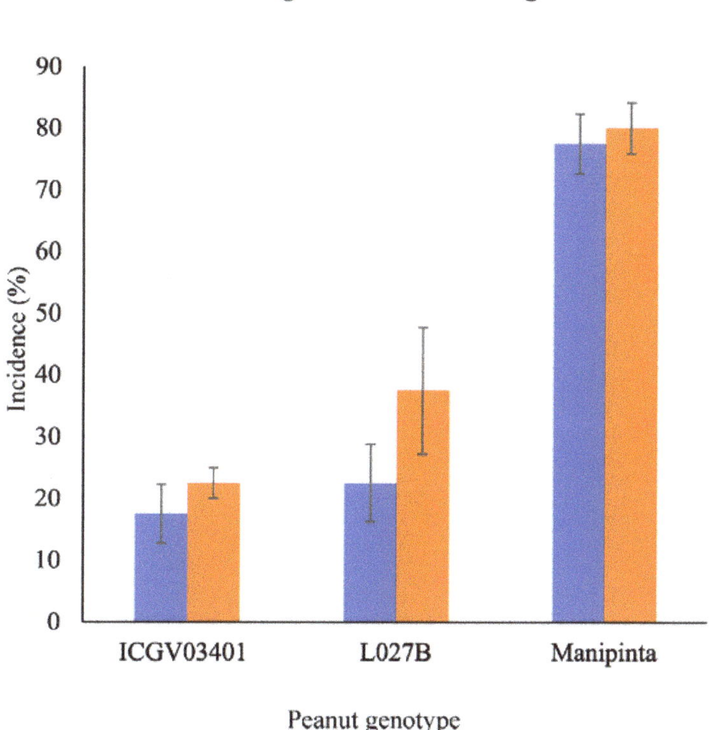

Figure 5. Incidence of aflatoxigenic and non-aflatoxigenic *A. flavus* on resistant peanut genotypes (ICGV–03401 and L027B) and susceptible peanut genotype (Manipinta). Error bars represent standard errors of means.

2.4. Severity of Aflatoxigenic and Non-Aflatoxigenic A. flavus

There were significant ($p < 0.001$) differences in severity of the three peanut genotypes inoculated with aflatoxigenic and non-aflatoxigenic *A. flavus*. The resistant genotypes (ICGV–03401 and L027B) had severities of less than 30% for aflatoxigenic and non-aflatoxigenic *A. flavus* whereas the susceptible genotype (Manipinta) had the highest (53%–63%). However, the incidence of non-aflatoxigenic *A. flavus* was higher in all the peanut genotypes than aflatoxigenic *A. flavus* (Figure 6).

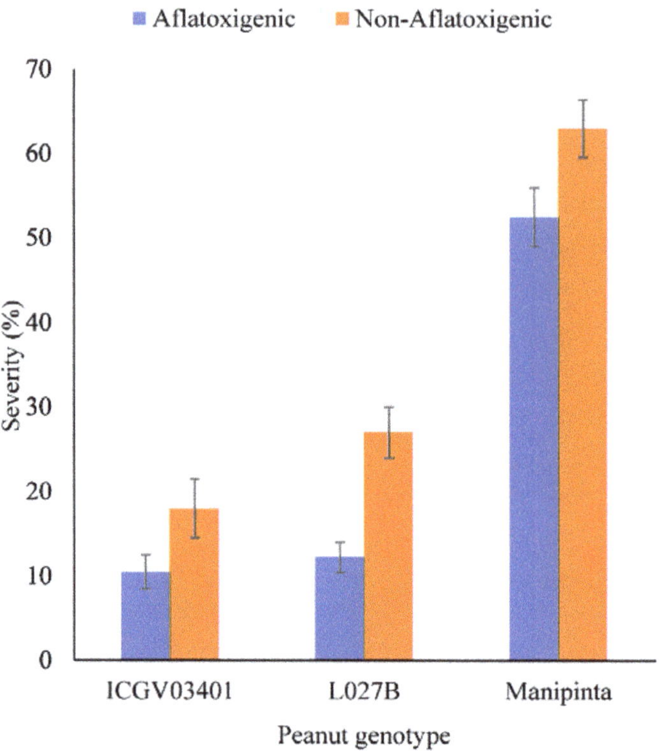

Figure 6. Severity of aflatoxigenic and non-aflatoxigenic *A. flavus* on resistant peanut genotypes (ICGV–03401 and L027B) and susceptible peanut genotype (Manipinta). Error bars represent standard errors of means.

2.5. Quantification of Aflatoxin

Extraction and estimation of aflatoxin accumulation in aflatoxigenic and non-aflatoxigenic isolates in peanut was quantified using HPLC (Figure 7) with aflatoxin standards (Figure 8). The chromatogram results revealed the peaks at retention times of about 3 min. Figure 7A,C confirmed the detection and accumulation of aflatoxin B_1, B_2, and G_1 in L027B and Manipinta, while only B_1 and G_1 was detected in ICGV−03402 Figure 7B following the inoculation of peanut seeds with the aflatoxigenic isolate A2. However, chromatogram in Figure 7D indicates non-detection of aflatoxins in peanut genotypes L027B, ICGV–03401, and Manipinta by non-aflatoxigenic isolate used in inoculation. Aflatoxin accumulation of the three peanut genotypes (ICGV–03401, L027B, and Manipinta) based on HPLC results is shown in Table 2.

Table 2. Concentration of accumulated aflatoxin following in vitro inoculation with the aflatoxigenic isolate A2 on peanut genotypes.

Genotype	Concentration of Aflatoxin (ppb)		
	AFB_1	AFB_2	G_1
ICGV–03401	0.86	Nil	0.76
L027B	2.10	0.85	0.40
Manipinta	0.61	1.09	0.65

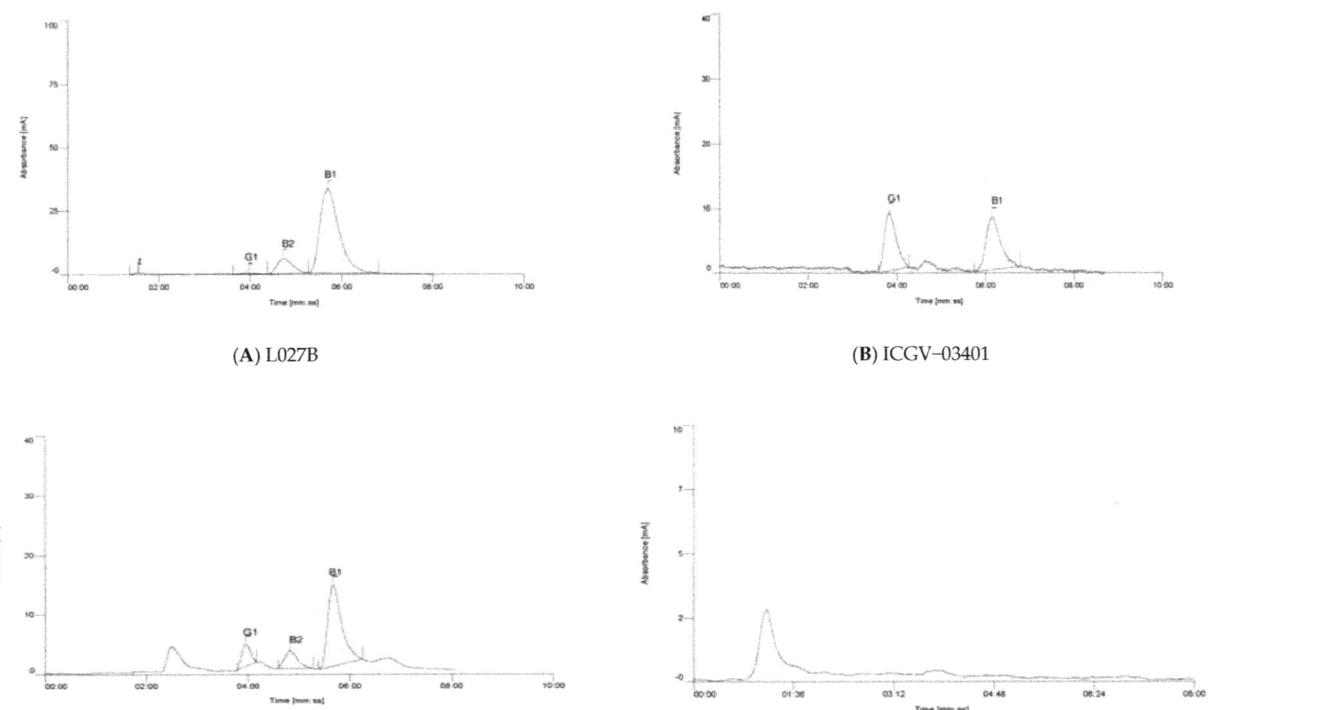

Figure 7. Chromatogram (**A**) shows the peaks of aflatoxin B_1, B_2, and G_1 in peanut genotype L027B, chromatogram (**B**) indicates the peaks of aflatoxin B_1 and G_1 in ICGV–03401, chromatogram (**C**) represents the peaks of aflatoxin B_1, B_2, and G_1 in Manipinta infested with aflatoxigenic isolate. The chromatogram (**D**) shows the non-detection of aflatoxin accumulation in peanut genotypes L027B, ICGV–03401, and Manipinta infested with non-aflatoxigenic isolate. Red line denotes the peaks, while black line refers to the baseline.

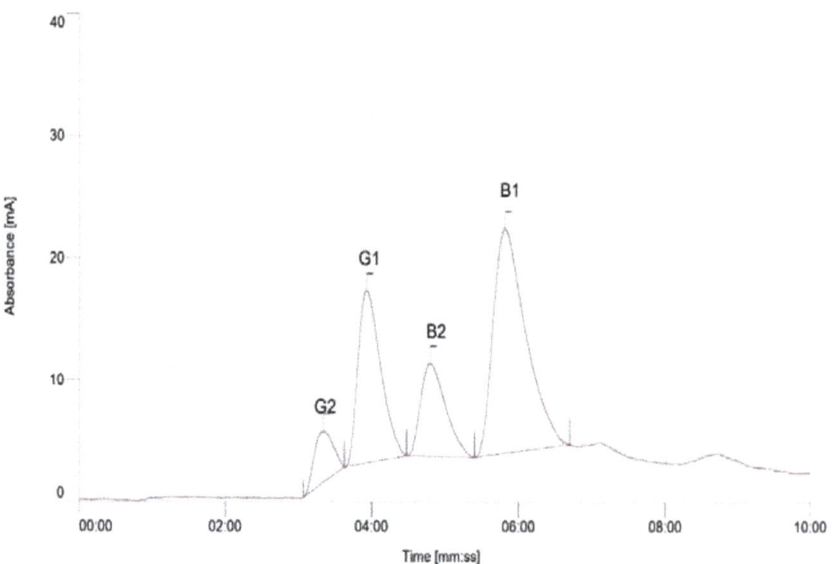

Figure 8. Chromatogram represents the aflatoxin standards used in the HPLC analysis. G_1—Aflatoxin G_1, G_2—Aflatoxin G_2, B_1—Aflatoxin B_1 and B_2—Aflatoxin B_2. Red line denotes the peaks, while black line refers to the baseline.

3. Discussion

Several interventions have been reviewed and recommended for aflatoxin management in peanuts; among these, host plant resistance have proven to be the most effective [29]. The use of resistant genotypes if available is the best option. Apart from being less cost-effective and easy to disseminate, it does not require any special expertise in its application and it is compatible with other control methods [29,30]. An understanding of the genetics and mechanisms of resistance to aflatoxin will help efficiently breed for resistance to this trait in peanut.

In order to elucidate the mechanisms of resistance in selected *A. flavus* resistant and susceptible peanut genotypes, it was hypothesised that (1) Non-aflatoxigenic *A. flavus* grows faster than aflatoxigenic *A. flavus* and could therefore limit the accumulation of aflatoxin and (2) *A. flavus* infection and colonisation is influenced by the genetic makeup of the peanut genotype and that subsequent aflatoxin production is dependent on the toxigenicity of the pathogen. These were investigated through multiple cultural methods of determining the aflatoxigenicity of the isolates followed by a radial growth bioassay to determine which of the isolates was fast growing. An in vitro seed colonisation (IVSC) assay was then conducted with the selected non-aflatoxigenic isolate and aflatoxigenic isolate, followed by detection and quantification of aflatoxins produced and accumulated in the infected tissues.

A. flavus was identified by its characteristic yellow green colour and with circular conidia which had a whitish colour around the edge [31].

The present study revealed that under the UV light observations, 81.25% of the isolates were detected by YESA+β-cyclodextrin medium as aflatoxigenic, whereas on PDA medium, 62.5% of the isolates were aflatoxigenic. A similar observation was made by Mahmoud et al. [32], who detected aflatoxigenic isolates under UV light using PDA amended with Sodium Chloride and YESA and the positive isolates observed were 20% and 53.33% respectively. This gives an indication that different culture media have different abilities in detecting aflatoxigenic and non-aflatoxigenic *A. flavus* isolates.

Again, by exposing the isolates to concentrated ammonia vapour, the study revealed that 13 (81.25%) out of the 16 isolates were observed to be aflatoxigenic whereas the remaining 3 (18.75%) isolates were non-aflatoxigenic *A. flavus*. Similar observations were made by Navya et al. [13], who identified 81.58% (31 out of 38 *A. flavus* isolates) to be aflatoxigenic *A. flavus* while the remaining 7 (18.42%) were non-aflatoxigenic in *A. flavus* culture plates exposed to ammonia vapour. This trend may mean majority of *A. flavus* isolates in most areas are aflatoxigenic as reported in previous studies [33–35]. The percentage of aflatoxigenic strains of *A. flavus* has been demonstrated to vary with the kind of substrate and environmental conditions, according to a study by Bharose et al. [19].

The AUDPC values and morphological variation in radial growth revealed that the non-aflatoxigenic *A. flavus* isolates grows faster than the aflatoxigenic isolates. This corroborates with reports that in nature the non-aflatoxigenic *A. flavus* grows faster than aflatoxigenic *A. flavus* since it does not require energy to produce aflatoxins and can invade and displace aflatoxigenic isolates [13,35].

The peanut seeds from the three genotypes reacted differently to aflatoxigenic and non-aflatoxigenic *A. flavus* infection as shown by their incidence and severity values. These differences could be associated with the differences in genetic makeup of the genotypes. ICGV–03401 and L027B had the lowest incidence and severity values, Manipinta had the highest. Among the two putative resistant lines, ICGV–03401 highly restricted fungal infection and colonisation, which could be attributed to the thickness and permeability of its seed coat [20,29,36] or seed coat biochemicals [37].

The selected aflatoxigenic and non-aflatoxigenic isolates infected and colonised the peanut genotypes with the non-aflatoxigenic isolate, resulting in the highest percentage incidence and severity. This means that the genotypes were not immune to neither aflatoxigenic nor non-aflatoxigenic *A. flavus* infection since both strains have an equal chance to survive and multiply. The genotypes also followed the same pattern (ICGV–03401 > L027B > Manipinta) in terms of their resistance to either aflatoxigenic or non-aflatoxigenic isolates. HPLC results, however, revealed that inoculation with a non-aflatoxigenic isolate did not result in the production and accumulation of aflatoxin but aflatoxins were detected in the genotypes inoculated with the aflatoxigenic strain. Resistance mechanisms to *A. flavus* infection are the same irrespective of the toxigenicity of the isolate. In terms of aflatoxin production, a pathogen factor (aflatoxigenic) and genotype factor (susceptibility) is involved. The toxigenicity of the isolate is very important in the production of aflatoxin. Peanut genotypes that exhibit strong resistance to any of these strains through IVSC may subsequently be resistant to aflatoxin production. IVSC could therefore be regarded as a quick way of screening for resistance to *A. flavus* infection and aflatoxin accumulation.

Aflatoxin types detected and identified in the three peanut genotypes were AFB_1, AFB_2, and AFG_1 with aflatoxin accumulation ranging from 0.4 ppb to 2.096 ppb. These values were obtained after 7 days incubation period following inoculation of a highly concentrated toxigenic strain. Increase in incubation period does not really influence aflatoxin concentration [38]; however, these values could increase with respect to favourable environmental factors such as temperature and water activity and storage methods, which makes the pathogen thrive. Most studies have a longer storage time and favourable environmental conditions associated with a higher aflatoxin accumulation [38–40].

The non-detection of AFB_2 in ICGV–03401 further boldens the resistance of this genotype, which could probably be due to the expression of genes that interfere with the aflatoxin biosynthetic pathway. This could mean apart from the concentration of specified aflatoxins, its absence could account for the resistance in a given genotype. Additionally, scenarios where non-aflatoxigenic *A. flavus* infection did not lead to aflatoxin production could be attributed to the presence of deleterious genes in the fungi that block the aflatoxin biosynthetic pathway. It is also possible that one or more aflatoxin biosynthetic genes may be absent, as reported by Commey et al. [37] when aflatoxin biosynthetic genes were compared between a non-aflatoxigenic *A. flavus* and aflatoxigenic *A. flavus* isolates. Non aflatoxigenic isolates have been used as biocontrol agents to reduce aflatoxin contamination in peanuts by 70%–100% in Ghana [41]. Similarly, Xu et al. [42] used non-aflatoxigenic strains to reduce aflatoxin B_1 accumulation in peanut kernels by 90%. In countries such as USA, Senegal, Nigeria, and Gambia, non-aflatoxigenic *A. flavus* are now serving as biocontrol agents to reduce aflatoxin production commercial in peanut production [43,44].

4. Conclusions

In conclusion, this study isolated and categorised *A. flavus* isolates into aflatoxin-producing isolates and non-aflatoxin-producing isolates from infected peanut genotypes using cultural techniques. Additionally, there were variations in the infection and colonisation of resistant and the susceptible peanut genotypes when inoculated with the aflatoxigenic and non-aflatoxigenic strains. The HPLC results confirmed that inoculations with the aflatoxigenic isolate resulted in the production and accumulation of aflatoxin B_1, B_2, and G_1, whereas inoculation with non-aflatoxigenic *A. flavus* did not produce any aflatoxin. The most resistant genotype lacked one of the detected aflatoxins under HPLC. Therefore, aflatoxigenic and non-aflatoxigenic strains all have equal ability to infect peanut seeds, but peanut genotypes demonstrated varied resistance to these isolates and varied aflatoxin accumulation. IVSC technology can be used as a quick method to screen for resistance to *A. flavus* infection and aflatoxin accumulation. The peanut genotypes ICGV–03401 and L027B could serve as sources of donors for resistance to *A. flavus* infection and aflatoxin accumulation. Future studies could also be carried out on the utility of non-aflatoxigenic isolates as a potential source of biocontrol agents.

5. Materials and Methods

5.1. Study Area

This experiment was carried out at the Plant pathology laboratory of the CSIR-Savanna Agricultural Research Institute, Nyankpala, Ghana.

5.2. Media Preparation for Fungal Isolation

5.2.1. Preparation of Potato Dextrose Agar (PDA)

Culturing of *A. flavus* was carried out using Potato Dextrose Agar (PDA) medium (Oxoid) according to the manufacturer's procedures. The prepared PDA was amended with chloramphenicol to inhibit the growth of bacteria and then dispensed into the sterilised disposable Petri dishes to solidify in an antiseptic environment [45].

5.2.2. Preparation of Yeast Extract Sucrose Agar (YESA)

A total amount of 40.50 g of Yeast extract sucrose agar made up of yeast extract of 4.0 g, sucrose of 20 g, KH_2PO_4 of 1.0 g, $MgSO_4$ of 0.5 g, and agar of 15 g was used and prepared according to the manufacture's procedures. It was amended with chloramphenicol to inhibit the growth of bacteria and beta-Cyclodextrin an aflatoxin inducing chemical respectively. This was dispensed into the sterilised Petri dishes in an antiseptic environment to solidify. Plates were exposed to Ultraviolet (UV) light to check for their fluorescence.

5.2.3. Source of Infected Peanuts for *A. flavus* Isolation

Infected peanut samples were obtained from a peanut processing centre at Nyankpala, Northern Region of Ghana. The infected seeds were collected and well packaged into a well labelled brown paper bag, and kept at room temperature until the time of isolation of *A. flavus*.

5.2.4. Isolation of *A. flavus* Isolates

With a modified isolation method by ref. [46], infected peanut seeds obtained were surface sterilised with 0.5% sodium hypochlorite solution for 1 min, rinsed in distilled water three times, and dried on tissue paper. Using sterilised forceps in the lamina flow hood cabinet, the surface-sterilised and dried seeds were placed onto the solidified PDA growth medium in the Petri dishes (10 seeds per plate) and incubated for 72 h at 27 °C.

The plates were visually observed and any visible mycelium on the seeds characterised by green-yellowish colouration was considered as the initial isolation criterion for *A. flavus*.

5.2.5. Obtaining Pure Cultures of *A. flavus*

Pure culture isolates were obtained by transferring mycelium with green-yellowish colour with sterilised inoculation needle from the previous plates and inoculating onto the centre of a fresh petri dish with PDA medium under aseptic environment in the lamina flow and incubating the plates for 10 days at 25 °C [46].

5.2.6. Identification of *A. flavus*

A. flavus was identified by observing morphological structures such as colony growth, conidia, texture, and colour [34,47] of the isolates obtained.

5.2.7. Detection of Aflatoxigenic and Non-Aflatoxigenic *A. flavus*
YESA Medium under UV Light

Pure *A. flavus* isolates were inoculated on YESA+β-cyclodextrin medium in three replicates for 7 days at room temperature. They were viewed under the UV light at 312 nm.

Exposure of Isolates on PDA to Ammonia Vapour

In 9 cm glass Petri plates, each isolate of *A. flavus* was inoculated in the centre of a solidified PDA medium and cultured for 7 to 8 days at 25 °C. The plates were subjected to 2 mL of concentrated Ammonia solution for 4 to 5 min and observed for change in colour of the under-side of colonies of aflatoxin-producing isolates. Those plates that changed colour from green-yellowish colour to orange-yellow pigmentation were classified as aflatoxigenic and those that did not change colour were noted as non-aflatoxigenic [48].

5.2.8. Radial Growth of Aflatoxigenic and Non-Aflatoxigenic Isolate on PDA

Four replicates of each isolate were plated on PDA and observed for their radial growth until the isolate with the fastest growth covers the entire plate. The area under the disease progress curve (AUDPC) was computed for each isolate growing on the PDA using the radial growth results for days 2, 4, 6, and 8 using the formula as described by Shaner and Finney (1977).

$$\text{AUDPC} = \sum_{i}^{n-1} \left(\frac{y_i + y_{i+1}}{2} \right) (t_{i+1} - t_i) \tag{1}$$

where n is the total number of observations, y_i is an assessment of a disease at the ith observation, and t_i is time at the ith observation.

5.2.9. Peanut Genotypes

Peanut genotypes less than 2 months old after harvest and free from any fungal contamination were used in this study. The three peanut genotypes, namely, ICGV–03401, L027B, and Manipinta, used in this study were obtained from the CSIR-Savanna Agricultural Research Institute, Nyankpala, Ghana. The ICGV–03401 have been reported to have resistance to *A. flavus* infection in vitro and similarly L027B have also been reported to have resistance to *A. flavus* infection in vitro (unpublished data). Manipinta is known to be susceptible to *A. flavus* infection in vitro.

5.3. In Vitro Seed Colonisation
5.3.1. Experimental Design

The experiment was laid out in a Completely Randomised Design (CRD) with three replications. The treatments were the three peanut genotypes and these were inoculated first with aflatoxigenic *A. flavus* in one experiment and in another experiment inoculated with non-aflatoxigenic *A. flavus*. Each plate served as a replicate with ten healthy peanut seeds.

5.3.2. Seed Sterilisation and Inoculation with Aflatoxigenic and Non-Aflatoxigenic Isolate

Preceding the inoculations, 30 peanut seeds with intact seed coats of each genotype were surface sterilised with 0.5% sodium hypochlorite and rinsed properly in sterile distilled water three times. The sterilised peanut seeds were placed in Petri dishes lined with moist Whatman No. 2 filter paper. Conidial suspensions were prepared from a 10-day-old culture of *A. flavus* isolate grown on PDA medium. Each seed was inoculated using 60 μL of suspension of *A. flavus* containing 1×10^6 spores per mL. They were incubated at room temperature for eight days [20,36].

After 8 days, the seeds were visually observed for seed infection by *A. flavus* by recording the percentage of seeds infested, which was shown by the presence of sporulating surface growth. Inoculations were performed using aflatoxigenic *A. flavus* and non-aflatoxigenic *A. flavus*.

5.3.3. Data Collection

Data were recorded on percent incidence and severity of aflatoxigenic and non-aflatoxigenic *A. flavus*. The incidence percent was calculated using the formulae below:

$$\text{Pathogen incidence} = \frac{\text{No. of seeds showing the } Aspergillus\ flavus\ \text{Colonisation}}{\text{Total number of seeds}} \times 100 \qquad (2)$$

Severity data was estimated using a modified scale of 0–5 [20]. The criterion for severity estimation is defined as follows: 0: non-infected peanut seeds (Highly resistant), 1: Less than 20% peanut seed surface covered (resistant), 2: 20% to 40% peanut seed surface covered (moderately resistant), 3: 40% to 60% peanut seed surface covered (susceptible), 4: 60% to 80% peanut seed surface covered (moderately susceptible), and 5: 80% to 100% peanut seed surface covered (Highly susceptible).

$$\text{Pathogen severity} = \frac{\text{Sum of pathogen rating}}{\text{Highest rating} \times \text{Total number of kernel rated}} \times 100 \qquad (3)$$

5.3.4. Detection of Aflatoxins Using High Performance Liquid Chromatography
Sample Extraction

Aflatoxin was extracted using methods described by ref. [49] with slight modifications of using acetonitrile: acetic acid v/v (9:1) as the extraction solution. Using a Preethi Mixer Grinder (Tamil Nadu, India), peanut samples were homogenised. A weight of 2 g of the homogenised sample was quantitatively transferred into a 15 mL centrifuge tube, 5 mL of distilled water was added, and the tube was vortexed for 1 min. After allowing the sample solution to stand for 5 min, 5 mL of the extraction solution was added. Using the Genie Vortex machine (New York, NY, USA), the resultant mixture was vortexed for 3 min. Next, 1.32 g of anhydrous $MgSO_4$ and 0.2 g of NaCl were added to the mixture and vortexed for 1 min. This was followed by centrifuging the tubes for 5 min at $3300 \times g$. The upper organic layer was then filtered through a 0.45 μm nylon syringe prior to injection. Finally, 50 μL of the filtered extract was injected into the high-performance liquid chromatographer (HPLC).

HPLC Analysis Technique

HPLC analysis was carried out based on AOAC Official Method 2005.08 (AOAC,2006) with Photochemical Reactor for Enhanced Detection (PHRED) for post-column derivatisation. A Cecil-Adept Binary Pump HPLC (Cambridge, UK) was joined with Shimadzu 10AxL fluorescence detector (Shimadzu Corporation, Tokyo, Japan) (Ex: 360 nm, Em: 440 nm) with Sunfire® C18 Column (150 × 4.60 mm, 5 um). The mobile phase used was methanol: water (40:60, v/v) at a flow rate of 1 mL/min with column temperature maintained at 40 °C. LCTech UVE (Dorfen, Germany) was used for post-colum photochemical derivatisation. The aflatoxin mix (G_1, G_2, B_1, B_2) standards (ng/g) of 5.02 ng/μL in acetonitrile from Romer Labs® was used for matrix-based calibration. Aflatoxins in the samples were detected by using the retentions of the standard solution run and quantification was conducted using the calibration of curves of each respective toxin. Quality assurance for established by checking for precision and trueness by spiking blank samples with aflatoxins standard (Table 3). Blank samples were run periodically confined to the absence of aflatoxins. The coefficient of variation obtained for replicates was less than 15%.

Table 3. Quality assurance for aflatoxin analysis.

Aflatoxins	LOD (ppb)	LOQ (ppb)	R^2	Recovery (%)
Aflatoxin B_1	0.2	0.4	0.999	98 ± 0.71
Aflatoxin B_2	0.1	0.2	0.999	98 ± 1.05
Aflatoxin G_1	0.2	0.4	0.999	99 ± 0.14
Aflatoxin G_2	0.1	0.2	0.995	99 ± 0.62

Aflatoxin calculation; Aflatoxin (ng/g) = A $\times \frac{T}{I} \times \frac{1}{W}$, where A = ng of aflatoxin as eluate injected, T = volume (μL) of eluate of the final test solution, I = volume eluate injected into LC (μL), W = mass (g) of commodity (final extract).

5.3.5. Data Analysis

Data were subjected to analysis of variance using GenStat (12th Edition) version 12.0.0.3033. Treatment means were separated using the Least Significance Difference (LSD) at 5% significant level.

Author Contributions: Conceptualisation, T.K.T.; methodology, D.N.N. and D.F.; validation, T.K.T., F.K. and D.N.N.; formal analysis, T.K.T., W.O.A. and D.N.N.; resources, T.K.T.; writing—original draft preparation, T.K.T. and D.N.N.; writing—review and editing, F.K. and T.K.T. and W.O.A.; supervision, F.K. and T.K.T.; funding acquisition, T.K.T. All authors have read and agreed to the published version of the manuscript.

Funding: This research was funded by USAID-NIFA, grant number 2021-67013-33943 and supported, in part, by the Bill & Melinda Gates Foundation [OPP1198373].

Institutional Review Board Statement: Not applicable.

Informed Consent Statement: Not applicable.

Data Availability Statement: Not applicable.

Acknowledgments: The authors thank the USDA-National Institute of Food and Agriculture (Award number 2021-67013-33943) awarded to Mark Burow. The CSIR-Savanna Agricultural Research Institute (SARI) and the Department of Crop Science, University for Development Studies are duly acknowledged for the partnership to ensure successful implementation of this work.

Conflicts of Interest: The authors declare no conflict of interest.

References

1. Raza, A.; Khan, Z.H.; Khan, K.; Anjum, M.M.; Ali, N.; Owais, M. Evaluation of Groundnut Varieties for the Agro-Ecological Zone of Malak and Division. *Int. J. Environ. Sci. Nat. Resour.* **2017**, *5*, 1–4. [CrossRef]
2. Onat, B.; Bakal, H.; Gulluoglu, L.; Arioglu, H. The effects of row spacing and plant density on yield and yield components of peanut grown as a double crop in Mediterranean environment in Turkey. *Turkish J. F. Crop.* **2017**, *22*, 71–80. [CrossRef]
3. Bakal, H.; Kenetli, A.; Arioglu, H. The effect of plant density on pod yield and some agronomic characteristics of different growthtype peanut varieties (*Arachis hypogaea* L.) grown as a main crop. *Turkish J. F. Crop.* **2020**, *25*, 92–99. [CrossRef]
4. Sylvanus, U.J. Effect of Defoliation on The Growth and Seed Yield of Four Groundnut (*Arachis Hypogaea* L.) Cultivars. Master's Thesis, Ahmadu Bello University, Faculty of Science, Department of Biological Sciences, Zaria, Nigeria, 2014.
5. Mwatawala, H.W.; Kyaruzi, P.P. An Exploration of Factors Affecting Groundnut Production in Central Tanzania: Empirical Evidence from Kongwa District, Dodoma Region. *Int. J. Progress. Sci. Technol.* **2019**, *14*, 122–130.
6. Danso-abbeam, G.; Dahamani, A.M.; Bawa, G.A.-S. Resource-use-efficiency among smallholder groundnut farmers in Northern Region, Ghana. *Am. J. Exp. Agric.* **2015**, *6*, 290–304. [CrossRef]
7. Tsigbey, F.; Brandenburg, R.L.; Clottey, V. Peanut Production Methods in Northern Ghana and Some Disease Perspectives. In *World Geography of the Peanut Knowledge Base Website*; 2003; Volume 9, pp. 33–38.
8. Jayaprakash, A.; Thanmalagan, R.R.; Roy, A.; Arunachalam, A.; Lakshmi, P. Strategies to understand *Aspergillus flavus* resistance mechanism in *Arachis hypogaea* L. Curr. *Plant Biol.* **2019**, *20*, 100123. [CrossRef]
9. Ajeigbe, H.A.; Waliyar, F.; Echekwu, C.A.; Kunihya, A.; Motagi, B.N.; Eniaiyeju, D.; Inuwa, A. *A Farmer's Guide to Profitable Groundnut Production in Nigeria*; International Crops Research Institute for the Semi-Arid Tropics (ICRISAT): Kano, Nigeria, 2015.

10. Guo, B.; Fedorova, N.D.; Chen, X.; Wan, C.-H.; Wang, W.; Nierman, W.C.; Bhatnagar, D.; Yu, J. Gene expression profiling and identification of resistance genes to *Aspergillus flavus* infection in peanut through EST and microarray strategies. *Toxins* **2011**, *3*, 737–753. [CrossRef] [PubMed]

11. Dube, M.; Maphosa, M. Prevalence of Aflatoxigenic *Aspergillus* spp and Groundnut Resistance in Zimbabwe. *J. Agric. Vet. Sci.* **2014**, *7*, 8–12. [CrossRef]

12. Korani, W.A.; Chu, Y.; Holbrook, C.; Clevenger, J.; Ozias-Akins, P. Genotypic regulation of aflatoxin accumulation but not *Aspergillus* fungal growth upon post-Harvest infection of peanut (*Arachis hypogaea* L.) seeds. *Toxins* **2017**, *9*, 218. [CrossRef]

13. Navya, H.M.; Hariprasad, P.; Naveen, J.; Chandranayaka, S.; Niranjana, S.R. Natural occurrence of aflatoxin, aflatoxigenic and non-aflatoxigenic *Aspergillus flavus* in groundnut seeds across India. *Afr. J. Biotechnol.* **2013**, *12*, 2587–2597. [CrossRef]

14. Bediako, K.A.; Ofori, K.; Offei, S.K.; Dzidzienyo, D.; Asibuo, J.Y.; Amoah, R.A. A flatoxin contamination of groundnut (*Arachis hypogaea* L.): Predisposing factors and management interventions. *Food Control* **2019**, *98*, 61–67. [CrossRef]

15. Bediako, K.A.; Dzidzienyo, D.; Ofori, K.; Offei, S.K.; Asibuo, J.Y.; Amoah, R.A.; Obeng, J. Prevalence of fungi and aflatoxin contamination in stored groundnut in Ghana. *Food Control* **2019**, *104*, 152–156. [CrossRef]

16. Pandey, M.K.; Kumar, R.; Pandey, A.K.; Soni, P.; Gangurde, S.S.; Sudini, H.K.; Fountain, J.C.; Liao, B.; Desmae, H.; Okori, P.; et al. Mitigating aflatoxin contamination in groundnut through a combination of genetic resistance and post-harvest management practices. *Toxins* **2019**, *11*, 315. [CrossRef]

17. Guchi, E.; Ayalew, A.; Dejene, M.; Ketema, M.; Asalf, B.; Fininsa, C. Occurrence of Aspergillus species in groundnut (*Arachis hypogaea* L.) along the value chain in different agro-ecological zones of Eastern Ethiopia. *J. Appl. Environ. Microbiol.* **2014**, *2*, 309–317. [CrossRef]

18. Mohammed, A.; Chala, A.; Dejene, M.; Fininsa, C.; Hoisington, D.A.; Sobolev, V.S.; Arias, R.S. *Aspergillus* and aflatoxin in groundnut (*Arachis hypogaea* L.) and groundnut cake in Eastern Ethiopia. *Food Addit. Contam.* **2016**, *9*, 290–298. [CrossRef]

19. Bharose, A.A.; Gajera, H.P.; Hirpara, D.G.; Kachhadia, V.H.; Golakiya, B.A. Morphological Credentials of Afla-Toxigenic and Non-Toxigenic *Aspergillus* Using Polyphasic Taxonomy. *Int. J. Curr. Microbiol. Appl. Sci.* **2017**, *6*, 2450–2465. [CrossRef]

20. Dieme, R.M.A.; Faye, I.; Zoclanclounon, Y.A.B.; Fonceka, D.; Ndoye, O.; Diedhiou, P.M. Identification of sources of resistance for peanut *Aspergillus flavus* colonization and aflatoxin contamination. *Int. J. Agron.* **2018**, *2018*, 1–7. [CrossRef]

21. Sudini, H.; Srilakshmi, P.; Kumar KV, K.; Njoroge, S.M.; Osiru, M.; Seetha, A.; Waliyar, F. Detection of aflatoxigenic *Aspergillus* strains by cultural and molecular methods: A critical review. *Afr. J. Agric. Res.* **2015**, *9*, 484–491.

22. El-Aziz, A.R.M.A.; Shehata, S.M.; Hisham, S.M.; Alobathani, A.A. Molecular profile of aflatoxigenic and non-aflatoxigenic isolates of *Aspergillus flavus* isolated from stored maize. *Saudi J. Biol. Sci.* **2021**, *28*, 1383–1391. [CrossRef]

23. Mahuku, G.; Warburton, M.L.; Makumbi, D.; Vicente, F.S. Managing aflatoxin contamination of maize: Developing host resistance. Aflatoxins Finding Solutions to Improve Food Safety 2020 Vision. 2013, p. 62. Available online: http://cdm15738.contentdm.oclc. org/utils/getfile/collection/p15738coll2/id/127887/filename/128098.pdf (accessed on 2 May 2022).

24. Widstrom, N.; Butron, A.; Guo, B.; Wilson, D.; Snook, M.; Cleveland, T.; Lynch, R. Control of preharvest aflatoxin contamination in maize by pyramiding QTL involved in resistance to ear-feeding insects and invasion by *Aspergillus* spp. *Eur. J. Agron.* **2003**, *19*, 563–572. [CrossRef]

25. Warburton, M.L.; Williams, W.P. Aflatoxin resistance in maize: What have we learned lately? *Adv. Bot.* **2014**, *2014*, 1–10. [CrossRef]

26. Guo, B.; Chen, Z.Y.; Lee, R.D.; Scully, B.T. Drought stress and preharvest aflatoxin contamination in agricultural commodity: Genetics, genomics and proteomics. *J. Integr. Plant Biol.* **2008**, *50*, 1281–1291. [CrossRef]

27. Brown, R.L.; Chen, Z.Y.; Cleveland, T.E.; Russin, J.S. Advances in the development of host resistance in corn to aflatoxin contamination by *Aspergillus flavus*. *Phytopathology* **1999**, *89*, 113–117. [CrossRef]

28. Williams, W.; Krakowsky, M.; Scully, B.; Brown, R.; Menkir, A.; Warburton, M.; Windham, G. Identifying and developing maize germplasm with resistance to accumulation of aflatoxins. *World Mycotoxin J.* **2015**, *8*, 193–209. [CrossRef]

29. Soni, P.; Gangurde, S.S.; Ortega-Beltran, A.; Kumar, R.; Parmar, S.; Sudini, H.K.; Lei, Y.; Ni, X.; Huai, D.; Fountain, J.C.; et al. Functional biology and molecular mechanisms of host-pathogen interactions for aflatoxin contamination in groundnut (*Arachis hypogaea* L.) and maize (*Zea mays* L.). *Front. Microbiol.* **2020**, *11*, 227. [CrossRef]

30. Garrido-Bazan, V.; Mahuku, G.; Bibbins-Martinez, M.; Arroyo-Bacerra, A.; Villalobos-López, M.Á. Dissection of mechanisms of resistance to *Aspergillus flavus* and aflatoxin using tropical maize germplasm. *World Mycotoxin J.* **2018**, *11*, 215–224. [CrossRef]

31. Khan, R.; Ghazali, F.M.; Mahyudin, N.A.; Samsudin, N.I.P. Morphological Characterization and Determination of Aflatoxigenic and Non-Aflatoxigenic *Aspergillus flavus* Isolated from Sweet Corn Kernels and Soil in Malaysia. *Agriculture* **2020**, *10*, 450. [CrossRef]

32. Mahmoud, M.A.; Ali, H.M.; El-Aziz, A.R.M.; Al-Othman, M.R.; Al-Wadai, A.S. Molecular characterization of aflatoxigenic and non-aflatoxigenic *Aspergillus flavus* isolates collected from corn grains. *Genet. Mol. Res.* **2014**, *13*, 9352–9370. [CrossRef]

33. Guezlane-Tebibel, N.; Bouras, N.; Mokrane, S.; Benayad, T.; Mathieu, F. Aflatoxigenic strains of *Aspergillus* section Flavi isolated from marketed peanuts (*Arachis hypogaea*) in Algiers (Algeria). *Ann. Microbiol.* **2013**, *63*, 295–305. [CrossRef]

34. Okayo, R.O.; Andika, D.O.; Dida, M.M.; K'Otuto, G.O.; Gichimu, B.M. Morphological and Molecular Characterization of Toxigenic *Aspergillus flavus* from Groundnut Kernels in Kenya. *Int. J. Microbiol.* **2020**, *2020*, 1–10. [CrossRef] [PubMed]

35. Ehrlich, K.C.; Wei, Q.; Brown, R.L.; Bhatnagar, D. Inverse correlation of ability to produce aflatoxin and *Aspergillus* colonization of maize seed. *Food Nutr. Sci.* **2011**, *2*, 486–489. [CrossRef]

36. Kasno, A.; Trustinah, T.; Purnomo, J.; Sumartini, S. Seed coat resistance of groundnut to *Aspergillus flavus* and their stability performance in the field. *AGRIVITA J. Agric. Sci.* **2011**, *33*, 53–62. [CrossRef]
37. Commey, L.; Tengey, T.K.; Cobos, C.J.; Dampanaboina, L.; Dhillon, K.K.; Pandey, M.K.; Sudini, H.K.; Falalou, H.; Varshney, R.K.; Burow, M.D.; et al. Peanut seed coat acts as a physical and biochemical barrier against *Aspergillus flavus* infection. *J. Fungi* **2021**, *7*, 1000. [CrossRef]
38. Yeboah, A.; Ahiakpa, K.; Adjei-Nsiah, S. Aflatoxin levels in seeds of commonly grown groundnut varieties (*Arachis hypogaea* L.) in Ghana as influenced by storage method. *Afr. J. Food Agric. Nutr. Dev.* **2020**, *20*, 15402–15414. [CrossRef]
39. Astoreca, A.; Vaamonde, G.; Dalcero, A.; Marin, S.; Ramos, A. Abiotic factors and their interactions influence on the co-production of aflatoxin B1 and cyclopiazonic acid by *Aspergillus flavus* isolated from corn. *Food Microbiol.* **2014**, *38*, 276–283. [CrossRef] [PubMed]
40. Lahouar, A.; Marin, S.; Crespo-Sempere, A.; Saïd, S.; Sanchis, V. Effects of temperature, water activity and incubation time on fungal growth and aflatoxin B1 production by toxinogenic *Aspergillus flavus* isolates on sorghum seeds. *Rev. Argent. Microbiol.* **2016**, *48*, 78–85. [CrossRef] [PubMed]
41. Agbetiameh, D.; Ortega-Beltran, A.; Awuah, R.T.; Atehnkeng, J.; Islam, M.S.; Callicott, K.A.; Bandyopadhyay, R. Potential of Atoxigenic *Aspergillus flavus* Vegetative Compatibility Groups Associated With Maize and Groundnut in Ghana as Biocontrol Agents for Aflatoxin Management. *Front. Microbiol.* **2019**, *10*, 1–15. [CrossRef]
42. Xu, J.; Wang, P.; Zhou, Z.; Cotty, P.J.; Kong, Q. Selection of Atoxigenic *Aspergillus flavus* for Potential Use in Aflatoxin Prevention in Shandong Province, China. *J. Fungi* **2021**, *7*, 773. [CrossRef] [PubMed]
43. Senghor, L.A.; Ortega-Beltran, A.; Atehnkeng, J.; Callicott, K.A.; Cotty, P.J.; Bandyopadhyay, R. The atoxigenic biocontrol product Aflasafe SN01 is a valuable tool to mitigate aflatoxin contamination of both maize and groundnut cultivated in Senegal. *Plant Dis.* **2020**, *104*, 510–520. [CrossRef]
44. Bandyopadhyay, R.; Ortega-Beltran, A.; Akande, A.; Mutegi, C.; Atehnkeng, J.; Kaptoge, L.; Senghor, A.; Adhikari, B.; Cotty, P. Biological control of aflatoxins in Africa: Current status and potential challenges in the face of climate change. *World Mycotoxin J.* **2016**, *9*, 771–789. [CrossRef]
45. Raed, N.K.A. Cultural and Molecular Detection of Aflatoxigenic Activity in Aspergillus flavus Isolated from Poultry Feed. Master's Thesis, University of Basrah-College of Veterinary Medicine, Basra, Iraq, 2016.
46. Abdi, M.; Alemayehu, C. Incidence of *Aspergillus* contamination of groundnut (*Arachis hypogaea* L.) in Eastern Ethiopia. *Afr. J. Microbiol. Res.* **2014**, *8*, 759–765. [CrossRef]
47. Klich, M.A. *Identification of Common Aspergillus Species*; Centraalbureau voor Schimmelcultures (CBS) Fungal Biodiversity Centre, Royal Netherlands Academy of Arts and Sciences: Utrecht, The Netherlands, 2002.
48. Saito, M.; Machida, S. A rapid identification method for aflatoxin-producing strains of *Aspergillus flavus* and *A. parasiticus* by ammonia vapor. *Mycoscience* **1999**, *40*, 205–208. [CrossRef]
49. Sirhan, A.Y.; Tan, G.H.; Al-Shunnaq, A.; Abdulra'uf, L.; Wong, R.C.S. QuEChERS-HPLC method for aflatoxin detection of domestic and imported food in Jordan. *J. Liq. Chromatogr. Relat. Technol.* **2014**, *37*, 321–342. [CrossRef]

Article

Elimination of Deoxynivalenol, Aflatoxin B1, and Zearalenone by Gram-Positive Microbes (*Firmicutes*)

Cintia Adácsi [1], Szilvia Kovács [2], István Pócsi [3] and Tünde Pusztahelyi [2,*]

[1] Doctoral School of Nutrition and Food Sciences, University of Debrecen, Böszörményi Str. 138, H-4032 Debrecen, Hungary

[2] Central Laboratory of Agricultural and Food Products, Faculty of Agricultural and Food Sciences and Environmental Management, University of Debrecen, Böszörményi Str. 138, H-4032 Debrecen, Hungary

[3] Department of Molecular Biotechnology and Microbiology, Institute of Biotechnology, Faculty of Science and Technology, University of Debrecen, Egyetem Tér 1, H-4032 Debrecen, Hungary

* Correspondence: pusztahelyi@agr.unideb.hu; Tel.: +36-20-210-9491

Abstract: Mycotoxin contaminations in the feed and food chain are common. Either directly or indirectly, mycotoxins enter the human body through the consumption of food of plant and animal origin. Bacteria with a high mycotoxin elimination capability can reduce mycotoxin contamination in feed and food. Four Gram-positive endospore-forming bacteria (*Bacillus thuringiensis* AMK10/1, *Lysinibacillus boronitolerans* AMK9/1, *Lysinibacillus fusiformis* AMK10/2, and *Rummeliibacillus suwonensis* AMK9/2) were isolated from fermented forages and tested for their deoxynivalenol (DON), aflatoxin B1 (AFB1), and zearalenone (ZEA) elimination potentials. Notably, the contribution of bacterial cell wall fractions to the observed outstanding ZEA elimination rates was demonstrated; however, the ZEA elimination differed considerably within the tested group of Gram-positive bacteria. It is worth noting that the purified cell wall of *L. boronitolerans* AMK9/1, *L. fusiformis* AMK10/2 and *B. thuringiensis* AMK10/1 were highly efficient in eliminating ZEA and the teichoic acid fractions of *B. thuringiensis* AMK10/1, and *L. fusiformis* AMK10/2 could also be successfully used in ZEA binding. The ZEA elimination capacity of viable *R. suwonensis* AMK9/2 cells was outstanding (40%). Meanwhile, *R. suwonensis* AMK9/2 and *L. boronitolerans* AMK9/1 cells produced significant esterase activities, and ZEA elimination of the cell wall fractions of that species did not correlate with esterase activity. DON and AFB1 binding capabilities of the tested bacterial cells and their cell wall fractions were low, except for *B. thuringiensis* AMK10/1, where the observed high 64% AFB1 elimination could be linked to the surface layer (S-layer) fraction of the cell wall.

Keywords: mycotoxins; cell wall; peptidoglycan; S-layer; zearalenone; elimination; esterase

Key Contribution: Mycotoxin elimination by microbes, bacterial cell fractions, and enzymes is highly desired. For bioadsorption, besides peptidoglycan, chemical structures in teichoic acid and S-layer fractions were successful in mycotoxin adsorption. Mycotoxin elimination-related studies were done on *R. suwonensis* and *L. boronitolerans* species for the first time. Purified cell wall fractions of *L. boronitolerans* AMK9/1, *L. fusiformis* AMK10/2 and *B. thuringiensis* AMK10/1 are valuable tools in eliminating ZEA, and the S-layer fraction of *B. thuringiensis* AMK10/1 was applicable in AFB1 binding and available for further cell-free bioadsorption system development.

Citation: Adácsi, C.; Kovács, S.; Pócsi, I.; Pusztahelyi, T. Elimination of Deoxynivalenol, Aflatoxin B1, and Zearalenone by Gram-Positive Microbes (*Firmicutes*). *Toxins* **2022**, *14*, 591. https://doi.org/10.3390/toxins14090591

Received: 22 July 2022
Accepted: 24 August 2022
Published: 27 August 2022

Publisher's Note: MDPI stays neutral with regard to jurisdictional claims in published maps and institutional affiliations.

1. Introduction

Mycotoxin production by molds is a global problem that cannot be solved easily in the field, even with the introduction of highly efficient proper agricultural practices. Environmental conditions affected by climate change and improper storage conditions may facilitate mold growth and secondary metabolite production [1,2]. These fungi can infect many plants and a wide array of food and agricultural products, and they can also produce a broad spectrum of harmful secondary metabolites, such as carcinogenic aflatoxins (AFs) [3,4]. Another dangerous mycotoxin, zearalenone (ZEA), is produced mainly by *Fusarium* species [5], and due to its estrogenic character, ZEA and its derivatives may cause severe reproductive and sexual dysfunctions [6–8]. The co-occurrence of ZEA with another *Fusarium* mycotoxin deoxynivalenol (DON) is common [9,10], and DON is a potent inhibitor of the protein synthesis in *Eukarya* [11], causing nausea, vomiting, diarrhoea, or anorexia in animals [12].

Not surprisingly, removing these harmful mycotoxins from feed and food is a crucial issue worldwide. In addition to the extensive range of available physical and chemical mycotoxin decontamination methods, biological technologies often based on food-grade microorganisms are gaining ground. The tested biological detoxification methodologies typically incorporate biodegradation and biosorption processes [13–16]. Mycotoxin elimination by bacteria and yeasts can happen through binding to biopolymers (e.g., peptidoglycan, PG), degradation by enzymes (e.g., esterase), or enzymatic conjugation to various molecules, giving rise to so-called masked or bound mycotoxins that make the original mycotoxin forms hidden to classic chemical analytical methods [17].

Bacteria are distinguished tools in the biological control of mycotoxigenic molds and the mycotoxins themselves. For example, bacterial cells can adsorb mycotoxins, e.g., aflatoxins are bound effectively by various cell wall components [18], or can detoxify these contaminants via enzymic degradation [15,19], or can inhibit the growth of molds by different antifungal metabolites, e.g., bacteriocins and organic acids [20].

Gram-positive lactic acid bacteria (LAB) typically have an excellent mycotoxin elimination capability, and both viable and nonviable LAB can eliminate aflatoxin B1 (AFB1). However, the elimination depends on the genus, pH, and bacterial density [21]. The binding of some major mycotoxins, including AFB1 [22], ZEA [23], and certain trichothecenes [24] by some probiotic LAB, was also demonstrated in vitro. Furthermore, some non-lactic acid bacteria, including *Bacillus* spp., were also investigated for their possible biotechnological application in mycotoxin elimination [18].

The bacterial cell wall consists of several biopolymers that can absorb different toxic substances, such as mycotoxins. Therefore, differences between Gram-negative and Gram-positive bacterial cell walls (Figure 1) would allow for predicting different adsorption capabilities in mycotoxin elimination due to the apparent differences in the thickness of the PG layer and the presence of the outer membrane. The surface layer, which is the outermost cell wall layer in many *Bacteria* and *Archaea* cells, attaches to the PG by electrostatic interactions and possesses inherent, entropy-driven affinities to self-assemble with each other [25] or, in the Gram-negative bacterial cell wall, to attach to outer membrane lipopolysaccharides (LPS) [26].

In this work, bacteria and bacterial cell wall preparations were screened for mycotoxin elimination using a unique collection of endospore-forming Gram-positive bacteria isolated from forages in Hungary. Novel mycotoxin-eliminating bacteria were aimed to find and shed light on cell wall fractions and other cell constituents, such as enzymes, which have the most critical role in mycotoxin elimination.

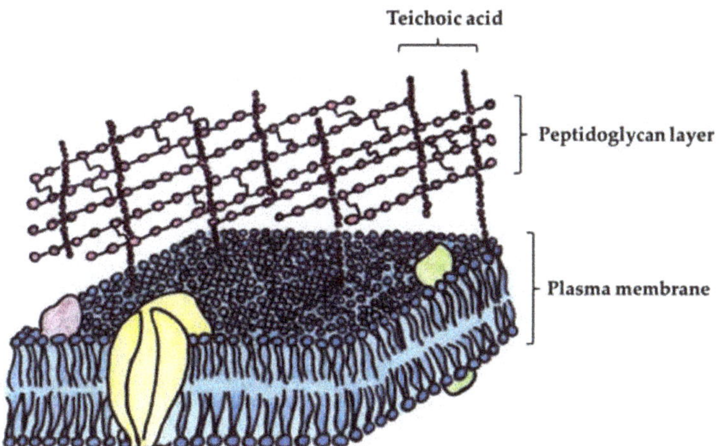

Figure 1. Main structural elements of the Gram-positive cell wall. The schematic figure did not show S-layer proteins present in the outermost part of the cell wall. Besides the main structural similarities, at the microstructural level, microbial cell walls are diverse and, therefore, highly differ in mycotoxin elimination capabilities.

2. Results

2.1. Identification of Isolated Bacteria

Bacterial isolates from fermented forages were identified with higher than 97% homology using 16S rRNA gene sequences (Table 1). All bacteria were Gram-positive endospore-forming organisms and taxonomically belonged to the phylum *Firmicutes*. Except for *Rummeliibacillus suwonensis* (*Planococcaceae*), all isolates were placed in the family *Bacillaceae*.

Table 1. Bacterium identification was based on 16S rRNA gene sequences. The sequences were submitted to the National Library of Medicine at National Center for Biotechnological Information (NCBI) under accession numbers OP183257-OP183263.

Description	Strain	Primer	Query Length	Homology	References
Rummeliibacillus suwonensis	AMK9/2	27F	1214	97.42%	[27]
	AMK9/2	1492R	1148	99.29%	
Bacillus thuringiensis	AMK10/1	1492R	1152	99.20%	[28]
Lysinibacillus boronitolerans	AMK9/1	27F	1165	98.37%	[29]
	AMK9/1	1492R	1156	98.93%	
Lysinibacillus fusiformis	AMK10/2	27F	1191	97.52%	[29]
	AMK10/2	1492R	1153	97.85%	

Source and homology search: NCBI database [30].

2.2. Mycotoxin Elimination

Viable cells of *R. suwonensis* AMK9/2, *L. boronitolerans* AMK9/1, *L. fusiformis* AMK10/2, and *B. thuringiensis* AMK10/1, as well as their cell wall fractions, were supplemented with zearalenone (ZEA), aflatoxin B1 (AFB1), and deoxynivalenol (DON), respectively, to study the mycotoxin elimination capability.

All samples showed negligible DON elimination under the tested experimental conditions. Furthermore, AFB1 elimination rates were low and typically under 20% by all tested bacteria and their cell wall fractions except the S-layer fraction of *B. thuringiensis* AMK10/1, which eliminated 64% of AFB1 (Figure 2).

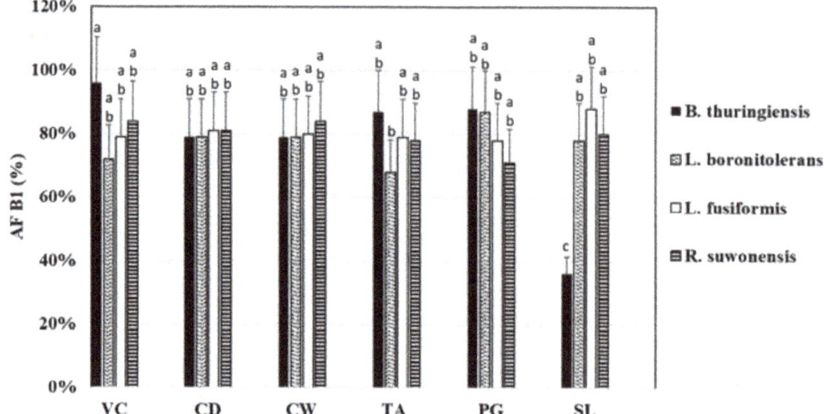

Figure 2. Aflatoxin B1 (AFB1) measured from the supernatants after elimination test with viable cells (*Bacillus thuringiensis* AMK10/1, *Lysinibacillus boronitolerans* AMK9/1, *Lysinibacillus fusiformis* AMK10/2, and *Rummeliibacillus suwonensis* AMK9/2) and their cell wall fractions. Viable cells (VC), cell debris (CD), purified cell wall (CW), teichoic acid fraction (TA), peptidoglycan fraction (PG), and S-layer fraction (SL) were tested. The letters above each column indicate the results of pairwise comparisons of all samples. Results that share the same letter do not differ significantly ($p \leq 0.05$) from one another in AFB1 elimination.

The elimination of ZEA by the bacteria and their cell wall preparations showed remarkable species-specific differences (Figure 3). In many cases, bacterial cells and their cell wall preparations showed remarkable ZEA elimination rates (Figure 3A), but the eliminated ZEA could only be partially recovered from the bacteria and cell wall pellets (Figure 3B). In addition, for some bacteria and cell wall preparations, the quantity of the recovered ZEA remained even below the detection limit (LOD, 2.6 µg/L).

Interestingly, ZEA elimination by viable *R. suwonensis* AMK9/2 cultures and its cell wall preparations showed approximately the same values: about 40% (Figure 3A). Significantly, purified cell wall, PG, and S-layer protein fractions of *L. boronitolerans* AMK 9/1 nearly eliminated ZEA, and the remaining ZEA concentrations in the supernatants were below LOD. ZEA concentrations recovered from the various cell wall fraction pellets were 1–12% of the starting values (cell wall fraction: 1%, PG: 12%, S-layer proteins: 6%) (Figure 3B).

L. fusiformis AMK10/2 cell debris, purified cell wall, PG fraction, and S-layer protein fraction were valuable for eliminating ZEA (Figure 3A). However, only small portions of the eliminated ZEA could be recovered from the fractions, except for the S-layer fraction of AMK10/2, where 25% of the original ZEA could be extracted (Figure 3B).

Considering *B. thuringiensis* AMK10/1, viable cells, cell debris, purified cell wall, and teichoic acid fractions were suitable for efficiently eliminating ZEA from the liquid phase (Figure 3A). Interestingly, the elimination of ZEA by PG fraction of the AMK10/1 strain was almost negligible, while, for the S-layer fraction, it was about 38%. Meanwhile, the teichoic acid fraction was the most important cell wall fraction for ZEA elimination (Figure 3A). Nevertheless, AMK10/1 cells and cell wall fractions released ZEA under extraction (Figure 3B) and the most significant release was achieved from the total cells.

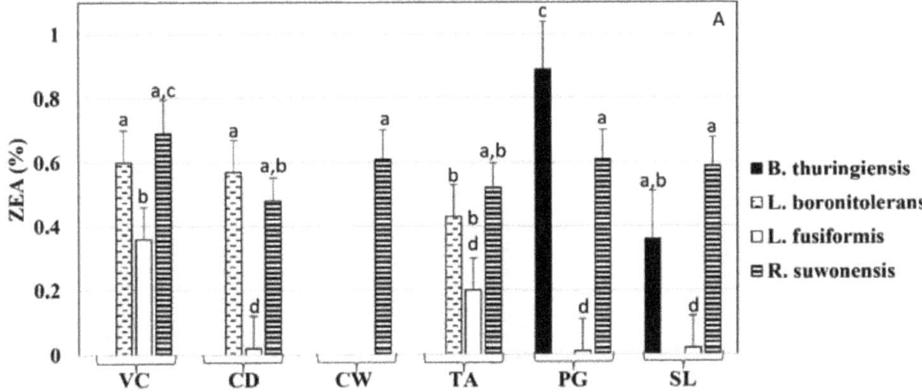

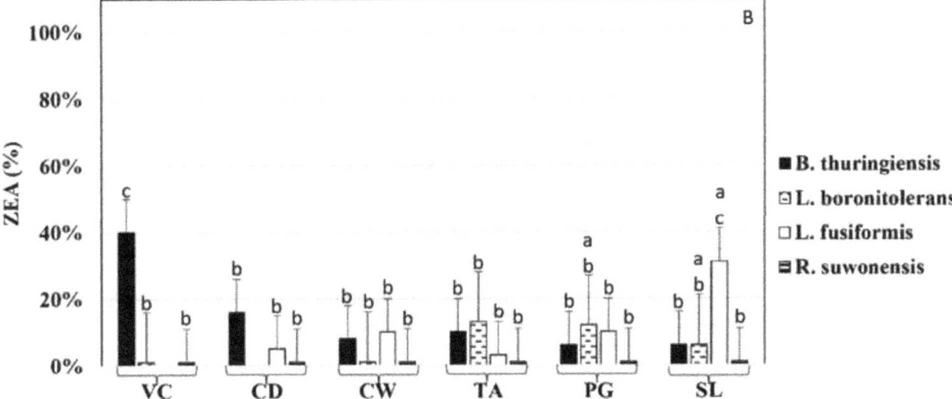

Figure 3. Zearalenone (ZEA) was measured from the supernatant (**A**) and the extract of pellets (**B**) after the elimination test with viable cells (*Bacillus thuringiensis* AMK10/1, *Lysinibacillus boronitolerans* AMK9/1, *Lysinibacillus fusiformis* AMK10/2, and *Rummeliibacillus suwonensis* AMK9/2) and their cell wall fractions. Viable cells (VC), cell debris (CD), purified cell wall (CW), teichoic acid fraction (TA), peptidoglycan fraction (PG), and S-layer fraction (SL) were tested. The letters above each column indicate the results of pairwise comparisons of all samples. Bars that share the same letter do not differ significantly ($p \leq 0.05$) from one another in ZEA content.

2.3. Esterase Activity

Since esterase activity is connected to ZEA degradation in living bacterial cells [19], this enzyme activity was measured and varied in a wide range. *R. suwonensis* AMK9/2 and *L. boronitolerans* AMK9/1 produced high esterase activities compared to *L. fusiformis* AMK10/2 and *B. thuringiensis* AMK10/1, the highest enzyme activity was measured in *R. suwonensis* AMK9/2 cell debris (1.98 ± 0.3 mM *p*-nitrophenol released/min) (Figure 4). Correlation analysis revealed a correlation coefficient of 0.618 between enzyme activity and extracted ZEA from cell debris and −0.748 between enzyme activity and the ZEA remaining in the cells' supernatant. AMK 9/2 cells and AMK 9/2 cell wall fractions eliminated ZEA with the same ratio. However, esterase activity could not be measured from the cell wall fractions because those samples were produced using different chemicals and heat treatment. Therefore, esterase activity could not explain ZEA elimination by AMK 9/2 cell wall fractions.

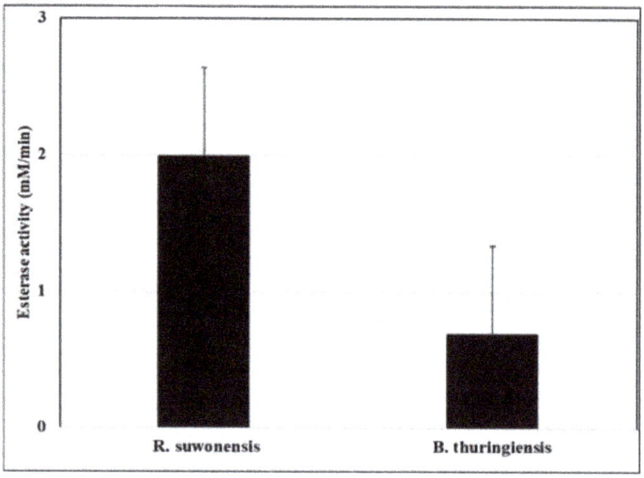

Figure 4. Esterase activity was measured from the lysed cell debris of *Rummeliibacillus suwonensis* AMK9/2 and *Bacillus thuringiensis* AMK10/1.

3. Discussion

Mycotoxin elimination technologies can use viable and dead bacterial biomasses and various cell fractions. The mycotoxin adsorption capability of bacteria depends primarily on cell wall components [23,24], where mainly polysaccharides, such as peptidoglycan (PG) and teichoic acids, play a critical role in mycotoxin binding [31]. Besides adsorption, enzymatic degradation (e.g., [14]) is possible, making a cell-free biotechnological process for mycotoxin elimination achievable.

Elimination of zearalenone (ZEA), deoxynivalenol (DON), and aflatoxin B1 (AFB1) using forage-based isolates of the Gram-positive *R. suwonensis, L. boronitolerans, L. fusiformis,* and *B. thuringiensis* was targeted. For *R. suwonensis* AMK9/2 and *L. boronitolerans* AMK9/1, no mycotoxin elimination-related studies are published thus far, to the best of our knowledge. For *L. fusiformis*, there was only one study on AFB1 elimination [32] but it was only performed with viable bacteria and not with cell fractions; meanwhile, for *B. thuringiensis*, there are several papers available on its antagonistic effects on patulin-producing fungi and its lactonase production, which degrades patulin mycotoxin (e.g., [33]). In the present study, viable cells, cell wall debris, purified cell walls, teichoic acid, PG, and S-layer fractions were produced with heat and different chemical treatments and tested in mycotoxin elimination experiments.

Both heat and organic acids affect the adsorption of mycotoxins by the cell walls because both factors reduce the PG's thickness, increasing the pore size of the structure [22,23]. In addition, heat treatment disrupts the glycosidic bonds of cell wall polysaccharides, such as PG, and organic acids break the amide ligaments in the structure of PG [22,23]. These treatments also denature proteins and split them into smaller peptides by breaking peptide bonds and exposing more binding sites [34]. Protein denaturation caused by high temperature is also a part of pore generation, resulting in enhanced permeability in living cells [35].

Under the experimental conditions, the trichothecene mycotoxin DON was not eliminated by bacterial cells and cell fractions, unlike in other Gram-positive cells [21,36,37]. Success in controlling DON-producing fungi by microbial antagonists and detoxifying DON with microbes and enzymes was reported in the past several years [16]. El-Nezami et al. [24] reported on significant differences in the ability of bacteria to bind trichothecenes in vitro. Despite heat and acid treatments, which significantly enhanced the ability of Gram-positive lactic acid bacteria to remove DON from the MRS medium [38], these parameters did not increase DON elimination yields. Results on presented Gram-positive bacteria or cell wall fractions suggest the lack of degrading enzymes or adsorption.

In *Lacticaseibacillus rhamnosus*, aflatoxin binding occurred through the carbohydrate and protein contents of the cell wall [22]. For *B. thuringiensis* AMK10/1, the S-layer fraction had the highest potential in AFB1 elimination. Lahtinen et al. [39] emphasized the significant role of the carbohydrate moieties of PG and other structures closely associated with PG in the binding process. Moreover, the teichoic acid fraction of *B. thuringiensis* showed a more significant AFB1 eliminating potential than the intact bacterial cells [39]. Similar observations were reported by Hernandez-Mendoza et al. [40] for *Lactobacillus reuteri* NRRL14171 and *Lactobacillus casei* Shirota [40], where the teichoic acid-deficient microbes bound significantly lower amounts of AFB1. Other lactic acid bacteria showed that the higher the D-alanine or teichoic acid contents, the higher level of AFB1 bound [41].

Teichoic acids may constitute up to 50% of the bacterial cell wall [42], but its amount is species-specific. Some reports indicated that the extraction of teichoic acids is likely influenced by the extraction time and the applied trichloroacetic acid (TCA) concentration [43]. Therefore, teichoic acid-based elimination of aflatoxins should be carefully interpreted when comparing various bacterium species. Adebo et al. [32] also suggested enzymatic degradation of AFB1 by an extracellular enzyme produced by *L. fusiformis*. However, the enzymatic reactions were carried out at an AFB1 concentration of 2200 ppb (3–24 h incubations), almost two orders of magnitude higher than the AFB1 concentration employed in this study.

Considering the elimination of ZEA, *L. fusiformis* AMK10/2 viable cells eliminated less mycotoxin than dead bacteria, indicating massive, heat-induced changes in the conformation of cell wall biopolymers, increasing the number and availability of ZEA binding sites [23]. Similarly, *L. rhamnosus* cell wall polysaccharide components showed good ZEA adsorption potential, and both heat and acid treatments significantly enhanced the ZEA adsorption capability of the fractions [24]. In addition, ZEA was bound predominantly by carbohydrate components of the cell wall of lactic acid bacteria [44]. Interestingly, the ZEA elimination capability of *R. suwonensis* AMK9/2 was identical by both viable and heat-treated bacterial cells and their fractions. Therefore, we can assume that the applied treatments did not affect cell wall structures to modify ZEA elimination yields.

Enzymatic degradation can also play a role in ZEA elimination. Tinyiro et al. [45] demonstrated that the quantities of ZEA bound by autoclaved and acid-treated cells of a *Bacillus* strain were identical, while suggesting that a metalloenzyme was responsible for ZEA degradation by a *Bacillus natto* strain. Several recent studies supported the view that bacterial esterase activities could contribute to ZEA degradation. For example, Wang et al. [19] selected ZEA-eliminating bacteria based on their esterase activities. In this study, *R. suwonensis* AMK9/2 showed the highest cell wall-bound esterase activity, but the esterase activities did not correlate to the eliminated ZEA in cell wall fractions. It was also demonstrated that even much higher esterase activities did not result in significantly improved ZEA eliminations (14–52%, depending on the strains) under ZEA expositions of *Lactiplantibacillus plantarum* [46]. Esterase activities of ruminal bacteria were also tested for deacetylating *Fusarium* T-2 toxin and showed varying success rates [47]. In *Lactobacillus*, ZEA elimination was suggested to occur through adsorption and not enzymatic degradation [23]. In the case of *B. thuringiensis* AMK10/1, the amounts of ZEA eliminated by the cell fractions were below those eliminated by the viable cells. The most significant ZEA release was also achieved from the viable cells, suggesting adsorption without enzymatic degradation.

4. Conclusions

Application of cell wall fractions for mycotoxin elimination can be more successful in some cases compared to viable cells. Cell-free technologies applying only cell fractions have a future in postharvest technologies. Industries can use bioadsorption techniques without producing metabolites with unknown physiological effects. Besides preharvest biocontrol techniques, bioadsorption can be a suitable method for decreasing mycotoxin presence. Species-specific elimination abilities could be originated from the diverse available adsorption surfaces of the microbial cell wall. However, the effect of the chemical treatment (which resulted in new surfaces and binding sites) should also be considered for its positive effect on making more potent microbial bioadsorption matrixes.

5. Materials and Methods

5.1. Isolation of Bacteria

Fermented forages were collected at the final stage of the fermentations (after 4–6 weeks) from different Hungarian dairy cattle farms in 2019–2020. The producers collected ten parallel samples per site from freshly opened silos and bales, combined them in sterile Velcro bags (min. 5 kg), and transferred them to the analytical laboratory for further analysis [48].

Fermented forage samples (100 g) placed in sterile homogenizing Stomacher bags were suspended in a 1:9 ratio buffered peptone water (BPW) solution (Scharlab, Barcelona, Spain) and homogenized with a Stomacher masticator homogenizer (IUL Instruments, Barcelona, Spain). Following that, decimal dilutions were made of the suspensions, and total microbial counts were determined on plate count agar (Scharlab, Barcelona, Spain) medium applying the pour plate method. Inoculated solid agar media were incubated at 30 °C for three days under either anaerobic or aerobic conditions as the standard method [49].

5.2. Identification of Bacteria

5.2.1. Isolation of Genomic DNA

Solitary colonies of bacterial cultures were isolated from the pour plates and inoculated in liquid nutrient broth (Scharlab, Barcelona, Spain). Following the DNA extraction protocol based on Wilson [50], 200 µL aliquots of 16 h cultures were mixed with a 1000 µL CTAB (Biochemica, Darmstadt, Germany) lysis buffer {2% (*w/v*) cetrimonium bromide (CTAB), 1.4 M NaCl, 100 mM Tris/HCl, 20 mM EDTA, pH 8.0} (Applichem Ltd. ITW Company, Darmstadt, Germany). The samples were incubated for 30 min at 65 °C and then put in 2 mL Lysing Matrix B tubes (MP Biomedicals Germany GmbH, Schwegel, Germany). Bead-based lysis was carried out for 20 s at 6500 rpm with Precellys 24 homogenizer (Peqlab Biotechnologie Ltd., Erlangen, Germany). Cell debris of lysed bacteria was separated by centrifugation (10 min at 14,000 rpm). To each 600 µL aliquot of the supernatant was added 240 µL of chloroform, stirred for 30 s, and centrifuged (20 min at 14,000 rpm). Aliquots (400 µL each) of the upper aqueous phase were mixed with equal volumes of isopropanol and were centrifuged again (10 min, 14,000 rpm). After discarding the supernatants, the pellets were washed with 500 µL aliquots of 70% ethanol, re-centrifuged, and left to dry at room temperature. Genomic DNA was re-suspended in 500 µL of water and was purified further using an Amicon Ultra filtration kit (Merck Millipore, Darmstadt, Germany) following the manufacturer's instructions.

5.2.2. PCR Method

In all PCR, the iProof high-fidelity PCR kit (BIO-RAD Ltd., Hercules, CA, USA Lithuania) was used with 27F (5′-AGAGTTTGATCCTGGCTCAG-3′) and 1492R (5′-TACGGTTAC CTTGTTACGACTT-3′) primers of 16S rRNA [51]. PCR was performed at 98 °C for 3 min; after that, 98 °C for 30 s, 54 °C for 30 s, 72 °C for 45 s, for 30 cycles, and 72 °C for 7 min on T100 thermal cycler (BIO-RAD Ltd., Lithuania).

DNA amplicons were cleaned in a 0.8% agarose gel (BIOLINE), Nucleo SpinGel, and PCR clean-up column (Macherey-Nagel, Düren, Germany) kit to the protocols of the manufacturers. The PCR products were sequenced by BIOMI Ltd. (Gödöllő, Hungary). The sequences were submitted to the National Library of Medicine at the National Center for Biotechnological Information (NCBI) under accession numbers OP183257-OP183263.

5.3. Bacterial Cell Fractions

Bacterium strains were grown in nutrient broth (Scharlab; 16 h at 30 °C), and exponential phase cultures were centrifuged (8000 rpm, 10 min, 4 °C), the supernatants were removed, and the cell pellets were washed three times with 200 µL aliquots of sterile phosphate-buffered saline solution (PBS). After washing, the aliquots of the pellets were exposed to various chemicals to get different cell wall fractions based on Niederkorn et al. [31] and Goh et al. [52]: H_2O (100 °C, 15 min), cell debris; 2% (w/v) sodium dodecyl sulphate (SDS) (100 °C, 15 min), purified cell wall fraction; 0.1 M HCl (100 °C, 15 min), teichoic acid fraction; 10% (w/v) trichloroacetic acid (TCA) (100 °C, 15 min), peptidoglycan fraction and 1M LiCl (100 °C, 15 min), S-layer proteins. After treatments, samples were centrifuged at 14,000 rpm 10 min, the supernatants were discarded, and the pellets were washed with 3×200 µL PBS. Bacterial cells and cell wall fractions were stored at -18 °C.

5.4. Mycotoxin Elimination

Deoxynivalenol (DON), zearalenone (ZEA), and aflatoxin B1 (AFB1) (Biopure, Romer Labs, Tulln, Austria) calibration solutions were purchased and used in appropriate dilution for mycotoxin elimination tests. Mycotoxins were diluted to the final concentration in phosphate-buffered saline (PBS) and were added at the following concentrations to living bacterial cells and different bacterial cell wall preparations: AFB1—24 µg/L, DON—700 µg/L, ZEA—100 µg/L. The toxin concentrations applied were based on literature search, the average toxin content of the forage samples, and preliminary mycotoxin tolerance studies [48]. All mycotoxin-supplemented samples were incubated in PBS for 1 h at 25 °C with shaking (250 rpm), centrifuged (8000 rpm, 10 min, 4 °C), and the supernatants were removed, extracted, and analyzed by HPLC. All assays were performed in triplicate, and positive controls (without cells or cell wall fractions) and negative controls (without mycotoxin) were included. HPLC detection of DON was carried out on Hitachi Elit LaChrom HPLC (San Jose, CA, USA) equipment. For deoxynivalenol (DON) measurement, the filtrated supernatant samples were loaded onto a Phenomenex (Torrance, CA, USA) RP-C18 column (125 × 4 mm, 5 µm) and detected with a diode array detector in UV 218 nm with acetonitrile: water (10:90) eluent.

For AFB1, the sample and methanol were mixed in 1:1 ratio and vortexed at high speed. HPLC detection of the tested mycotoxins was carried out in Dionex Ultimate 3000 (Thermo Scientific, Waltham, MA, USA) equipment [48]. The diluted extract was filtered and loaded onto a Phenomenex (Torrance, CA, USA) RP-C18 column (150 × 4.6 mm, 5 µm) with a Romer UV derivatization unit (Romer Labs Ltd., Tulln, Austria) and a fluorescence detector (ex360 nm, em440 nm) with methanol: water (45:55) eluent.

In samples exposed to ZEA, the mycotoxin contents of the supernatants and the pellets were also measured. Supernatants were treated with methanol in a 1:1 ratio and vortexed at high speed. Pellet-adsorbed ZEA was extracted by a mixture of acetonitrile–water–methanol (46:46:8) based on ZearalaTest™ (VICAM, Watertown, USA). HPLC detection of the tested mycotoxins was carried out in Dionex Ultimate 3000 (Thermo Scientific) equipment [48]. The extracts were filtered and loaded onto a Phenomenex (Torrance, CA, USA) RP-C18 column (150 × 4.6 mm, 5µm) and a fluorescence detector (ex274 nm, em440 nm) with acetonitrile–water–methanol (46:46:8) eluent.

For AFB1, DON, and ZEA, the limits of detection (LOD) were 0.02 µg/L, 0.05 mg/L, and 2.6 µg/L, respectively. The retention times were 10.3 min, 3.2 min, and 8.9 min, respectively. The coefficient of variation within test repetitions was calculated and found to be under 15% in all cases.

5.5. Esterase Activity

Enzyme activity was measured spectrophotometrically based on the method of Castilo et al. [53]. The reaction mixture contained 800 μL 50 mM Tris-HCl buffer, pH 7.5, 100 μL *p*-nitrophenyl butyrate (Sigma-Aldrich, Saint Louis, USA) as substrate (8.1 mM in acetone), and 100 μL lysed (Precellys 24 homogenizer, Peqlab Biotechnologie Ltd., Erlangen, Germany) and PBS-washed cell debris or the heat-treated cell wall fraction samples. Enzyme activity was detected as *p*-nitrophenol deliberated after 10 min at 37 °C incubation (λ = 346 nm). The esterase activity was expressed as mM *p*-nitrophenyl released per min.

5.6. Statistical Analysis

Data analyses were done in Windows Excel Analysis ToolPac (Microsoft), where a t-probe (at $p \leq 0.05$) was performed for the significance analysis. A correlation (Pearson) test of esterase activity and mycotoxin elimination was also done in Windows Excel data Analysis ToolPac (Microsoft).

Author Contributions: Conceptualization and supervision, I.P.; methodology, T.P.; investigation, C.A. and S.K.; writing—original draft preparation, C.A.; writing—review and editing, T.P.; visualization, C.A. All authors have read and agreed to the published version of the manuscript.

Funding: This research was funded by NATIONAL RESEARCH, DEVELOPMENT, AND INNOVATION FUND OF HUNGARY project no. 2018-1.2.1-NKP-2018-00002 and 2019-2.1.13-TÉT_IN-2020-00056.

Institutional Review Board Statement: Not applicable.

Informed Consent Statement: Not applicable.

Data Availability Statement: Not applicable.

Acknowledgments: We thank Mrs. Istvan Sőrés for her technical assistance.

Conflicts of Interest: The authors declare no conflict of interest.

References

1. Martins, H.M.; Guerra, M.M.M.; Bernardo, F.M.D. Presencia de aflatoxina B1 en piensos para ganado lechero en Portugal durante el periodo 1995–2004. *Rev. Iberoam. Micol.* **2007**, *24*, 69–71. [CrossRef]
2. Medina, Á.; Rodríguez, A.; Magan, N. Climate change and mycotoxigenic fungi: Impacts on mycotoxin production. *Curr. Opin. Food Sci.* **2015**, *5*, 99–104. [CrossRef]
3. Yu, J. Current Understanding on aflatoxin biosynthesis and future perspective in reducing aflatoxin contamination. *Toxins* **2012**, *4*, 1024–1057. [CrossRef] [PubMed]
4. Gnonlonfin, G.J.; Hell, K.; Adjovi, Y.; Fandohan, P.; Koudande, D.O.; Mensah, G.A.; Sanni, A.; Brimer, L.A. Review on aflatoxin contamination and its implications in the developing world: A Sub-Saharan African perspective. *Crit. Rev. Food Sci. Nutr.* **2013**, *53*, 349–365. [CrossRef] [PubMed]
5. Dänicke, S.; Winkler, J. Invited Review: Diagnosis of Zearalenone (ZEN) exposure of farm animals and transfer of its residues into edible tissues (carry over). *Food Chem. Toxicol.* **2015**, *84*, 225–249. [CrossRef]
6. Fink-Gremmels, J.; Malekinejad, H. Clinical Effects and biochemical mechanisms associated with exposure to the mycoestrogen zearalenone. *Anim. Feed Sci. Technol.* **2007**, *137*, 326–341. [CrossRef]
7. Zinedine, A.; Soriano, J.M.; Moltó, J.C.; Mañes, J. Review on the toxicity, occurrence, metabolism, detoxification, regulations and intake of zearalenone: An oestrogenic mycotoxin. *Food Chem. Toxicol.* **2007**, *45*, 1–18. [CrossRef]
8. Gromadzka, K.; Waskiewicz, A.; Chelkowski, J.; Golinski, P. Zearalenone and its metabolites: Occurrence, detection, toxicity and guidelines. *World Mycotoxin J.* **2008**, *1*, 209–220. [CrossRef]
9. Kushiro, M. Effects of milling and cooking processes on the deoxynivalenol content in wheat. *Int. J. Mol. Sci.* **2008**, *9*, 2127–2145. [CrossRef]
10. Grenier, B.; Applegate, T. Modulation of intestinal functions following mycotoxin ingestion: Meta-analysis of published experiments in animals. *Toxins* **2013**, *5*, 396–430. [CrossRef]
11. Döll, S.; Dänicke, S. The Fusarium toxins deoxynivalenol (DON) and zearalenone (ZON) in animal feeding. *Prev. Vet. Med.* **2011**, *102*, 132–145. [CrossRef]
12. Mishra, S.; Srivastava, S.; Dewangan, J.; Divakar, A.; Kumar Rath, S. Global occurrence of deoxynivalenol in food commodities and exposure risk assessment in humans in the last decade: A survey. *Crit. Rev. Food Sci. Nutr.* **2022**, *60*, 1346–1374. [CrossRef] [PubMed]

13. Abbès, S.; Salah-Abbès, J.B.; Sharafi, H.; Jebali, R.; Noghabi, K.A.; Oueslati, R. Ability of *Lactobacillus rhamnosus* GAF01 to remove AFM1 in vitro and to counteract AFM1 immunotoxicity in vivo. *J. Immunotoxicol.* **2013**, *10*, 279–286. [CrossRef] [PubMed]

14. Ji, C.; Fan, Y.; Zhao, L. Review on biological degradation of mycotoxins. *Anim. Nutr.* **2016**, *2*, 127–133. [CrossRef]

15. Li, Z.; Wang, Y.; Liu, Z.; Jin, S.; Pan, K.; Liu, H.; Liu, T.; Li, X.; Zhang, C.; Luo, X.; et al. Biological detoxification of fumonisin by a novel carboxylesterase from *Sphingomonadales* bacterium and its biochemical characterization. *Int. J. Biol. Macromol.* **2021**, *169*, 18–27. [CrossRef] [PubMed]

16. Tian, Y.; Zhang, D.; Cai, P.; Lin, H.; Ying, H.; Hu, Q.N.; Wu, A. Elimination of fusarium mycotoxin Deoxynivalenol (DON) via microbial and enzymatic strategies: Current status and future perspectives. *Trends Food Sci. Technol.* **2022**, *124*, 96–107. [CrossRef]

17. Freire, L.; Sant'Ana, A.S. Modified mycotoxins: An updated review on their formation, detection, occurrence, and toxic effects. *Food Chem. Toxicol.* **2018**, *111*, 189–205. [CrossRef] [PubMed]

18. Peles, F.; Sipos, P.; Kovács, S.; Győri, Z.; Pócsi, I.; Pusztahelyi, T. Biological control and mitigation of aflatoxin contamination in commodities. *Toxins* **2021**, *13*, 104. [CrossRef]

19. Wang, G.; Yu, M.; Dong, F.; Shi, J.; Xu, J. Esterase activity inspired selection and characterization of zearalenone degrading bacteria *Bacillus pumilus* ES-21. *Food Cont.* **2017**, *77*, 57–64. [CrossRef]

20. Kim, D.H.; Lee, K.D.; Choi, K.C. Role of LAB in silage fermentation: Effect on nutritional quality and organic acid production—an overview. *AIMS Agric. Food* **2021**, *6*, 216–234. [CrossRef]

21. Niderkorn, V.; Boudra, H.; Morgavi, D.P. Binding of fusarium mycotoxins by fermentative bacteria in vitro. *J. Appl. Microbiol.* **2006**, *101*, 849–856. [CrossRef] [PubMed]

22. Haskard, C.; Binnion, C.; Ahokas, J. Factors Affecting the sequestration of aflatoxin by *Lactobacillus rhamnosus* Strain GG. *Chem.-Biol. Interact.* **2000**, *128*, 39–49. [CrossRef]

23. El-Nezami, H.; Polychronaki, N.; Salminen, S.; Mykkanen, H. Binding rather than metabolism may explain the interaction of two food-grade *Lactobacillus* strains with zearalenone and its derivative-zearalenol. *Appl. Environ. Microbiol.* **2002**, *68*, 3545–3549. [CrossRef] [PubMed]

24. El-Nezami, H.S.; Chrevatidis, A.; Auriola, S.; Salminen, S.; Mykkänen, H. Removal of common Fusarium toxins in vitro by strains of *Lactobacillus* and *Propionibacterium*. *Food Addit. Contam.* **2002**, *19*, 680–686. [CrossRef]

25. Legg, M.S.G. Advancing Understanding of Secondary Cell Wall Polymer Binding and Synthesis in S-Layers of Gram-Positive Bacteria. Ph.D. Thesis, University of Victoria, Victoria, BC, Canada, 2022.

26. Beeby, M. Toward organism-scale structural biology: S-layer reined in by bacterial LPS. *Trends Biochem. Sci.* **2020**, *45*, 549–551. [CrossRef]

27. Her, J.; Kim, J. *Rummeliibacillus suwonensis* sp. nov. isolated from soil collected in a mountain area of South Korea. *J. Microbiol.* **2013**, *51*, 268–272. [CrossRef]

28. Joung, K.B.; Côté, J.C. A single phylogenetic analysis of *Bacillus thuringiensis* strains and bacilli species inferred from 16S rRNA gene restriction fragment length polymorphism is congruent with two independent phylogenetic analyses. *J. Appl. Microbiol.* **2002**, *93*, 1075–1082. [CrossRef]

29. Ahmed, I.; Yokota, A.; Yamazoe, A.; Fujiwara, T. Proposal of *Lysinibacillus boronitolerans* gen. nov. sp. nov. and transfer of *Bacillus fusiformis* to *Lysinibacillus fusiformis* comb. nov. and *Bacillus sphaericus* to *Lysinibacillus sphaericus* comb. nov. *Int. J. Syst. Evol. Microbiol.* **2007**, *57*, 1117–1125. [CrossRef]

30. Schoch, C.L.; Ciufo, S.; Domrachev, M.; Hotton, C.L.; Kannan, S.; Khovanskaya, R.; Leipe, D.; Mcveigh, R.; O'Neill, K.; Robbertse, B.; et al. NCBI Taxonomy: A comprehensive update on curation, resources and tools. *Database* **2020**, *2020*, baaa062. [CrossRef]

31. Niderkorn, V.; Morgavi, D.P.; Aboab, B.; Lemaire, M.; Boudra, H. Cell wall component and mycotoxin moieties involved in the binding of Fumonisin B1 and B2 by lactic acid bacteria. *J. Appl. Microbiol.* **2009**, *106*, 977–985. [CrossRef]

32. Adebo, O.A.; Njobeh, P.B.; Mavumengwana, V. Degradation and detoxification of AFB1 by *Staphylococcus warneri*, *Sporosarcina* sp. and *Lysinibacillus fusiformis*. *Food Cont.* **2016**, *68*, 92–96. [CrossRef]

33. Chalivendra, S.; Ham, J.H. Bacilli in the biocontrol of mycotoxins. In *Bacilli and Agrobiotechnology: Phytostimulation and Biocontrol*; Islam, M.T., Rahman, M.M., Pandey, P., Boehme, M.H., Haesaert, G., Eds.; Springer Nature Switzerland: Cham, Switzerland, 2019; pp. 49–62.

34. Lili, Z.; Junyan, W.; Hongfei, Z.; Baoqing, Z.; Bolin, Z. Detoxification of cancerogenic compounds by lactic acid bacteria Strains. *Crit. Rev. Food Sci. Nutr.* **2018**, *58*, 2727–2742. [CrossRef] [PubMed]

35. Abedi, E.; Mousavifard, M.; Hashemi, S.M.B. Ultrasound-assisted detoxification of ochratoxin a: Comparative study of cell wall structure, hydrophobicity, and toxin binding capacity of single and co-culture lactic acid bacteria. *Food Bioproc. Technol.* **2022**, *15*, 539–560. [CrossRef]

36. Yao, Y.; Long, M. The biological detoxification of deoxynivalenol: A review. *Food Chem. Toxicol.* **2020**, *145*, 111649. [CrossRef]

37. Luo, Y.; Liu, X.; Yuan, L.; Li, J. Complicated interactions between bio-adsorbents and mycotoxins during mycotoxin adsorption: Current research and future prospects. *Trends Food Sci. Technol.* **2020**, *96*, 127–134. [CrossRef]

38. Zou, Z.Y.; He, Z.F.; Li, H.J.; Han, P.F.; Meng, X.; Zhang, Y.; Zhou, F.; Ouyang, K.P.; Chen, X.Y.; Tang, J. In vitro removal of deoxynivalenol and T-2 toxin by lactic acid bacteria. *Food Sci. Biotechnol.* **2012**, *21*, 1677–1683. [CrossRef]

39. Lahtinen, S.J.; Haskard, C.A.; Ouwehand, A.C.; Salminen, S.J.; Ahokas, J.T. Binding of Aflatoxin B1 to cell wall components of Lactobacillus rhamnosus strain GG. *Food Addit. Contam.* **2004**, *21*, 158–164. [CrossRef]

40. Hernandez-Mendoza, A.; Guzman-De-Peña, D.; Garcia, H.S. Key role of teichoic acids on aflatoxin B1 binding by probiotic bacteria. *J. Appl. Microbiol.* **2009**, *107*, 395–403. [CrossRef]

41. Serrano-Niño, J.C.; Cavazos-Garduño, A.; Cantú-Cornelio, F.; González-Córdova, A.F.; Vallejo-Córdoba, B.; Hernández-Mendoza, A.; García, H.S. In vitro reduced availability of aflatoxin B1 and acrylamide by bonding interactions with teichoic acids from *Lactobacillus* strains. *LWT-Food Sci. Technol.* **2015**, *64*, 1334–1341. [CrossRef]

42. Brown, S.; Santa Maria, J.P.; Walker, S. Wall teichoic acids of Gram-positive bacteria. *Ann. Rev. Microbiol.* **2013**, *67*, 313–336. [CrossRef]

43. Wicken, A.J.; Gibbens, J.W.; Knox, K.W. Comparative studies on the isolation of membrane lipo-teichoic acid from *Lactobacillus fermenti* 6991. *J. Bacteriol.* **1972**, *113*, 365–372. [CrossRef] [PubMed]

44. El-Nezami, H.; Polychronaki, N.; Lee, Y.K.; Haskard, C.; Juvonen, R.; Salminen, S.; Mykkänen, H. Chemical moieties and interactions involved in the binding of zearalenone to the surface of *Lactobacillus rhamnosus* strains GG. *J. Agric. Food Chem.* **2004**, *52*, 4577–4581. [CrossRef] [PubMed]

45. Tinyiro, S.E.; Yao, W.; Sun, X.; Wokadala, C.; Wang, S. Scavenging of zearalenone by *Bacillus* strains in vitro. *Res. J. Microbiol.* **2011**, *6*, 304–309. [CrossRef]

46. Chen, S.-W.; Hsu, J.-T.; Chou, Y.-A.; Wang, H.-T. The application of digestive tract lactic acid bacteria with high esterase activity for zearalenone detoxification. *J. Sci. Food Agric.* **2018**, *98*, 3870–3879. [CrossRef] [PubMed]

47. Westlake, K.; Mackie, R.I.; Dutton, M.F. T-2 toxin metabolism by ruminal bacteria and its effect on their growth. *Appl. Environ. Microbiol.* **1987**, *53*, 587–592. [CrossRef]

48. Adácsi, C.; Kovács, S.; Pócsi, I.; Győri, Z.; Dombrádi, Z.; Pusztahelyi, T. Microbiological and toxicological evaluation of fermented forages. *Agriculture* **2022**, *12*, 421. [CrossRef]

49. *ISO 4833-1:2013*; Microbiology of the Food Chain—Horizontal Method for the Enumeration of Microorganisms—Part 1: Colony Count at 30 °C by the Pour Plate Technique. International Organization for Standardization: Geneva, Switzerland, 2013.

50. Wilson, K. Preparation of genomic DNA from bacteria. *Curr. Protoc. Mol. Biol.* **2001**, *56*, 2.4.1–2.4.5. [CrossRef]

51. Pradhan, P.; Tamang, J.P. Phenotypic and genotypic identification of bacteria isolated from traditionally prepared dry starters of the Eastern Himalayas. *Front. Microbiol.* **2019**, *10*, 2526. [CrossRef]

52. Goh, Y.J.; Azcarate-Peril, M.A.; O'Flaherty, S.; Durmaz, E.; Valence, F.; Jardin, J.; Lortal, S.; Klaenhammer, T.R. Development and application of a upp-based counterselective gene replacement system for the study of the S-Layer protein Slpx of *Lactobacillus acidophilus* NCFM. *Appl. Environ. Microbiol.* **2009**, *75*, 3093–3105. [CrossRef]

53. Castillo, I.; Requena, T.; De Palencia, P.F.; Fontecha, J.; Gobbetti, M. Isolation and characterization of an intracellular esterase from *Lactobacillus casei* subsp. *casei* IFPL731. *J. Appl. Microbiol.* **1999**, *86*, 653–659. [CrossRef]

 *toxins*

Communication

Multiple Mycotoxin Contamination in Medicinal Plants Frequently Sold in the Free State Province, South Africa Detected Using UPLC-ESI-MS/MS

Julius Ndoro [1], Idah Tichaidza Manduna [2,*], Makomborero Nyoni [3] and Olga de Smidt [2]

[1] Department of Life Sciences, Faculty of Health and Environmental Sciences, Central University of Technology, Free State, Private Bag X20539, Bloemfontein 9300, South Africa

[2] Centre for Applied Food Sustainability and Biotechnology (CAFSaB), Central University of Technology, Free State, Bloemfontein 9300, South Africa

[3] Research, Development and Innovation Department, National Biotechnology Authority, 21 Princess Drive Newlands, Harare, Zimbabwe

* Correspondence: imanduna@cut.ac.za

Abstract: Medicinal plants are important in the South African traditional healthcare system, the growth in the consumption has led to increase in trade through *muthi* shops and street vendors. Medicinal plants are prone to contamination with fungi and their mycotoxins. The study investigated multiple mycotoxin contamination using Ultra High Pressure Liquid Chromatography–Tandem Mass Spectrometry (UPLC-ESI-MS/MS) for the simultaneous detection of Aflatoxin B1 (AFB1), Deoxynivalenol (DON), Fumonisins (FB$_1$, FB$_2$, FB$_3$), Nivalenol (NIV), Ochratoxin A (OTA) and Zearalenone (ZEN) in frequently sold medicinal plants. Medicinal plant samples (n = 34) were purchased and analyzed for the presence of eight mycotoxins. DON and NIV were not detected in all samples analyzed. Ten out of thirty-four samples tested positive for mycotoxins —AFB$_1$ (10.0%); OTA (10.0%); FB1 (30.0%); FB2 (50.0%); FB3 (20.0%); and ZEN (30.0%). Mean concentration levels ranged from AFB$_1$ (15 μg/kg), OTA (4 μg/kg), FB$_1$ (7–12 μg/kg), FB$_2$ (1–18 μg/kg), FB$_3$ (1–15 μg/kg) and ZEN (7–183 μg/kg). Multiple mycotoxin contamination was observed in 30% of the positive samples with fumonisins. The concentration of AFB$_1$ reported in this study is above the permissible limit for AFB1 (5 μg/kg). Fumonisin concentration did not exceed the limits set for raw maize grain (4000 μg/kg of FB$_1$ and FB$_2$). ZEN and OTA are not regulated in South Africa. The findings indicate the prevalence of mycotoxin contamination in frequently traded medicinal plants that poses a health risk to consumers. There is therefore a need for routine monitoring of multiple mycotoxin contamination, human exposure assessments using biomarker analysis and establishment of regulations and standards.

Keywords: medicinal plants; mycotoxins; Ultra High Pressure Liquid Chromatography–Tandem Mass Spectrometry; contamination; street vendors; *muthi* shops; fungi; plant trade

Key Contribution: This is the first report on multiple mycotoxin contamination in Free State Province, South Africa in marketed medicinal plants using UPLC-ESI-MS/MS. The study shows the contribution of commercially traded medicinal plants in human mycotoxin exposure. The study findings help in advocating for consumer safety and quality monitoring as well as development of strategies to maintain the safety of medicinal plants sold in markets.

Citation: Ndoro, J.; Manduna, I.T.; Nyoni, M.; de Smidt, O. Multiple Mycotoxin Contamination in Medicinal Plants Frequently Sold in the Free State Province, South Africa Detected Using UPLC-ESI-MS/MS. *Toxins* **2022**, *14*, 690. https://doi.org/10.3390/toxins14100690

Received: 9 August 2022
Accepted: 3 October 2022
Published: 8 October 2022

Publisher's Note: MDPI stays neutral with regard to jurisdictional claims in published maps and institutional affiliations.

1. Introduction

There has been a steady increase in the demand for medicinal plants, herbs and preparations as complementary and alternative medicine (CAM) and in traditional medicine both in developing and developed countries [1]. In developed countries, between 25 and 70% of the population rely on complementary and/or alternative medicine (CAM) [2]. In

Africa, the use of medicinal plants and consumption is significantly higher, around 80%, due to economic, social, and cultural factors. Medicinal plants play a vital role in disease prevention and their promotion and use compliments current prevention strategies under the Primary Health Care Approach [3,4].

Due to a complex supply chain involving different players and conditions from pre-harvesting, harvesting, storage and trade, medicinal plants are prone to infestation by pests, microbes and toxins [5,6]. Mycotoxins are toxic fungal secondary metabolites and are common contaminants of both human food and animal feed. Contamination is more common in developing countries with poor crop storage and production technologies, and climatic conditions which promote fungal growth and toxin production [7]. There are over 400 mycotoxins known today. Aflatoxins, ochratoxins, fumonisins and trichothecenes are the major classes of mycotoxins that have been recognized as being of public health significance due to their high occurrence and associated carcinogenic properties [6,8].

Mycotoxin exposures occur via various routes of entry such as oral, dermal, respiratory and parenteral. The oral/ingestion route is the major route of entry for mycotoxin exposures. A potential chain reaction can occur when contaminated animal feed results in infected meat, milk and eggs [9] which in turn, can affect human health. Acute and chronic mycotoxicosis can be developed depending on an individual's susceptibility, the type of mycotoxin and dosage [7,10]. For example, approximately a third of all cases of liver cancer in Africa are due to chronic exposures to mycotoxins [11]. Table 1 shows the adverse effects of some mycotoxins on animal and human health [9,12–15]. Additive or synergistic harmful effects may also be a result of co-occurring mycotoxins [12]

Table 1. Mycotoxigenic effects on animal and human health.

Mycotoxin	Fungal Source (Genus)	Health Effects
Aflatoxin B1	*Aspergillus*	Teratogenic, hepatotoxic, immunosuppressive, carcinogenic and mutagenic.
Deoxynivalenol	*Fusarium*	Gastrointestinal damage, reproductive effects toxicosis, genotoxicity and immunosuppressive.
Fumonisins	*Fusarium*	Teratogenic, carcinogenic, hepatotoxic, nephrotoxic, immunosuppressive and neurotoxic.
Nivalenol	*Fusarium*	Anorexic, immunotoxic, hematotoxic and genotoxic.
Ochratoxin A	*Aspergillus Penicillium*	Carcinogenic, teratogenic, immunosuppressive and nephrotoxic.
Zearalenone	*Fusarium*	Carcinogenic, hormonal imbalance (hyperestrogenism) and reproductive effects.

Despite the reported and potential impacts of mycotoxins including their relation to many diseases, they are poorly studied in South African medicinal plants sold in markets which are prone to contamination [16,17]. Furthermore, the control of mycotoxins is inadequately funded, and many African governments do not give priority to mycotoxin control in medicinal plants [18]. However, the occurrence of mycotoxins has been reported in South Africa [16,17]; Kenya [19–21]; Nigeria [22–25]; and Egypt [26–29].

Most of these studies have been limited in scope focusing mainly on aflatoxin and fumonisin contamination. In view of the increasing demand for and trade in medicinal plants and the health risks from fungal contamination and their toxins, there is a need to have a broad understanding of the prevalence of mycotoxins in commercially traded medicinal plants. Regrettably, there is limited information on mycotoxins in medicinal plants in South Africa which is not commensurate with the escalating economic value of the trade. As mentioned earlier, previous studies have been completed in South Africa, but no studies have been published on mycotoxin contamination in medicinal plants sold in the Free State Province *muthi* (traditional medicine) shops and by street vendors, hence there is no information available. The aim of this study was to assess the safety of medicinal plants with respect to multiple mycotoxin contamination namely Aflatoxin B_1 (AFB$_1$), Ochratoxin A (OTA), Zearalenone (ZEN), Deoxynivalenol (DON), Nivalenol (NIV) and Fumonisins (FB$_1$, FB$_2$, FB$_3$) as supported by Keter et al. [20].

2. Results

2.1. Mycotoxin Extraction

Whilst other studies have employed sample clean up, the present study directly injected mycotoxin extracts into the LC-MS/MS without any simple clean up. This is in line with a study on simultaneous LC/MS/MS determination of aflatoxins, fumonisins, OTA and patulin, type A and B trichothecenes and Zearalenone, with no sample clean-up [30] The method employed in this study is also supported by an HPLC-ESI-MS/MS method which was developed for simultaneous determination of 33 mycotoxins in various products. The mycotoxins were extracted with acetonitrile/water and then directly injected into a LC-MS/MS system without any clean-up [31]. Other studies have also conducted mycotoxin analysis with no clean-up step in various matrices [32]. Therefore, the extraction method was quite efficient in isolating the targeted mycotoxins under investigation.

2.2. Mycotoxin Analysis

A total of 34 samples from commonly sold medicinal plants were analyzed for the presence of multiple mycotoxins. None of the plant samples contained detectable levels of DON and NIV. Of the 34 samples, 10 (29%) were positive for OTA (1); AFB$_1$ (1); FB$_1$(3); FB$_2$(5); FB$_3$ (2); and ZEN (3) as illustrated in Figure 1. Multi-mycotoxin contamination was observed in 30% of the positive samples with fumonisin derivatives (FB$_1$, FB$_2$, FB$_3$).

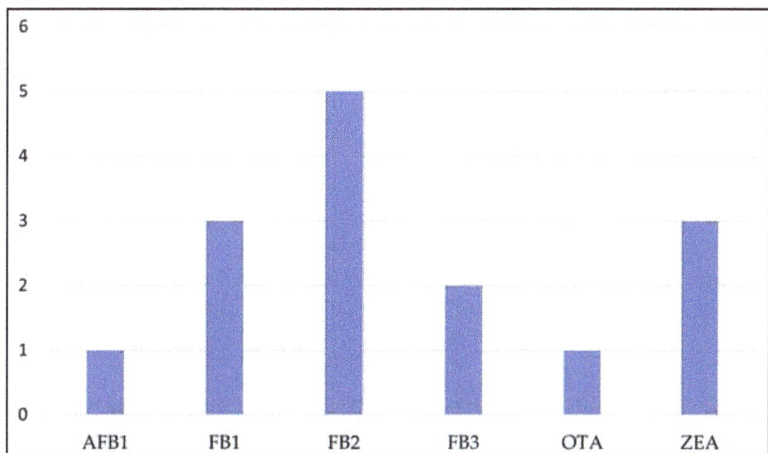

Figure 1. Frequency of occurrence of mycotoxins in medicinal plants.

The occurrence of mycotoxins and their concentrations in medicinal plants are presented in Table 2. AFB1 was only found in *Dicoma anomala* at a concentration of 15 µg/kg and OTA (4 µg/kg) was found in *Aloe ferox*. FB1 ranged from 1 µg/kg to 12 µg/kg while FB$_2$ was detected in five different pants from five different locations. FB$_2$ concentrations ranged between 1 µg/kg and 18 µg/kg. FB$_3$ (1–15 µg/kg) with a mean of 4.5 µg/kg and ZEN (7–183 µg/kg) with a mean of 81.3 µg/kg. The highest mycotoxin contamination level in the study was recorded for ZEN at 183 µg/kg. A sample contaminated with ZEN had the highest total mycotoxin levels whilst the least contaminated had a concentration of 15 µg/kg.

Table 2. Mycotoxin contamination in medicinal plants.

Plant Name	Trader: Location	AFB_1	FB_1	FB_2	FB_3	OTA	ZEN
Bulbine narcissifolia Salm-Dyck	MS: Thaba 'Nchu	-	10.0	18.0	1.0	-	-
Helichrysum odoratissimum (L.) Sweet.	SV: Zastron	-	12.0	15.0	-	-	-
Hypoxis hemerocallidea Fisch., C.A.Mey. & Avé-Lall.	MS: Dewetsdorp	-	-	-	-	-	183.0
Adenia gummifera (Harv.) Harms	MS: Sasolburg	-	-	-	-	-	54.0
Aloe ferox Mill.	MS: Senekal	-	-	-	-	4.0	-
Galium capense Thunb	MS: Winburg	-	-	2.0	-	-	-
Siphonochilus aethiopicus (Schweif.) B.L. Burt	SV: Kroonstad	-	-	-	-	-	7.0
Helichrysum odoratissimum (L.) Sweet.	SV: Kroonstad	-	-	6.0	-	-	-
Dicoma anomala Sond.	SV: Bloemfontein	15.0	-	-	-		
Pentanisia prunelloides (Klotzsch ex Eckl. & Zeyh.) Walp.	SV: Bloemfontein	-	7.0	1.0	1.0	-	-
Mean of positive samples ± standard deviation		15.0	9.6 ± 2.5	8.4 ± 7.7	1.0	4.0	81.3 ± 91.1

Only positive sample results have been shown; Concentrations in µg/kg; Not detected (-). SV-Street Vendor; MS-*Muthi* Shop.

3. Discussion

3.1. Aflatoxin (AFB₁)

Aflatoxins have been reported as mycotoxins of human importance [33]. There are various worldwide reports on aflatoxin contamination in medicinal plants. The current study findings seem to be consistent with a report by Aiko and Mehta, in their study of 63 Indian medicinal herbs samples only one sample tested positive for aflatoxin B [34]. In another study of African traditional herbal medicines sold in South Africa (Tshwane-Pretoria and Cape Town), all 16 samples were not contaminated with aflatoxins [17]. A study in Italy found that all samples of medicinal plants, aromatic herbs and herbal infusions were not contaminated with aflatoxins [29].

Another study of 500 herbal plants in Poland reported all samples to be safe from aflatoxins [35]. This is quite similar to the current study findings which only found one sample to be contaminated with AFB_1. In contrast, the authors of [36] reported aflatoxin contamination in 58.9% of herbal tea samples from Moroccan market. Tassaneeyakul et al., reported aflatoxin contamination in herbal medicinal plant products in Thailand, in the range of 1.7–0.0000143 µg/kg in 5 out of 28 samples, which is lower than the present findings [37]. The AFB_1 concentration reported in our study was above the results reported by Yang et al., with AFs (up to 32 µg/kg) in 3 of 19 samples of Chinese herbal medicines [38]. Commonly used Nigerian indigenous crude herbal preparations tested positive for aflatoxin contamination in the range of 0.004–0.345 µg/kg, which is also lower than the current study findings [18]. AFB_1 has also been reported in kava kava at a concentration of 0.0005 µg/kg. In the same study, other botanical roots' samples tested negative for aflatoxins [39] In China, the authors of [40] reported one sample of medicinal materials of radix and rhizome to be contaminated with AFB_1 (5 µg/kg).

The optimum conditions for aflatoxin production by *Aspergillus flavus* and *A. parasiticus* species are at (0.94–0.99 a_w) and temperatures (25–37 °C) [41,42]. The climate of the Free State province, especially the summer temperatures, might contribute to aflatoxin contamination in medicinal plants whilst another factor might be the climatic conditions where the plants are collected: Gauteng as well as the KwaZulu Natal major markets. The concentrations reported in this study are above the permissible limit for AFB_1 of 5 µg/kg and total AFs (10 µg/kg) [43]. The AFB_1 level reported in the study is also above maximum limits of 2 µg/kg for AFB$_1$ and 4 µg/kg for total aflatoxins in herbal drugs set by the European Pharmacopoeia [44] as well as Liu et al. [45], who proposed maximum limits of 5 µg/kg and 10 µg/kg for AFB$_1$ and total aflatoxins, respectively. The presence of AFB$_1$ sheds light on the possibility of contamination of medicinal plants by aflatoxins. Therefore, consumers of medicinal plants sold in the markets might be at risk of mycotoxicosis due to aflatoxin contamination.

3.2. Deoxynivalenol (DON) and Nivalenol (NIV)

DON and NIV were not detected in all the samples that were analyzed in the current study. This outcome contrasts with a study conducted in Spain which found 62% of the 84 types of aromatics and/or medicinal herb samples analyzed were contaminated with DON [46]. A study of Chinese medicinal herbs and related products reported DON (17.2–50.5 µg/kg) contamination in 3 out of 58 samples [47]. A most recent study by Darra et al. [48] in Lebanon on multi-mycotoxin occurrence in commercial spices and herbs found DON (12% in spices, 3% in herbs) but NIV was not detected.

In Latvia, DON was detected in 45% of marketed herbal tea samples at concentrations of 129 µg/kg in the herbal blend and 5.463 µg/kg in wormwood tea [49]. Whilst previous studies have detected the presence of DON and NIV, the absence of these mycotoxins may be attributed to the environmental conditions which do not favor the production of these mycotoxins by fungi species. The European Commission has set limits of DON at maximum of 200 µg/kg for processed cereal-based food and 1250 µg/kg for unprocessed cereals [50]. In South Africa, DON is regulated and for maize or barley ready for human consumption they may not contain more than 1000 ug/kg of deoxynivalenol [43]. However, regulatory limits have not yet been provided for NIV.

3.3. Fumonisins (FB$_1$, FB$_2$, FB$_3$)

In accordance with the present results, previous studies have demonstrated that medicinal plants can be contaminated with fumonisins. In Turkey, 2% of 115 medicinal plants and herbal tea samples tested positive for FB$_1$ at levels of 0.00016 and 0.001487 µg/kg [51]. The current study findings were lower than previous studies in South Africa [16], where it was reported in the Eastern Cape that only 4 out of 30 medicinal wild plants samples tested were contaminated with FB$_1$ (8–1553 µg/kg. In another study of African traditional herbal medicines sold in Tshwane and Cape Town, 81% of 16 samples were found contaminated with FB1 (14–139 µg/kg) [17]. These earlier studies concluded that FB$_1$ contamination was more common in South African medicinal herbs whilst the current study found FB$_2$ to be the predominant fumonisin derivative.

Samples of black tea and medicinal plants sold in Lisbon supermarkets in Portugal 65% tested positive for FB$_1$ (range, 20 to 700 µg/kg) whilst none were contaminated with FB$_2$ [52]. Han et al. [53] reported that more than 50% of 35 samples of traditional Chinese medicines tested positive for fumonisins' contamination (0.58–88.95 µg/kg). In a recent study of mycotoxin contamination in Menthae haplocalycis, Luo et al., reported FB$_1$ and FB$_2$ in the samples analyzed [54]. The presence of fumonisins' contamination has been reported before which demonstrates the risk to consumers, the need for continuous monitoring and in vitro studies in exposure risk assessments.

In South Africa, fumonisin is regulated for raw maize grain intended for further processing, that may not contain more than 4000 µg/kg of FB$_1$ and FB$_2$, whereas for maize flour, maize meal ready for human consumption has a limit of 200 µg/kg for FB$_1$ and FB$_2$ whole commodity [43] The present study findings for total fumonisins were below the regulatory limits, but this does not absolve the consumer from fumonisin-mycotoxin health risks.

3.4. Ochratoxin (OTA)

In Poland, 49% of the 79 samples of herbs analyzed for natural occurrence of OTA contamination tested positive whilst 22.3% exceeded OTA acceptable limits [55]. The

OTA concentration from this study was higher than the results of Roy and Kumar [56] who reported that 44% of 129 herbal samples destined for Ayurvedic medicines were contaminated with OTA (range 0.3–0.00234 µg/kg). Aziz et al., reported higher OTA levels compared to our findings; in their study, 3 out of 17 medicinal plant samples were contaminated by OTA at a mean concentration of (20–80 µg/kg [25]).

In addition, Bresch et al., reported OTA contamination in 50% of the 19 licorice samples (range, 0.3 to 216 µg/kg) [57]. A study on Chinese medicinal plants reported OTA presence in 44% of the 57 samples analyzed (range 1.2–158.7 µg/kg) [38]. In a recent study Ochratoxin A (OTA) was detected in 10% of herbal teas marketed in Latvia at concentrations that ranged between 2.99–30.3 µg/kg [49].

According to EFSA, OTA has been found in breast milk, which could represent a possible health concern for breast-fed infants [58]. Shim et al., reported an OTA transfer rate of (12.72–61.33%) from herbal medicines to decoctions indicating that the use of mycotoxin-contaminated medicinal plants presents a health risk to the consumers after consumption of such products [59]. OTA is not regulated in South Africa, however, according to the European Union Commission Regulation [50], the maximum residue level (MRL) for OTA in nutmeg, ginger, turmeric, black and white pepper, licorice root and its extract, the legislative limit varies from 15 µg/kg to 80 µg/kg. In the current study, the OTA concentration was below the set limit as well as the European regulatory standard (5 µg/kg in unprocessed cereals) [50].

3.5. Zearalenone (ZEN)

The mean concentration for zearalenone recorded in this study was lower than the one reported in China wherein all nine samples of the coix seed medicinal herb tested positive for ZEN (range, 18.7–211.4 µg/kg) [60]. Similarly, another study in China [61] also reported ZEN contamination of coix seeds (68.9 to 119.6 µg/kg). ZEN was not detected in an earlier study of 84 medicinal plant samples using direct determination methods [26]. Different countries have set a maximum limit for ZEN ranging from 20 to 1000 µg/kg in raw and processed food items [50]. In our study, ZEN contamination levels (7–183 µg/kg) did not exceed the permissible limits. ZEN-advanced pubertal changes in young children have been reported in Puerto Rico and gynecomastia with testicular atrophy has been reported in rural males in Southern Africa [62,63].

3.6. Multiple Mycotoxin Contamination

In this study, the mycotoxin occurrence was mainly extracted from the fumonisin derivatives (FB_1, FB_2, FB_3). Mycotoxin co-occurrence has been reported in previous studies of medicinal plants. In Spain, all 84 samples of medicinal and aromatic herbs analyzed showed multi-contamination with AFs, OTA, ZEN, FBs, DON, T-2 toxin and citrinin [46]. Another study reported contamination in 20.58% of the powdered herbal samples with mycotoxins (total aflatoxins, sterigmatocystin, citrinin) [64]. In an analysis of ginger products, aflatoxins and OTA were detected in 67% and 74% samples, respectively, with a range of 0.001–0.03 ng/kg [65]. A study by Koul and Sumbali, [66], found the presence of ZEN and DON in 13.07% and 6.92% of 130 samples of medicinally important dried rhizomes and root tubers. Veprikova et al., analyzed herbal-based dietary supplements for the presence of 57 mycotoxins. The study reported Fusarium trichothecenes, ZEN and ENs and Alternaria as the main mycotoxins and mycotoxin co-occurrence of ENs, HT-2, T-2 and Alternaria toxins [67].

A simultaneous analysis of multiple mycotoxins in 44 samples of *Alpinia oxyphylla* by UPLC-MS/MS detected AFB_1, ZEN, OTA, FB_1 and FB_2 in four moldy samples [68]. Another study of multiclass mycotoxins in Chinese medicinal and edible lotus seeds found three of the ten batches of samples tested positive for AFB_1, FB_2, T-2 and ZEN [69]. An investigation into the presence of multi-class mycotoxins in 40 batches of *Menthae haplocalycis* samples found the most common mycotoxin was tentoxin, followed by alternariol, alternariol monomethyl ether, ZEN, FB_2, FB_3, OTA, AFB_1, AFB_2, AFG_1 and T-2 toxin [54].

Reinholds et al. [49] analyzed 60 samples of herbal teas from Latvia drugstores for the presence of 12 mycotoxins. Among the dry tea samples, 90% were positive for at least one–eight mycotoxins. A study on teas and medicinal plants used to prepare infusions in Portugal reported that 84% of the analyzed samples tested positive for at least one of the mycotoxins [70]. Narvaez, in the analysis of the presence of 16 mycotoxins in botanical nutraceuticals, reported a co-occurrence in 4 out of 10 samples (EN B1, EN A and EN A1). Meanwhile, the prevalent mycotoxins were ZEN (60%) and EN B1 (30%) in samples analyzed [71]. A recent study by Caldeirão et al. [72], analyzed 58 herbs from Brazil for the presence of 14 mycotoxins by LC-MS/MS. Mycotoxin multiple contamination (range 1–8) was reported in 72% of the samples. The most prevalent mycotoxins were enniatins (EN), beauvericin (BEA), sterigmatocystin (STE) and HT-2 toxin, whilst FB_1, FB_2, and T-2 were not detected in any of the samples. Furthermore, the concentration of mycotoxins in the herbal infusions was 88% lower than in the raw herbs.

The percentage of positive samples and mycotoxin co-occurrence of mycotoxins varied among the different studies. This can be attributed to the sensitivity of the methods used and the wide range of mycotoxins analyzed as compared to the current study which only investigated the presence of eight mycotoxins. The reports from preceding studies further indicate that multiple mycotoxin contamination in medicinal plants, herbs and herbal products is cause of concern. Therefore, there is need for a comprehensive analysis of other emerging mycotoxins in routine monitoring of medicinal plants and their products.

The presence of mycotoxins in human food and animal feed increases the risk of endemic diseases such as malaria, hepatitis and HIV with consequent acute and chronic effects [8]. The lack of epidemiological studies, focusing on co-exposure to multi-class mycotoxins and associated health outcomes, is partly attributable to the absence of valid biomarkers [8,73]. A study of 53 South African women, found eight single or combined mycotoxins in urine samples including: DON; FB_1; OTA; and ZEN [73]. In another study conducted in Cameroon, the authors reported the detection of 11 single or combined mycotoxins and their metabolites in 63% of 175 urine samples including AFM_1, OTA and DON [74]. The presence of more than one mycotoxin demonstrates the possibility of mycotoxin exposures from single or multiple sources not limited to food but also from other non-food sources as reported earlier. Therefore, the contribution of medicinal plants as source of mycotoxin exposures should not be underestimated. Whilst co-occurrence has been reported in medicinal plants by several previous studies, present study findings warrant further research to analyze for a wide range of mycotoxins, especially the ones not frequently studied/reported.

4. Conclusions

The study evaluated the presence of mycotoxins (AFB_1, DON, FBs, NIV, OTA, and ZEN) as they have been reported to be the major mycotoxins of public health importance. The findings indicate the prevalence of mycotoxin contamination in frequently traded medicinal plants in South Africa. Mycotoxins pose a health risk to consumers due to the additive or synergistic effects of mycotoxins. Taking into consideration the frequency of use, dietary intakes and individual susceptibility among other factors, consumers are at an increased risk from mycotoxins and their adverse health effects. The current study's findings, supported by previous urinary biomarkers' assessment reports, demonstrate that consumers of medicinal plants are at risk despite the low concentration levels recorded for some mycotoxins in medicinal plants and their products. This is the first study in the Free State Province, South Africa to investigate multiple mycotoxin contamination in marketed medicinal plants. There is therefore a need for routine monitoring of multiple mycotoxins and regulations as well as human exposure assessments using biomarker analysis. Inspections of storage and trading conditions, including regulation of trade, are required to ensure that the trade in medicinal plants is conducted in environments that do not favor the growth of fungi and mycotoxin production.

5. Materials and Methods

5.1. Standards and Reagents

The mycotoxin standards comprising of Aflatoxin B_1 (AFB$_1$), Zearalenone (ZEN), Nivalenol (NIV), Deoxynivalenol (DON) and Ochratoxin A (OTA), were obtained from Sigma-Aldrich (Bornem, Belgium). Fumonisin (B_1, B_2, B_3) was purchased from Promec Unit (Tynberg, South Africa). Acetonitrile (VWR International, Zaventem, Belgium) and methanol (Biosolve, Valkenswaard, The Netherlands), Formic acid ($\geq$98%) (Merck, Darmstadt, Germany). All reagents were of analytical grade.

5.2. Sample Collection and Preparation

A survey was carried out throughout the Free State province, South Africa where 48 vendors were asked to list their top ten selling medicinal plants. Participants listed 165 medicinal plants. The plants which had a Frequency Index (percentage frequency of mention for a single species by informants) $\geq$ 10 were selected for further analysis. A total of 34 samples from 32 plant species (Table S1) were randomly selected and purchased from *muthi* shops and street vendors. The samples that were procured from the *muthi* shops (16) and street vendors (18) comprised of roots, bark, leaves, stems and bulbs. Samples were collected in a dry state in sterile zip-lock plastic bags and immediately transported to the CAFSaB laboratory at the Central University of Technology. Samples were further dried to reduce moisture content in a laminar air flow dryer (Lasec). All dried samples were milled using a Kinematica Polymix PX-MFC90D (Kinematica AG, Luzern, Switzerland) to less than 0.5 mm particle size. Homogenized samples of 30 g were divided into two, for mycotoxin and microbial analysis. Samples were stored in sterile zip-lock bags at 4 °C to inhibit mycotoxin production and fungal growth until analysis.

5.3. Mycotoxin Extraction

A simple solvent extraction method with no sample clean-up was used as described by Spanjer [27]. Homogenized samples were accurately weighed (approx. 5 g) using an analytical balance (3 dp) into a 50 mL tube. Extraction solvent of 20 mL of water/methanol/acetonitrile (2/1/1, v/v) was added and sonicated for 60 min. Then, 1 mL of sample was aliquoted into a 2 mL Eppendorf tube and double diluted with 75% water; 25% methanol solvent and centrifuged for 5 min at 13,000 rpm. A total of 1 mL of the diluted sample was aliquoted into an analysis vial for analysis.

5.4. Equipment Calibration

A calibration graph was created by plotting the obtained peak area or peak height for each standard working solution against the mass of each mycotoxin injected. Each mycotoxin peak in the chromatogram was identified by comparing the retention times with those of corresponding reference standards. The quantity of mycotoxins in injected eluate was determined by comparison to the respective standard curves of each mycotoxin standard.

5.5. Liquid Chromatography Tandem Mass Spectrometry

A Waters Acquity Ultra Performance Liquid Chromatography (UPLC) apparatus coupled to a Xevo Triple Quadrupole Tandem Mass Spectrometer (TQMS) (Waters, Milford, MA, USA) was used for high resolution UPLC/MS/MS for the detection and quantification of mycotoxins. A symmetry Waters column UPLC BEH -C18 (100 mm × 2.1 id; 1.7 μm particle size) attached to a guard column (10 mm × 2.1 mm i.d.) (Waters, Zellik, Belgium) was used. A fixed sample injection volume of 2 μL was used. Mobile phase solvent A consisted of acidified water with 0.1% formic acid (10/1, v/v) and mobile phase solvent B of 0.1% formic acid in acetonitrile acidified with 0.1% formic acid (10/1, v/v). Multiple mycotoxins were separated in the mass spectrometer operated using selected multiple reaction monitoring channels (MRM) in positive electrospray ionization mode (ESI+). ESI conditions were optimized as follows; capillary voltage, 3.5 V; cone voltage range, 15–50 V; collision energy range, 10–40 eV; source temperature, 140 °C. Nitrogen was used as the desolvation gas, desolvation temperature of 400 °C; desolvation gas, 800 L/h and cone gas, 50 L/h.

The gradient elution program (illustrated in Figure S1) at initial conditions of 98% A (Water + 0.1% formic acid), held for 0–0.5 min at a flow rate of 0.35 mL/min was used. This was followed by a slow gradient change of solvent A to 60% from 0.5–7 min. From 7–10 min there was another gradient change of solvent A to 30%. A rapid gradient change ensued for solvent A to 5% from 10–11 min. There was another gradient change of solvent A to 0% from 11–12 min. This was followed by a quick gradient change to initial conditions of 98% solvent A from 12–12.1 min. After that, an isocratic period of 98% of solvent A was kept for 12.1–14 min. The column was reconditioned with solvent B (Acetonitrile + 0.1% formic acid) for 5 min before the next injection. The total analytical run time was 14 min through a linear decrease of mobile phase.

5.6. Data Acquisition and Analysis

The frequency index (FI) used to select the plants used in this study was calculated using the formula FI = (FC ÷ N) × 100. FC is the number of informants who mentioned the use of the species, and N is the total number of informants. N = 48 in this study. Plant names were documented in the local languages, mostly Sotho and Zulu, and scientific names identified from literature.

MassLynx and QuanLynx software's version 4.1 (Micromass, Manchester, UK) were used for data acquisition and processing. Descriptive statistics (mean, range, maximum and the frequency of the data obtained in this study) were calculated using Microsoft Office Excel 2016.

Supplementary Materials: The following supporting information can be downloaded at: https://www.mdpi.com/article/10.3390/toxins14100690/s1, Figure S1: Gradient elution program; Table S1: Medicinal plants screened for mycotoxins.

Author Contributions: Research design: J N., I.T.M., M.N. and O.d.S.; Funding acquisition: I.T.M.; Fieldwork and sample collection: J.N.; laboratory analysis of samples: J.N. and O.d.S.; Data analysis and writing original draft: J.N.; Supervision and reviewing article drafts: I.T.M., M.N. and O.d.S. All authors have read and agreed to the published version of the manuscript.

Funding: The study was funded by the National Research Foundation of South Africa (Grant No 118587) and the Central University of Technology Free State Research and Innovation Grant.

Institutional Review Board Statement: The study protocol was approved by the University Research and Innovation Committee (URIC) of the Central University of Technology, Free State. Ethical clearance was obtained from the University of Free State's Environment and Biosafety Ethics Committee (UFS-ESD2019/01831504) and the Health Sciences Research Ethics Committee (UFS-HSD2019/1226/2605) on 15 April 2020 and 4 May 2020 respectively.

Informed Consent Statement: Participation in the study was voluntary, informed consent was obtained from all participants involved in the study.

Data Availability Statement: Not applicable.

Acknowledgments: We acknowledge the Centre for Applied Food Sustainability and Biotechnology Central University of Technology, and Stellenbosch University Central Analytical facilities for providing laboratory facilities for this study; B Ngobeni and Agnes Liako Mohale are acknowledged for assisting with sample collection.

Conflicts of Interest: The authors declare no conflict of interest.

References

1. Ekor, M. The growing use of herbal medicines: Issues relating to adverse reactions and challenges in monitoring safety. *Front. Pharmacol.* **2014**, *4*, 177. [CrossRef] [PubMed]
2. Egan, B.; Hodgkins, C.; Shepherd, R.; Timotijevic, L.; Raats, M. An overview of consumer attitudes and beliefs about plant food supplements. *Food Funct.* **2011**, *2*, 747–752. [CrossRef]

3. Khan, M.S.; Ahmad, I. Herbal medicine: Current trends and future prospects. In *New Look to Phytomedicine*; Khan, M.S.A., Ahmad, I., Chattopadhyay, D., Eds.; Academic Press: London, UK, 2019; pp. 3–13. [CrossRef]

4. Sofowora, A.; Ogunbodede, E.; Onayade, A. The role and place of medicinal plants in the strategies for disease prevention. *Afr. J. Tradit. Complement. Altern. Med.* **2013**, *10*, 210–229. [CrossRef]

5. Sitole, P.; Ndhlala, A.R.; Kritzinger, Q.; Truter, M. Mycoflora contamination of South African medicinal plants. *S. Afr. J. Bot.* **2017**, *100*, 369–370. [CrossRef]

6. Yu, J.; Yang, M.; Han, J.; Pang, X. Fungal and mycotoxin occurrence, affecting factors, and prevention in herbal medicines: A review. *Toxin Rev.* **2021**, *41*, 976–994. [CrossRef]

7. Agriopoulou, S.; Stamatelopoulou, E.; Varzakas, T. Advances in occurrence, importance, and mycotoxin control strategies: Prevention and detoxification in foods. *Foods* **2020**, *9*, 137. [CrossRef]

8. Wild, C.P.; Gong, Y.Y. Mycotoxins and human disease: A largely ignored global health issue. *Carcinogenesis* **2010**, *31*, 71–82. [CrossRef] [PubMed]

9. Silva, J.V.B.D.; Oliveira, C.A.F.D.; Ramalho, L.N.Z. An overview of mycotoxins, their pathogenic effects, foods where they are found and their diagnostic biomarkers. *Food Sci. Technol.* **2021**, *42*, 1–9. [CrossRef]

10. Zain, M.E. Impact of mycotoxins on humans and animals. *J. Saudi Chem. Soc.* **2011**, *15*, 129–144. [CrossRef]

11. Gibb, H.; Devleesschauwer, B.; Bolger, P.M.; Wu, F.; Ezendam, J.; Zeilmaker, M.; Verger, P.; Pitt, J.; Baines, J.; Adegoke, G. World Health Organization estimates of the global and regional disease burden of four foodborne chemical toxins, 2010: A data synthesis. *F1000Research* **2015**, *4*, 1393. [CrossRef]

12. Ma, R.; Zhang, L.; Liu, M.; Su, Y.T.; Xie, W.M.; Zhang, N.Y.; Dai, J.F.; Wang, Y.; Rajput, S.A.; Qi, D.S.; et al. Individual and combined occurrence of mycotoxins in feed ingredients and complete feeds in China. *Toxins* **2018**, *10*, 113. [CrossRef] [PubMed]

13. Egbuta, M.A.; Mwanza, M.; Babalola, O.O. Health risks associated with exposure to filamentous fungi. *Int. J. Environ. Health Res.* **2017** *14*, 719. [CrossRef]

14. Zhao, L.; Zhang, L.; Xu, Z.; Liu, X.; Chen, L.; Dai, J.; Karrow, N.A.; Sun, L. Occurrence of Aflatoxin B1, deoxynivalenol and zearalenone in feeds in China during 2018–2020. *J. Anim. Sci. Biotechnol.* **2021**, *12*, 74. [CrossRef]

15. Liu, M.; Zhao, L.; Gong, G.; Zhang, L.; Shi, L.; Dai, J.; Han, Y.; Wu, Y.; Khalil, M.M.; Sun, L. Invited review: Remediation strategies for mycotoxin control in feed. *J. Anim. Sci. Biotechnol.* **2022**, *13*, 9. [CrossRef] [PubMed]

16. Sewram, V.; Shephard, G.S.; van der Merwe, L.; Jacobs, T.V. Mycotoxin contamination of dietary and medicinal wild plants in the Eastern Cape Province of South Africa. *J. Agric. Food Chem.* **2006**, *54*, 5688–5693. [CrossRef] [PubMed]

17. Katerere, D.R.; Stockenström, S.; Thembo, K.M.; Rheeder, J.P.; Shephard, G.S.; Vismer, H.F. A preliminary survey of mycological and fumonisin and aflatoxin contamination of African traditional herbal medicines sold in South Africa. *Hum. Exp. Toxicol.* **2008**, *27*, 793–798. [CrossRef]

18. Ezekiel, C.; Ortega-Beltran, A.; Bandyopadhyay, R. The need for integrated approaches to address food safety risk: The case of mycotoxins in Africa. In Proceedings of the First FAO/WHO/AU International Food Safety Conference, Addis Ababa, Ethiopia, 12–13 February 2019; Available online: https://www.fao.org/3/CA3313EN/ca3313en.pdf (accessed on 20 July 2022).

19. Mukundi, J.W. Bacteria, Aflatoxins and Fluoride Levels in Locally Processed Herbal Medicines from Nairobi County, Kenya. Master's Thesis, University of Nairobi, Nairobi, Kenya, 2015.

20. Keter, L.; Too, R.; Mwikwabe, N.; Mutai, C.; Orwa, J.; Mwamburi, L.; Ndwigah, S.; Bii, C.; Korir, R. Risk of fungi associated with aflatoxin and fumonisin in medicinal herbal products in the Kenyan market. *Sci. World J.* **2017**, *2017*, 1892972. [CrossRef]

21. Korir, R.K. Microbial and Heavy Metal Contaminations in Selected Herbal Medicinal Products Sold in Nairobi, Kenya. Ph.D. Thesis, University of Nairobi, Nairobi, Kenya, 2017.

22. Efuntoye, M.O. Mycotoxins of fungal strains from stored herbal plants and mycotoxin contents of Nigerian crude herbal drugs. *Mycopathologia* **1999**, *147*, 43–48. [CrossRef]

23. Oyero, O.G.; Oyefolu, A.O. Fungal contamination of crude herbal remedies as a possible source of mycotoxin exposure in man. *Asian Pac. J. Trop. Med.* **2009**, *2*, 38–43.

24. Ezekwesili-Ofili, J.O.; Onyemelukwe, N.F.; Agwaga, P.; Orji, I. The bioload and aflatoxin content of herbal medicines from selected states in Nigeria. *Afr. J. Tradit. Complement. Altern. Med.* **2014**, *11*, 143–147. [CrossRef]

25. Ikeagwulonu, R.C.; Onyenekwe, C.C.; Ukibe, N.R.; Ikimi, C.G.; Ehiaghe, F.A.; Emeje, I.P.; Ukibe, S.N. Mycotoxin contamination of herbal medications on sale in Ebonyi State, Nigeria. *Int. J. Biol. Chem. Sci.* **2020**, *14*, 613–625. [CrossRef]

26. Aziz, N.H.; Youssef, Y.A.; El-Fouly, M.Z.; Moussa, L.A. Contamination of some common medicinal plant samples and spices by fungi and their mycotoxins. *Bot. Bull. Acad. Sin.* **1998**, *39*, 279–285.

27. Abou-Arab, A.A.; Kawther, M.S.; El Tantawy, M.E.; Badeaa, R.I.; Khayria, N. Quantity estimation of some contaminants in commonly used medicinal plants in the Egyptian market. *Food Chem.* **1999**, *67*, 357–363. [CrossRef]

28. Allam, N.G.; El-Shanshoury, A.E.; Emara, H.A.; Zaky, A.Z. Decontamination of ochratoxin-A producing Aspergillus niger and ochratoxin A in medicinal plants by gamma irradiation and essential oils. *J. Int. Environ. Appl. Sci.* **2012**, *7*, 161–169.

29. Abol-Ela, M.F.; AbdeL-Ghany, Z.M.; Atwa, M.A.; El-Melegy, K.M. Fungal Load and Mycotoxinogenesis of some Egyptian Medicinal Plants. *J. Agric. Chem. Biotechnol.* **2011**, *2*, 217–228. [CrossRef]

30. Rudrabhatla, M.; George, J.E.; Faye, T. Multicomponent mycotoxin analysis by LC/MS/MS. In Israel Analytical Chemistry Society editor. In Proceedings of the 10th Annual Meeting of the Israel Analytical Chemistry Society Conference and Exhibition, Tel-Aviv, Israel, 23–24 January 2007.

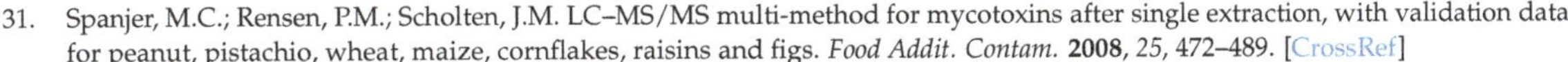

31. Spanjer, M.C.; Rensen, P.M.; Scholten, J.M. LC–MS/MS multi-method for mycotoxins after single extraction, with validation data for peanut, pistachio, wheat, maize, cornflakes, raisins and figs. *Food Addit. Contam.* **2008**, *25*, 472–489. [CrossRef]

32. Sadhasivam, S.; Britzi, M.; Zakin, V.; Kostyukovsky, M.; Trostanetsky, A.; Quinn, E.; Sionov, E. Rapid detection and identification of mycotoxigenic fungi and mycotoxins in stored wheat grain. *Toxins* **2017**, *9*, 302. [CrossRef]

33. Pickova, D.; Ostry, V.; Toman, J.; Malir, F. Aflatoxins: History, Significant Milestones, Recent Data on Their Toxicity and Ways to Mitigation. *Toxins* **2021**, *13*, 399. [CrossRef]

34. Aiko, V.; Mehta, A. Prevalence of toxigenic fungi in common medicinal herbs and spices in India. *3 Biotech* **2016**, *6*, 159. [CrossRef]

35. Ledzion, E.; Rybińska, K.; Postupolski, J.; Kurpińska-Jaworska, J.; Szczesna, M. Studies and safety evaluation of aflatoxins in herbal plants. *Rocz. Panstw. Zakl. Hig.* **2011**, *62*, 377–381.

36. Mannani, N.; Tabarani, A.; Zinedine, A. Assessment of aflatoxin levels in herbal green tea available on the Moroccan market. *Food Control* **2020**, *108*, 106882. [CrossRef]

37. Tassaneeyakul, W.; Razzazi-Fazeli, E.; Porasuphatana, S.; Bohm, J. Contamination of aflatoxins in herbal medicinal products in Thailand. *Mycopathologia* **2004**, *158*, 239–244. [CrossRef] [PubMed]

38. Yang, M.H.; Chen, J.M.; Zhang, X.H. Immunoaffinity column clean-up and liquid chromatography with post-column derivatization for analysis of aflatoxins in traditional Chinese medicine. *Chromatographia* **2005**, *62*, 499–504. [CrossRef]

39. Weaver, C.M.; Trucksess, M.W. Determination of aflatoxins in botanical roots by a modification of AOAC official method SM 991.31: Single-laboratory validation. *J. AOAC Int.* **2010**, *93*, 184–189. [CrossRef] [PubMed]

40. Hu, S.; Dou, X.; Zhang, L.; Xie, Y.; Yang, S.; Yang, M. Rapid detection of aflatoxin B1 in medicinal materials of radix and rhizome by gold immunochromatographic assay. *Toxicon* **2018**, *150*, 144–150. [CrossRef] [PubMed]

41. Lahouar, A.; Marin, S.; Crespo-Sempere, A.; Saïd, S.; Sanchis, V. Effects of temperature, water activity and incubation time on fungal growth and aflatoxin B1 production by toxinogenic Aspergillus flavus isolates on sorghum seeds. *Rev. Argent. Microbiol.* **2016**, *48*, 78–85. [CrossRef]

42. Gizachew, D.; Chang, C.H.; Szonyi, B.; De La Torre, S.; Ting, W.T. Aflatoxin B1 (AFB1) production by Aspergillus flavus and Aspergillus parasiticus on ground Nyjer seeds: The effect of water activity and temperature. *Int. J. Food Microbiol.* **2019**, *296*, 8–13. [CrossRef] [PubMed]

43. Government Gazette. Foodstuffs, S.A. Cosmetics and Disinfectant Act 1972 (ACT No 54 of 1972) Regulations Governing Tolerance for Fungusproduced Toxins in Foodstuffs Amendment No. 26849. R1145. 2004. Available online: https://www.gov.za/sites/default/files/gcis_document/201409/26849b0.pdf (accessed on 20 July 2022).

44. Artiges, A. The European directorate for the quality of medicines. *Ann. Pharm. Fr.* **2001**, *59*, 63–68. [PubMed]

45. Liu, L.; Jin, H.; Sun, L.; Ma, S.; Lin, R. Determination of Aflatoxins in Medicinal Herbs by High-performance Liquid Chromatography–Tandem Mass Spectrometry. *Phytochem. Anal.* **2012**, *23*, 469–476. [CrossRef]

46. Santos, L.; Marín, S.; Sanchis, V.; Ramos, A.J. Screening of mycotoxin multicontamination in medicinal and aromatic herbs sampled in Spain. *J. Sci. Food Agric.* **2009**, *89*, 1802–1807. [CrossRef]

47. Yue, Y.T.; Zhang, X.F.; Pan, J.; Ou-Yang, Z.; Wu, J.; Yang, M.H. Determination of deoxynivalenol in medicinal herbs and related products by GC–ECD and confirmation by GC–MS. *Chromatographia* **2010**, *71*, 533–538. [CrossRef]

48. El Darra, N.; Gambacorta, L.; Solfrizzo, M. Multimycotoxins occurrence in spices and herbs commercialized in Lebanon. *Food Control* **2019**, *95*, 63–70. [CrossRef]

49. Reinholds, I.; Bogdanova, E.; Pugajeva, I.; Bartkevics, V. Mycotoxins in herbal teas marketed in Latvia and dietary exposure assessment. *Food Addit. Contam. Part B* **2019**, *12*, 199–208. [CrossRef]

50. European Commission. Commission Regulation (EC) No 1881/2006 of 19 December 2006 setting maximum levels for certain contaminants in foodstuffs. *Off. J. Eur. Union* **2006**, *364*, 5–24.

51. Omurtag, G.Z.; Yazicioğilu, D. Determination of fumonisins B1 and B2 in herbal tea and medicinal plants in Turkey by high-performance liquid chromatography. *J. Food Prot.* **2004**, *67*, 1782–1786. [CrossRef] [PubMed]

52. Martins, M.L.; Martins, H.M.; Bernardo, F. Fumonisins B1 and B2 in black tea and medicinal plants. *J. Food Prot.* **2001**, *64*, 1268–1270. [CrossRef]

53. Han, Z.; Ren, Y.; Liu, X.; Luan, L.; Wu, Y. A reliable isotope dilution method for simultaneous determination of fumonisins B1, B2 and B3 in traditional Chinese medicines by ultra-high-performance liquid chromatography-tandem mass spectrometry. *J. Sep. Sci.* **2010**, *33*, 2723–2733. [CrossRef]

54. Luo, J.; Zhou, W.; Dou, X.; Qin, J.; Zhao, M.; Yang, M. Occurrence of multi-class mycotoxins in Menthae haplocalycis analyzed by ultra-fast liquid chromatography coupled with tandem mass spectrometry. *J. Sep. Sci.* **2018**, *41*, 3974–3984. [CrossRef]

55. Waśkiewicz, A.; Beszterda, M.; Bocianowski, J.; Goliński, P. Natural occurrence of fumonisins and ochratoxin A in some herbs and spices commercialized in Poland analyzed by UPLC–MS/MS method. *Food Microbiol.* **2013**, *36*, 426–431. [CrossRef] [PubMed]

56. Roy, A.K.; Kumar, S. Occurrence of ochratoxin A in herbal drugs of Indian origin—A report. *Mycotoxin Res.* **1993**, *9*, 94–98. [CrossRef] [PubMed]

57. Bresch, H.; Urbanek, M.; Nusser, M. Ochratoxin A in food containing liquorice. *Nahrung* **2000**, *44*, 276–278. [CrossRef]

58. EFSA Panel on Contaminants in the Food Chain (CONTAM); Schrenk, D.; Bodin, L.; Chipman, J.K.; del Mazo, J.; Grasl-Kraupp, B.; Hogstrand, C.; Hoogenboom, L.R.; Leblanc, J.C.; Nebbia, C.S.; et al. Risk assessment of ochratoxin A in food. *EFSA J.* **2020**, *18*, e06113. [CrossRef]

59. Shim, W.B.; Ha, K.S.; Kim, M.G.; Kim, J.S.; Chung, D.H. Evaluation of the transfer rate of ochratoxin a to decoctions of herbal medicines. *Food Sci. Biotechnol.* **2014**, *23*, 2103–2108. [CrossRef]
60. Zhang, X.; Liu, W.; Logrieco, A.F.; Yang, M.; Ou-yang, Z.; Wang, X.; Guo, Q. Determination of zearalenone in traditional Chinese medicinal plants and related products by HPLC–FLD. *Food Addit. Contam. Part A* **2011**, *28*, 885–893. [CrossRef]
61. Kong, W.J.; Shen, H.H.; Zhang, X.F.; Yang, X.L.; Qiu, F.; Ou-yang, Z.; Yang, M.H. Analysis of zearalenone and α-zearalenol in 100 foods and medicinal plants determined by HPLC-FLD and positive confirmation by LC-MS-MS. *J. Sci. Food Agric.* **2013**, *93*, 1584–1590. [CrossRef]
62. Shephard, G.S. Impact of mycotoxins on human health in developing countries. *Food Addit. Contam.* **2008**, *25*, 146–151. [CrossRef]
63. Milićević, D.R.; Škrinjar, M.; Baltić, T. Real and perceived risks for mycotoxin contamination in foods and feeds: Challenges for food safety control. *Toxins* **2010**, *2*, 572–592. [CrossRef] [PubMed]
64. Gautam, A.K.; Bhadauria, R. Mycoflora and mycotoxins in some important stored crude and powdered herbal drugs. *Biol-An Int. J.* **2009**, *1*, 1–7.
65. Trucksess, M.; Weaver, C.; Oles, C.; D'Ovidio, K.; Rader, J. Determination of aflatoxins and ochratoxin A in ginseng and other botanical roots by immunoaffinity column cleanup and liquid chromatography with fluorescence detection. *J. AOAC Int.* **2006**, *89*, 624–630. [CrossRef]
66. Koul, A.; Sumbali, G. Detection of zearalenone, zearalenol and deoxynivalenol from medicinally important dried rhizomes and root tubers. *Afr. J. Biotechnol.* **2008**, *7*, 4136–4139.
67. Veprikova, Z.; Zachariasova, M.; Dzuman, Z.; Zachariasova, A.; Fenclova, M.; Slavikova, P.; Vaclavikova, M.; Mastovska, K.; Hengst, D.; Hajslova, J. Mycotoxins in plant-based dietary supplements: Hidden health risk for consumers. *J. Agric. Food Chem.* **2015**, *63*, 6633–6643. [CrossRef] [PubMed]
68. Zhao, X.S.; Kong, W.J.; Wang, S.; Wei, J.H.; Yang, M.H. Simultaneous analysis of multiple mycotoxins in Alpinia oxyphylla by UPLC-MS/MS. *World Mycotoxin J.* **2017**, *10*, 41–51. [CrossRef]
69. Huang, P.; Kong, W.; Wang, S.; Wang, R.; Lu, J.; Yang, M. Multiclass mycotoxins in lotus seeds analysed by an isotope-labelled internal standard-based UPLC-MS/MS. *J. Pharm. Pharmacol.* **2018**, *70*, 1378–1388. [CrossRef]
70. Duarte, S.C.; Salvador, N.; Machado, F.; Costa, E.; Almeida, A.; Silva, L.J.; Pereira, A.M.; Lino, C.; Pena, A. Mycotoxins in teas and medicinal plants destined to prepare infusions in Portugal. *Food Control* **2020**, *115*, 107290. [CrossRef]
71. Narváez, A.; Rodríguez-Carrasco, Y.; Castaldo, L.; Izzo, L.; Ritieni, A. Ultra-high-performance liquid chromatography coupled with quadrupole Orbitrap high-resolution mass spectrometry for multi-residue analysis of mycotoxins and pesticides in botanical nutraceuticals. *Toxins* **2020**, *12*, 114. [CrossRef]
72. Caldeirão, L.; Sousa, J.; Nunes, L.C.; Godoy, H.T.; Fernandes, J.O.; Cunha, S.C. Herbs and herbal infusions: Determination of natural contaminants (mycotoxins and trace elements) and evaluation of their exposure. *Food Res. Int.* **2021**, *144*, 110322. [CrossRef]
73. Shephard, G.S.; Burger, H.M.; Gambacorta, L.; Gong, Y.Y.; Krska, R.; Rheeder, J.P.; Solfrizzo, M.; Srey, C.; Sulyok, M.; Visconti, A.; et al. Multiple mycotoxin exposure determined by urinary biomarkers in rural subsistence farmers in the former Transkei, South Africa. *Food Chem. Toxicol.* **2013**, *62*, 217–225. [CrossRef]
74. Abia, W.A.; Warth, B.; Sulyok, M.; Krska, R.; Tchana, A.; Njobeh, P.B.; Turner, P.C.; Kouanfack, C.; Eyongetah, M.; Dutton, M.; et al. Bio-monitoring of mycotoxin exposure in Cameroon using a urinary multi-biomarker approach. *Food Chem. Toxicol.* **2013**, *62*, 927–934. [CrossRef]

Article

Mycotoxins in Wheat Flours Marketed in Shanghai, China: Occurrence and Dietary Risk Assessment

Haiyan Zhou [1], Anqi Xu [1], Meichen Liu [1], Zheng Yan [1], Luxin Qin [2], Hong Liu [2], Aibo Wu [1] and Na Liu [1,*]

[1] SIBS-UGENT-SJTU Joint Laboratory of Mycotoxin Research, CAS Key Laboratory of Nutrition, Metabolism and Food Safety, Shanghai Institute of Nutrition and Health, University of Chinese Academy of Sciences, Chinese Academy of Sciences, Shanghai 200030, China

[2] Shanghai Municipal Center for Disease Control and Prevention, Shanghai 200336, China

* Correspondence: liuna@sinh.ac.cn; Tel.: +86-21-54920716

Abstract: The risk of exposure to mycotoxins through the consumption of wheat flours has long been a concern. A total of 299 wheat flours marketed in Shanghai Province of China were surveyed and analyzed for the co-occurrence of 13 mycotoxins through an ultra-high performance liquid chromatography-tandem mass spectrometry (UPLC-MS/MS) method. The detection rates of mycotoxins in wheat flours ranged from 0.7~74.9% and their average contamination levels in wheat flours (0.2~57.6 µg kg^{-1}) were almost lower than the existing regulations in cereals. However, their co-contamination rate was as high as 98.1%, especially Fusarium and Alternaria mycotoxins. Comparative analysis of different types of wheat flours showed that the average contamination levels in refined wheat flours with low-gluten were lower. Based on these contamination data and the existing consumption data of Shanghai residents, point evaluation and the Monte Carlo assessment model were used to preliminarily evaluate the potential dietary exposure risk. The probable daily intakes of almost all mycotoxins, except for alternariol, were under the health-based guidance values for 90% of different consumer groups. Health risks of dietary exposure to alternariol should be a concern and further studied in conjunction with an internal exposure assessment.

Keywords: wheat flours; mycotoxins; contamination characteristics; risk assessment

Key Contribution: We investigated and analyzed the occurrence and co-occurrence of mycotoxins in wheat flours used for human consumption from Shanghai, China, 2020–2021, through a modified and validated UPLC-MS/MS method for the simultaneous determination of 13 mycotoxins. The obtained data were used to compare distribution characteristics of mycotoxins in different types of wheat flours. The chronic dietary intake risks for Shanghai residents were profiled and the approximate cumulative exposure risks of major co-contamination patterns in wheat flours were also considered for the first time.

Citation: Zhou, H.; Xu, A.; Liu, M.; Yan, Z.; Qin, L.; Liu, H.; Wu, A.; Liu, N. Mycotoxins in Wheat Flours Marketed in Shanghai, China: Occurrence and Dietary Risk Assessment. *Toxins* **2022**, *14*, 748. https://doi.org/10.3390/toxins14110748

Received: 14 September 2022
Accepted: 22 October 2022
Published: 31 October 2022

Publisher's Note: MDPI stays neutral with regard to jurisdictional claims in published maps and institutional affiliations.

1. Introduction

Wheat flour, which is primary processing food from wheat, is widely used to produce staple foods (breads, biscuits, cakes, pastries, pasta, and noodles). It can be divided into whole wheat flour and refined wheat flour for their differences in processing technology. Whole wheat flour includes the bran and germ, in addition to the basic endosperm, which are important sources of dietary fibers, vitamins, minerals, or healthy phytochemicals. The sales of whole cereal foods exceeded 23.68 billion dollars in China in 2015 [1]. It is also classified as low-gluten, medium-gluten, and high-gluten wheat flour based on protein content. This quality difference will affect the viscoelasticity and rheology of the dough, resulting in the chewiness and gumminess of its derived foods. Recently, organic wheat flours from special agricultural practice have also received more and more attention [2]. Differences in the choice of various types of wheat flours depend on various requirements in terms of food composition and quality for consumers.

Due to differences in climates, resistance, and other factors [3], organic or conventional wheat is still susceptible to mycotoxin contamination by toxigenic fungi, such as *Aspergillus* spp., *Fusarium* spp., *Penicillium* spp., and *Alternaria* spp., especially for their outer layer [4]. Preexisting mycotoxins in wheat grain with spatial localization [5] cannot be destroyed or eliminated by milling, but can be redistributed in different milling fractions, such as germ, bran, grits, or flour [6,7]. Multiple factors mentioned above have confounding effects on differences in mycotoxin concentrations in wheat flours [2,8]. Therefore, wheat flours may be contaminated by individual or multiple mycotoxins concurrently for some adverse factors during the whole industrial chain from agriculture practices to the table [9,10]. Mycotoxin contamination of wheat flour in various markets has also been reported continuously [11,12]. Mycotoxins in terms of deoxynivalenol (DON), ochratoxin A (OTA), zearalenone (ZEN), fusarenon-X (FUS-X), or neosolaniol (NEO) were detected in wheat and wheat flours from China [13–16]. DON was frequently detected in wheat flours marked in Hungary and the metabolic forms of DON were also found in spelt or durum flour [17]. OTA-contaminated wheat flours of 8% and 11.5% in Lebanon and Poland were also reported [18,19]. Fumonisins (FBs) were also common in wheat-based foods [10]. Recently, more focus has been placed on Alternaria mycotoxins (ATs) with the detection in wheat flour samples, which may be related to growth capacity at low temperature for *Alternaria* spp. [20,21]. The contamination rates of DON, alternariol (AOH), tenuazonic acid (TEA) and FB_1 in 54 wheat flour samples from China were 90.7%, 16.7%, 3.7% and 9.3%, respectively [22]. Considering the public health concerns arising from their acute or chronic toxicity, the maximum levels of DON, ZEN, FB_1, and OTA have already been proposed, not only in cereals and cereal flour, bran and germ intended for direct human consumption (750, 75, 800, 3 $\mu g\ kg^{-1}$), but also in cereal-based processing foods and baby foods for infants and young children (200, 20, 200, and 0.5 $\mu g\ kg^{-1}$) in the European Union [23]. Moreover, the tolerable daily intakes (TDIs) for DON and its derivatives, ((15-acetyl-deoxynivalenol (15-AcDON), 3-acetyl-deoxynivalenol (3-AcDON), deoxynivalenol-3-glucoside (D_3G)), nivalenol (NIV), ZEN, OTA, and FB1, are 1.0, 0.7, 0.25, 0.01, and 2 $\mu g\ kg^{-1}$ bw day^{-1}, respectively [24–27]. The European Food Safety Authority (EFSA) also released the thresholds of toxicological concern (TTC) for the genotoxic compounds AOH at 0.0025 $\mu g\ kg^{-1}$ bw day^{-1} and for the non-genotoxic tentoxin (TEN) and TEA at 1500 $\mu g\ kg^{-1}$ bw day^{-1} [28]. The probabilistic dietary risk exposure exceeded the safe chronic exposure levels at the 95P of DON through the intake of foods made from contaminated-wheat flours for teenagers in Brazil, which indicate differences in eating habits and body functions that present various dietary risks [29].

Wheat flours remain the predominant dietary source of mycotoxin exposure in grain consumption patterns [30]. However, studies on the co-occurrence of Alternaria and Fusarium toxins in wheat flours, their contamination differences between different types of wheat flour, and the exposure and toxicity of the Alternaria toxin are still limited. Targeted and continuous surveillance of contamination is necessary to explore contamination patterns, ensure minimal contamination and early prediction of their dietary exposure risks and their possible cumulative risks. The first objective of the present study was to survey and analyze the occurrence and co-occurrence of mycotoxins in wheat flours used for human consumption from Shanghai, China, 2020–2021, through a modified and validated UPLC-MS/MS method for the simultaneous determination of multiple mycotoxins (DON, 15-AcDON, 3-AcDON, D_3G, NIV, NEO, FUS-X, AOH, TEA, TEN, OTA, ZEN, and FB_1). The investigated data were used to compare distribution characteristics of mycotoxins in different types of wheat flours and to assess the chronic dietary intake risks for Shanghai residents in terms of age and gender. In particular, the approximate cumulative exposure risks of co-occurred mycotoxins in the same sample (>5%) were also considered for the first time.

2. Results and Discussion

2.1. Method Modification and Validation

The co-occurrence of Alternaria and Fusarium toxins in wheat flours has rarely been explored, and the method for simultaneous determination has been limited until now. After optimization, 75% acetonitrile with 1% formic acid and ACQUITY UPLC BEH C_{18} VanGuard pre-column with ACQUITY UPLC BEH C_{18} column were finally selected as the extraction agent and chromatography column with the acceptable extraction recoveries for most mycotoxins in wheat flour (60~120%) (Figure S1). Analytical parameters in the modified UPLC-MS/MS method were analyzed and validated. Values for the limit of detection (LOD) and limit of quantification (LOQ) for 13 mycotoxins were in the range of 0.08~62.5 µg kg^{-1} and 0.18~125 µg kg^{-1}, which are far below the available MLs in foodstuffs. Acceptable linearities ($R^2 > 0.9$) within the tested range were obtained (Table 1). Mycotoxin recovery in this method of fortified wheat flour samples at three levels ranged from 67.7% to 120.0% and the intra- and inter-day precisions were less than 20% (1.2~20.0%). The results of these validation parameters indicated that the modified method could be applied in the quantitative detection of mycotoxins in wheat flour samples.

Table 1. Overview of the accuracy and precision of the developed LC-MS/MS method.

Mycotoxin	Spike (µg kg^{-1})	Recovery $\pm$ RSD r (%, n = 6)	Recovery $\pm$ RSD$_R$ (%, n = 6)	Linearity Equation (R^2)	LOD (µg kg^{-1})	LOQ (µg kg^{-1})
NIV	50	76.4 ± 19.8	85.7 ± 17.5	Y = 1.9X (0.9891)	7.5	25
	100	106.4 ± 19.8	96.2 ± 18.5			
	200	71.5 ± 17.8	110.9 ± 15.2			
DON	50	115.0 ± 4.5	73.6 ± 7.4	Y = 17.1X (0.9258)	1.6	3.1
	100	109.4 ± 7.1	67.7 ± 5.1			
	200	105.4 ± 5.9	100.0 ± 6.5			
15-AcDON	50	96.1 ± 15.4	119.9 ± 3.6	Y = 36.8X (0.9971)	1.6	3.1
	100	109.6 ± 10.1	118.0 ± 5.5			
	200	91.5 ± 9.5	120.0 ± 1.2			
3-AcDON	50	111.9 ± 4.6	114.5 ± 5.9	Y = 27.0X (0.9949)	3.1	6.2
	100	111.2 ± 6.1	114.6 ± 2.7			
	200	116.6 ± 3.3	118.6 ± 3.3			
D$_3$G	50	85.4 ± 19.3	119.0 ± 6.6	Y = 9.7X (0.9956)	3.1	6.2
	100	98.7 ± 8.1	117.1 ± 5.9			
	200	81.7 ± 12.7	108.9 ± 14.3			
FUS-X	50	105.7 ± 6.3	102.9 ± 9.5	Y = 43.1X (0.9972)	7.5	25
	100	109.5 ± 7.2	110.7 ± 5.3			
	200	80.7 ± 4.4	118.3 ± 5.3			
OTA	10	92.2 ± 5.7	113.6 ± 2.6	Y = 1271.2X (0.9889)	0.3	0.6
	20	108.7 ± 7.0	102.0 ± 12.6			
	40	83.8 ± 6.3	115.9 ± 9.8			
NEO	1	102.0 ± 14.8	119.5 ± 10.4	Y = 529.7X (0.9850)	0.2	0.5
	2	84.9 ± 18.7	115.8 ± 6.7			
	4	83.1 ± 16.8	118.7 ± 6.7			
ZEN	10	92.3 ± 6.5	85.6 ± 6.7	Y = 530.9X (0.9976)	0.3	0.6
	20	112.4 ± 5.5	117.4 ± 5.6			
	40	98.4 ± 3.3	104.5 ± 9.2			
AOH	30	82.3 ± 20.0	106.8 ± 9.9	Y = 38.4X (0.9964)	0.5	1.0
	60	91.8 ± 15.4	115.2 ± 11.4			
	120	74.6 ± 11.8	118.0 ± 5.1			

Table 1. *Cont.*

Mycotoxin	Spike (μg kg^{-1})	Recovery $\pm$ RSD r (%, n = 6)	Recovery $\pm$ RSD$_R$ (%, n = 6)	Linearity Equation (R^2)	LOD (μg kg^{-1})	LOQ (μg kg^{-1})
TEA	50 100 200	95.4 $\pm$ 12.8 91.1 $\pm$ 8.7 79.3 $\pm$ 10.0	113.2 $\pm$ 7.8 108.3 $\pm$ 10.1 104.3 $\pm$ 4.7	Y = 52.7X (0.9934)	1.6	3.1
TEN	5 10 20	99.6 $\pm$ 5.8 105.4 $\pm$ 4.9 100.8 $\pm$ 4.2	111.2 $\pm$ 6.4 91.2 $\pm$ 13.8 106.2 $\pm$ 10.5	Y = 1875.8X (0.9844)	0.08	0.2
FB$_1$	250 500 1000	109.1 $\pm$ 7.4 107.0 $\pm$ 7.9 110.1 $\pm$ 5.0	98.4 $\pm$ 14.1 87.5 $\pm$ 19.5 105.4 $\pm$ 7.2	Y = 5.4X (0.9949)	62.5	125

Low, Middle, and High represent the spiked low, middle, and high concentrations of mycotoxins respectively. RSD r: intraday precision (repeatability) in mycotoxin-fortified samples; RSD$_R$: inter-day precision (reproducibility) in mycotoxin-fortified samples.

2.2. Mycotoxins Occurrence in Wheat Flour Samples

2.2.1. Mycotoxin Presence in Wheat Flour Samples

Detailed data on mycotoxin occurrence are shown in Table 2 and Figure 1A. The contamination levels for mycotoxins in wheat flour samples were lower than the existing regulations in cereals (95.6~100.0%), except for one sample. Among these mycotoxins, DON (74.9%), TEA (73.2%), TEN (55.2%), and ZEN (40.1%) had a higher detection rate in the wheat flours analyzed. The average contamination levels of DON, TEA, and FB$_1$ were higher. The contamination levels of DON (4.4%) or OTA (0.3%) in some samples exceeded the limits of infant flour-based food. The detection rate and contamination level for OTA in this study were also lower than those in other results [18,19]. Only one of the 299 samples (1260.0 μg kg^{-1}) exceeded the maximum limit of FB$_1$ in maize and corn-based foods for direct human consumption, which contains 20% corn flour after further detailed examination. Exogenous food ingredients, such as corn or buckwheat, have possible effects on mycotoxin contamination in wheat flour.

Table 2. Concentration levels of mycotoxins in wheat flour samples from Shanghai, China, 2020–2021 (n = 299, μg kg^{-1}).

Mycotoxins	MRLs	Positive Rate (%)	Below MRLs (%)	Mean $\pm$ SD	Range
DON	750 [a], 200 [b]	74.9	100.0, 95.6	57.6 $\pm$ 65.3	0.8–371.4
15-Ac DON	NF	37.8	**	12.0 $\pm$ 19.8	0.8–140.6
3-Ac DON	NF	4.0	**	1.7 $\pm$ 0.8	1.6–10.8
D$_3$G	NF	32.1	**	8.6 $\pm$ 15.0	1.6–96.3
FUS-X	NF	11.7	**	8.5 $\pm$ 18.8	3.6–191.7
NIV	NF	13.7	**	6.6 $\pm$ 9.8	3.8–96.7
AOH	NF	18.4	**	6.6 $\pm$ 18.7	0.2–140.8
TEN	NF	55.2	**	0.5 $\pm$ 1.2	0.04–14.8
TEA	NF	73.2	**	23.1 $\pm$ 27.0	0.8–161.6
ZEN	75 [a], 20 [b]	40.1	100.0, 100.0	0.6 $\pm$ 0.7	0.2–5.7
FB$_1$	800 [a], 200 [b]	0.7	99.7, 99.7	35.5 $\pm$ 71.1	31.2–1260.4
OTA	3 [a], 0.5 [b]	2.7	100.0, 99.7	0.2 $\pm$ 0.06	0.2–1.0
NEO	NF	2.3	**	0.6 $\pm$ 0.2	0.1–2.6

MRLs, Maximum Regulation Limits; SD, Standard Deviation; NF, Not Found; **, No Data. [a] MRLs for DON, ZEN, FB1, and OTA in cereal flour, maize-based breakfast cereals, or processed cereals as end product marketed for direct human consumption (EC Regulation No 1881/2006). [b] MRLs for DON, ZEN, FB1, and OTA in processed cereal-based foods and baby foods for infants and young children (EC Regulation No 1881/2006).

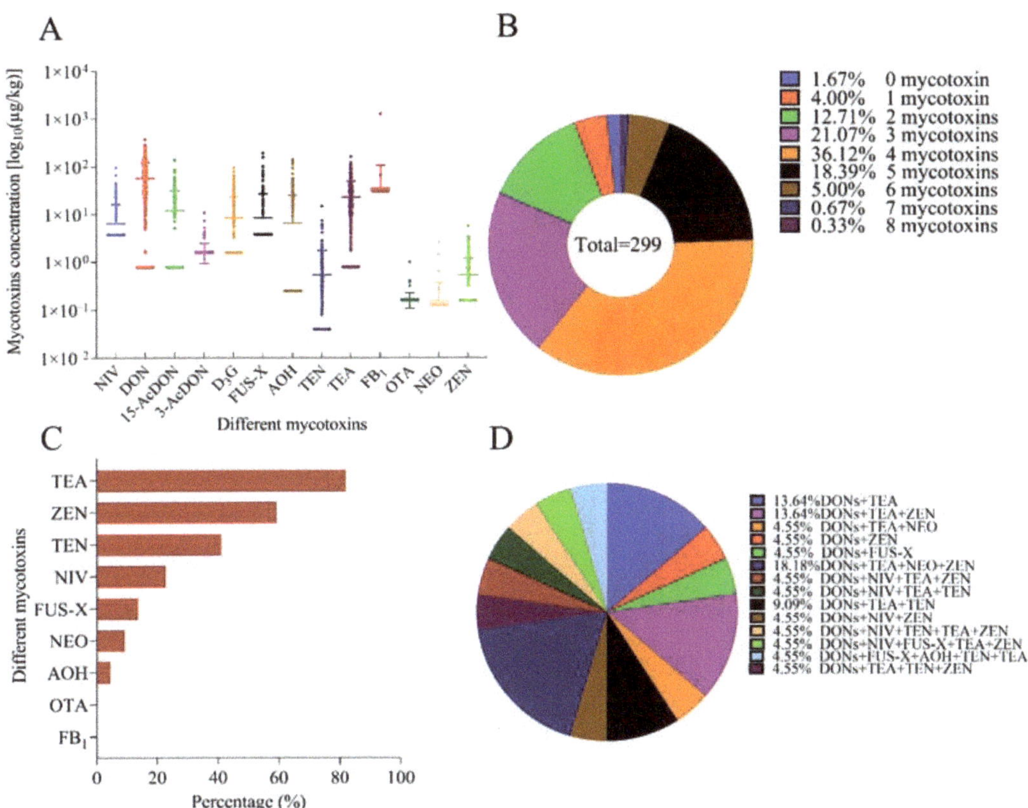

Figure 1. Mycotoxins occurrence in wheat flour samples. (**A**) occurrence of mycotoxins in wheat flour samples; (**B**) analytical results of mycotoxin coexistence in 299 wheat flour samples; (**C**) co-occurrence of mycotoxins in wheat flour samples ($\geq$200 µg kg^{-1}) contaminated with DONs; and (**D**) different combinations of the co-occurrence of mycotoxins in positive wheat flours for DONs (*n* = 22).

The co-contamination rate of mycotoxins in wheat flours was as high as 98.1%, and more than half of the samples (57.2%) contained three to four mycotoxins (Figure 1B). The detailed combination of the co-occurrence of mycotoxins in wheat flours are shown in Table S1. There were four main contamination patterns: DON+15-AcDON+TEA+TEN (12.0%), DON + TEA + TEN (8.4%), NIV+DON+D$_3$G+TEA+TEN (5.0%), and AOH+ZEN (5.0%). Among these samples with a total DON greater than 200, the detection rate of TEA, ZEN, TEN, NIV, FUS-X, NEO, or AOH in wheat flours was 81.8%, 59.1%, 40.9%, 22.7%, 13.6%, 9.1%, or 4.6%, respectively (Figure 1C,D). The co-contamination of *Fusarium* and *Alternaria* mycotoxin was relatively common, especially for the co-contamination of DON and TEA, which was also reported as the predominant contaminant pattern in the previous study [22].

2.2.2. Distribution Characteristics and Differences of Mycotoxins in Wheat Flour Samples

We have carried out a comparative analysis for mycotoxin contamination in wheat flour involving wheat flour refining processing technology and organic agriculture production, especially from the perspective of whole vs. refined, low-gluten vs. medium-gluten vs. high-gluten, and organic vs. conventional wheat flours. The results showed that the average levels of tested mycotoxins in different types of wheat flours with different significance, except for NIV, FB$_1$, NEO, and OTA (Figures 2 and S2).

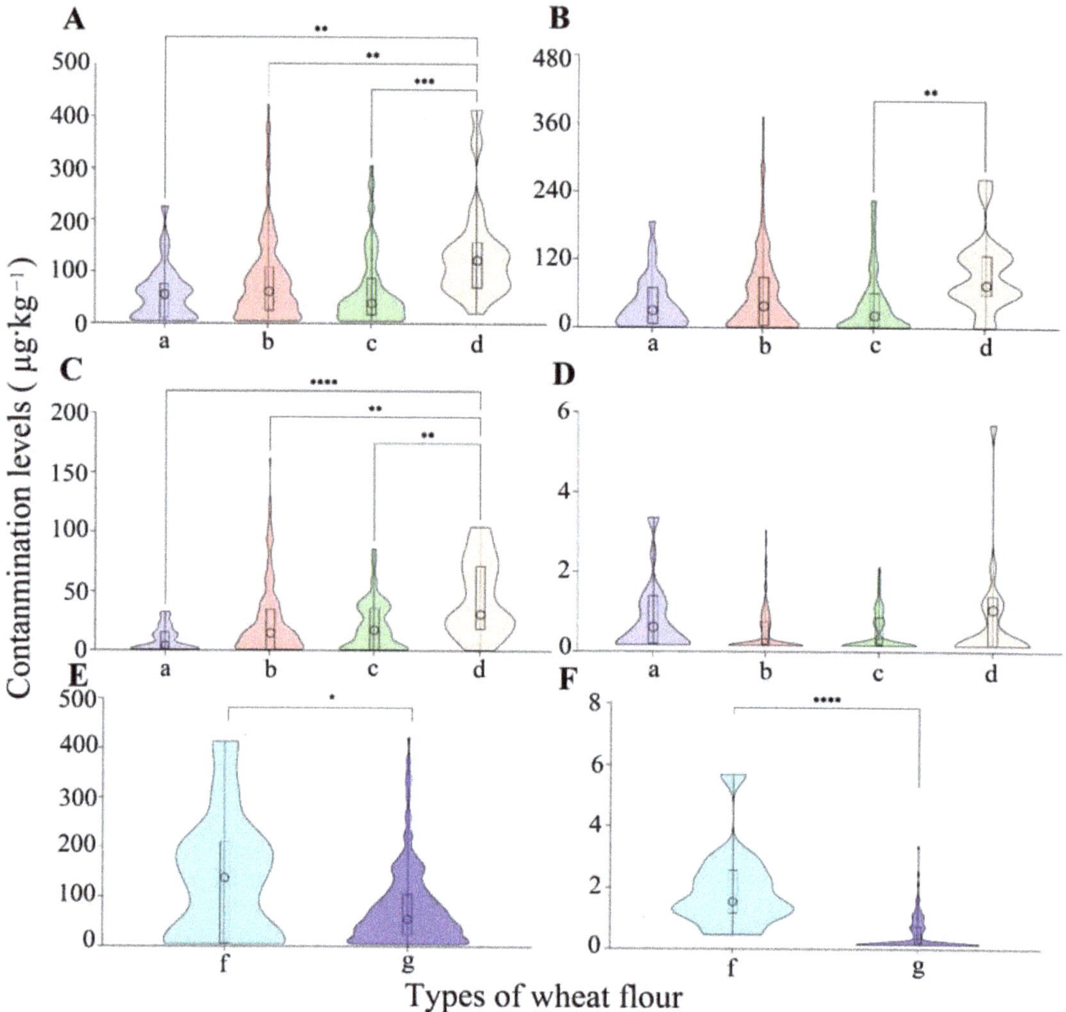

Figure 2. The distribution characteristics and differences of contamination levels for mycotoxins in various types of wheat flour samples (* $p < 0.05$, ** $p < 0.01$, *** $p < 0.001$ and **** $p < 0.0001$): (**A**) Occurrence of DONs in different wheat flour samples; (a) low-gluten wheat flours, (b) medium-gluten wheat flours, (c) high-gluten wheat flours, and (d) whole wheat flours; (**B**) occurrence of DON in different wheat flour samples; (a) low-gluten wheat flours, (b) medium-gluten wheat flours, (c) high-gluten wheat flours, and (d) whole wheat flours; (**C**) occurrence of TEA in different wheat flour samples; (a) low-gluten wheat flours, (b) medium-gluten wheat flours, (c) high-gluten wheat flours, and (d) whole wheat flours; (**D**) occurrence of ZEN in different wheat flour samples; (a) low-gluten wheat flours, (b) medium-gluten wheat flours, (c) high-gluten wheat flours, and (d) whole wheat flours; (**E**) occurrence of DONs in different wheat flour samples; (f) organic wheat flours, and (g) conventional wheat flours; and (**F**) occurrence of ZEN in different wheat flour samples; (f) organic wheat flours, and (g) conventional wheat flours.

The total amount of DON in refined wheat flours with different protein content (56.1~80.3 µg kg^{-1}) were significantly lower than that in whole wheat flour (138.4 µg kg^{-1}), and the average level of DON in whole wheat flours (96.5 µg kg^{-1}) was also higher than that in high-gluten wheat flours (43.4 µg kg^{-1}) (Figure 2A,B). The similar contamination difference was also found for TEA, ZEN, FUS-X, or TEN between whole wheat flour and refined wheat flour (Figures 2C,D and S2A,B). This distribution characteristic of DON or ZEN in whole wheat flours and refined wheat flours had also been found in a previous study, but the levels detected in this study were lower, which may be due to good management control [1]. Mycotoxin contamination levels in wheat flour were related to the proportions of different milling fractions because the epidermis, bran or germ of wheat grain as the natural medium of toxigenic fungi may concentrate mycotoxins. In other words, mycotoxins in wheat flours could be diminished by removing or adjusting grinding fractions with a high risk of contamination. Compared to low-gluten wheat flour, medium-gluten and high-gluten wheat flour had a higher degree and a wider range of contamination (Figures 2 and S2). The similar patterns of contamination have been found in wheat flour related foods. Oueslati et al. reported that the contamination levels of combination (DON + ENB) in the Tunisian whole bread samples were higher than that in white bread samples [31]. The mycotoxin contamination (DON, TEA, and TEN) in noodles derived from high- or medium-gluten wheat flours were also found to be more serious than that in biscuits derived from low-gluten wheat flours [22]. In short, the monitoring of primary processed wheat flour should be regarded with more concern, even if contamination levels are below the limits, minimizing health risk as early as possible, especially for whole wheat flours, medium-gluten, and high-gluten wheat flours.

Meanwhile, there is still a controversy that organic wheat flour, without use of fertilizers and fungicides, contain more mycotoxins than conventional products, thereby involving another risk for human health because of limited studies on the comparison of organic and conventional wheat flours [8]. In this study, significant differences were only found in some mycotoxins and the average level of DON, ZEN, or AOH in organic wheat flours (135.9, 2.0, or 13.9 µg kg^{-1}) was significantly higher than that in conventional wheat flours (77.3, 0.5, or 6.2 µg kg^{-1}), while the mean contamination for TEN in organic wheat flours (0.05 µg kg^{-1}) was exactly the opposite in conventional wheat flours (0.6 µg kg^{-1}) (Figure 2E,F). DON, NIV, and FUS-X were reported to occur frequently in organic cereals from Italy [32]. Crop rotation, good agricultural, or harvest practices in terms of the proper transport and storage conditions was associated with mycotoxins contamination of organic wheat flours [9]. Therefore, mycotoxin of organic wheat flours was not always higher than that of conventional wheat flours, which needs further research considering the relationship of fungi and mycotoxin for variable environmental conditions.

2.3. Risk Assessment and Uncertainty of Ingestion or Exposure

The deterministic and probabilistic assessment of exposure to each group of mycotoxins through wheat flour consumption for the local population were obtained (Table 3) by Equations (2) and (3) and are found in Section 4.4. The PDI values from the estimation of almost all mycotoxins were lower than the available *HBGVn*. In particular, the PDI values of AOH in the entire population ranged from 9.00×10^{-3} to 7.88×10^{-2} µg kg^{-1} bw day^{-1}, which were higher than the recommended TTC values for AOH at 0.0025 µg kg^{-1} bw day^{-1}. The %TDI ranking results of assessments were almost the same as AOH > DONs > FB$_1$ > OTA > TEA > NIV > ZEN > TEN. The %TDI or sum of %TDI from the single exposure assessment of almost all mycotoxins, except for AOH, were less than 100% in all populations (Table 3 and Figure 3), indicating that wheat flour consumption for Shanghai residents contributed little to the exposure risk of mycotoxins. The margin of exposure values were the ratios of BMDL$_{10}$ values of 4.7 and 14.5 µg kg^{-1} bw day^{-1} to *PDI*, respectively, reflecting the nonneoplastic and neoplastic effects of OTA. The results showed that they both exceeded 200 and 10,000, indicating no health problems (Table S2).

Table 3. Exposure levels of mycotoxins in wheat flour samples with the upper bound deterministic and probabilistic estimation for different consumer groups ($\mu g\ kg^{-1}\ bw\ day^{-1}$).

Mycotoxins / Poulation	Deterministic Estimation	Probabilistic Estimation	
		Median	P90
NIV (TDI, 0.7 $\mu g\ kg^{-1}\ bw\ day^{-1}$)			
Total population	1.40×10^{-2}	1.40×10^{-2}	3.04×10^{-2}
Adult men	1.49×10^{-2}	1.49×10^{-2}	3.24×10^{-2}
Adult women	1.31×10^{-2}	1.31×10^{-2}	2.84×10^{-2}
7–10-year-old boys	2.52×10^{-2}	2.52×10^{-2}	5.47×10^{-2}
7–10-year-old girls	2.20×10^{-2}	2.20×10^{-2}	4.78×10^{-2}
DONs (TDI, 1 $\mu g\ kg^{-1}\ bw\ day^{-1}$)			
Total population	1.18×10^{-1}	9.58×10^{-2}	2.62×10^{-1}
Adult men	1.26×10^{-1}	1.02×10^{-1}	2.79×10^{-1}
Adult women	1.11×10^{-1}	8.94×10^{-2}	2.45×10^{-1}
7–10-year-old boys	2.13×10^{-1}	1.72×10^{-1}	4.71×10^{-1}
7–10-year-old girls	1.86×10^{-1}	1.50×10^{-1}	4.12×10^{-1}
AOH (TTC, 0.0025 $\mu g\ kg^{-1}\ bw\ day^{-1}$)			
Total population	9.64×10^{-3}	9.64×10^{-3}	4.38×10^{-2}
Adult men	1.03×10^{-2}	1.03×10^{-2}	4.66×10^{-2}
Adult women	9.00×10^{-3}	9.00×10^{-3}	4.09×10^{-2}
7–10-year-old boys	1.73×10^{-2}	1.73×10^{-2}	7.88×10^{-2}
7–10-year-old girls	1.51×10^{-2}	1.51×10^{-2}	6.88×10^{-2}
TEN (TTC, 1.5 $\mu g\ kg^{-1}\ bw\ day^{-1}$)			
Total population	7.69×10^{-4}	5.37×10^{-4}	1.65×10^{-3}
Adult men	8.18×10^{-4}	5.72×10^{-4}	1.75×10^{-3}
Adult women	7.18×10^{-4}	5.01×10^{-4}	1.54×10^{-3}
7–10-year-old boys	1.38×10^{-3}	9.66×10^{-4}	2.96×10^{-3}
7–10-year-old girls	1.21×10^{-3}	8.44×10^{-4}	2.59×10^{-3}
TEA (TTC, 1.5 $\mu g\ kg^{-1}\ bw\ day^{-1}$)			
Total population	3.31×10^{-2}	2.60×10^{-2}	7.00×10^{-2}
Adult men	3.52×10^{-2}	2.76×10^{-2}	7.45×10^{-2}
Adult women	3.08×10^{-2}	2.42×10^{-2}	6.53×10^{-2}
7–10-year-old boys	5.94×10^{-2}	4.67×10^{-2}	1.26×10^{-1}
7–10-year-old girls	5.19×10^{-2}	4.08×10^{-2}	1.10×10^{-1}
FB$_1$ (TDI, 2 $\mu g\ kg^{-1}\ bw\ day^{-1}$)			
Total population	9.53×10^{-2}	8.95×10^{-2}	9.88×10^{-2}
Adult men	1.01×10^{-1}	9.52×10^{-2}	1.05×10^{-1}
Adult women	8.89×10^{-2}	8.35×10^{-2}	9.21×10^{-2}
7–10-year-old boys	1.71×10^{-1}	1.61×10^{-1}	1.78×10^{-1}
7–10-year-old girls	1.50×10^{-1}	1.41×10^{-1}	1.55×10^{-1}
OTA (TDI, 0.01 $\mu g\ kg^{-1}\ bw\ day^{-1}$)			
Total population	4.64×10^{-4}	4.58×10^{-4}	4.67×10^{-4}
Adult men	4.93×10^{-4}	4.88×10^{-4}	4.97×10^{-4}
Total population	4.33×10^{-4}	4.28×10^{-4}	4.36×10^{-4}
7–10-year-old boys	8.34×10^{-4}	8.24×10^{-4}	8.40×10^{-4}
7–10-year-old girls	7.28×10^{-4}	7.20×10^{-4}	7.33×10^{-4}
ZEN (TDI, 0.25 $\mu g\ kg^{-1}\ bw\ day^{-1}$)			
Total population	7.88×10^{-4}	6.17×10^{-4}	1.59×10^{-3}
Adult men	8.38×10^{-4}	6.57×10^{-4}	1.69×10^{-3}
Adult women	7.35×10^{-4}	5.76×10^{-4}	1.48×10^{-3}
7–10-year-old boys	1.42×10^{-3}	1.11×10^{-3}	2.85×10^{-3}
7–10-year-old girls	1.24×10^{-3}	9.70×10^{-4}	2.49×10^{-3}

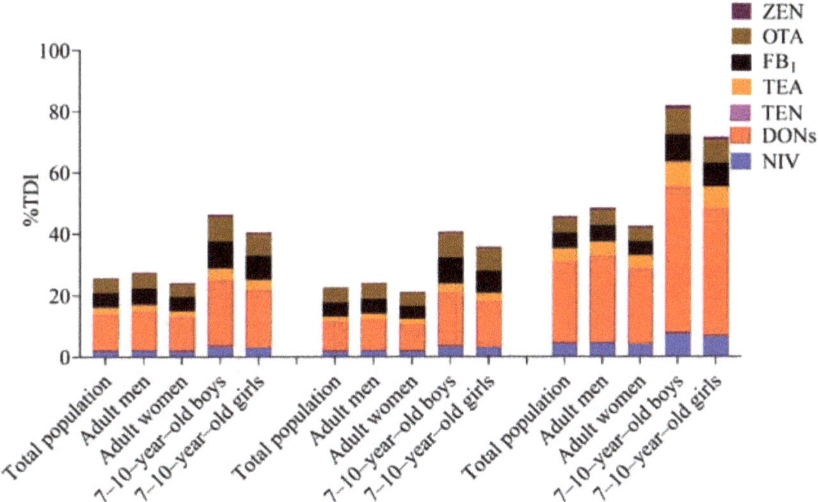

Figure 3. The health risk of various mycotoxins through wheat flour consumption were described in the upper bound deterministic and probabilistic estimation (Median and P90) for different consumer groups from left to right.

Considering mycotoxins co-contamination, the approximate cumulative exposure risks were also assessed in positive samples contaminated with multiple mycotoxins concurrently and their incidence greater than 5%. Among these four main contamination patterns, only the %TDI min or %TDI max from one of patterns (AOH + ZEN) was more than 100% (Table S3). It is important to note that limited knowledge is available on the transfer or fate of mycotoxins during food processing and digestion. It is reported that 60~80% of OTA and AFs in wheat flour can be retained after cooking [33]. The fates of mycotoxins, particularly DON and D$_3$G, in bread making are affected by fermentation or other complex factors [2,6,34]. Some externally additive food compositions can also affect the risk of eventual exposure to mycotoxins. Recently, bread enriched with pumpkin extract and fermented whey individually and in combination were reported to reduce the bio-accessibilities of mycotoxins and alleviate their associated neurotoxicity [35–37]. Therefore, the associated health risks of AOH needs to be further studied, in combination with internal exposure, especially in children. A recent study also pointed to the need to pay attention to the risks of infant exposure to AOH through cereal-based foods consumption [22].

3. Conclusions

Wheat flour is one of the important dietary sources of mycotoxin exposure and its safety deserves attention. We inspected 13 mycotoxins in wheat flours marketed in Shanghai by UPLC-MS/MS and profiled their external exposure risk. Of the wheat flour samples with low concentration levels, 95.6~100.0% met the regulations. The average contamination levels of DON and TEA in wheat flours were higher than other mycotoxins. Of the wheat flour samples, 57.2% were contaminated simultaneously by three to four mycotoxins. We found the co-occurrence of Fusarium and Alternaria mycotoxin in wheat flours. Particularly, four main co-contamination patterns in wheat flours were also found in this study, including DON+15-AcDON+TEA+TEN, DON+TEA+TEN, NIV+DON+D$_3$G+TEA+TEN, and AOH+ZEN. Lower average contamination levels and fewer types of mycotoxins were detected in refined flours than whole wheat flours. Combined with the relevant wheat food consumption data of Shanghai residents, chronic dietary intake risk assessments of mycotoxins were performed by point evaluation and Monte Carlo assessment model. Dietary exposure risks of DONs, ZEN, NIV, TEA, TEN, FB$_1$, and OTA by wheat flours intakes were considered to be acceptable (%TDI < 100%). However, the exposure risk of AOH and the approximate cumulative exposure risks of AOH and ZEN should be considered and further studied in conjunction with internal exposure assessments.

4. Materials and Methods

4.1. Materials and Reagents

The ZEN (Z2125), DON (D0156), 3-AcDON (A6166), 15-AcDON (A1556), D_3G (32911), OTA (O1877), AOH (A1312), TEN (35977), NIV (34131), and FB_1 (F1147) analytical standards are Sigma-Aldrich products (St. Louis, MO, USA). The TEA (ab142764) analytical standards were purchased from Abcam (Cambridge, MA, USA). The FUS-X (10003647) and NEO (10003640) analytical standards are Romer Lab Biopure™ products (Union, MO, USA). The certified wheat flour (Fusarium mycotoxins, ERMBC600) as the reference material for quality control was also obtained from Sigma-Aldrich (St. Louis, MO, USA). HPLC-grade acetonitrile was purchased from Merck (Darmstadt, Germany). Milli-Q quality water (Millipore, Billerica, MA, USA) was used throughout the experiments.

4.2. Sampling and Samples for Analysis

Two hundred and ninety-nine wheat flour samples were randomly taken from retail stores or supermarkets located in different zones of Shanghai from December 2020 to October 2021. These samples originated from twelve different districts representing central and suburban Shanghai, including Congming ($n = 21$), Changning ($n = 38$), Huangpu ($n = 29$), Hongkou ($n = 8$), Jingan ($n = 25$), Minhang ($n = 47$), Pudong ($n = 29$), Putuo ($n = 30$), Qingpu ($n = 10$), Songjiang ($n = 10$), Xuhui ($n = 46$), and Yangpu ($n = 6$). Of the samples, 13 out of 299 were bulk wheat flours, while the remaining wheat flour samples were categorized as whole wheat flours and refined wheat flours according to labels. According to the labeled varying content of the protein, refined wheat flours involved low-gluten wheat flours ($n = 29$), medium-gluten wheat flours ($n = 166$), and high-gluten wheat flours ($n = 71$). Almost all of them were packaged in plastic food containers at least 500 g/sample. 13 of 299 samples were organic wheat flours after a subdivision of the collected samples in terms of agricultural practice. The geographic origin, food composition, and other information about these samples used for humans were also recorded. All the collected samples were divided and stored in plastic cans at -20 °C before analysis.

4.3. Analytical Method and Validation

A total of 13 mycotoxins, including DON, 3-AcDON, 15-AcDON, D_3G, NIV, FUS-X, NEO, ZEN, OTA, FB_1, AOH, TEA, and TEN, were simultaneous determined in wheat flours following the UPLC-MS/MS method reported by previous studies [22,38] with some modifications. First, more Fusarium toxins and Alternaria toxins were monitored in the full-scan mode. The extraction recoveries for targeted mycotoxins were compared under different concentrations of organic reagents after adding 13 mycotoxins simultaneously in wheat flour samples at middle concentration levels. The equation for calculating their recoveries, where *Ca* is the calculated concentration in the mycotoxins-spiked sample, *Cb* is the calculated concentration in the non-spiked sample, and CA is the theoretical concentration of the analyte that was added into the sample.

$$\text{Recovery (\%)} = (Ca - Cb)/\text{CA} \times 100 \tag{1}$$

In brief, 1.0 g of sample was vortexed vigorously for 5 min using 4 mL acetonitrile: water (75:25, *v/v*) with 1% formic acid solution, and then ultrasound-assisted extracted for 40 min. The extraction mixture was centrifuged at 4000 rpm for 5 min. After centrifugation, 2 mL of the supernatant extract was filtered with 0.22 μm organic filter membrane and injected into the ultimate 3000 UPLC system (Thermo Fisher Scientific, San Jose, CA, USA). Mycotoxins were separated by the ACQUITY UPLC BEH C_{18} VanGuard pre-column (1.7 μm, 2.1 mm × 5 mm) and ACQUITY UPLC BEH C_{18} (1.7 μm, 2.1 mm × 100 mm) column with mobile phase A (water containing 5 mM ammonium acetate) and mobile phase B (methanol) gradient elution following: 0–1.0 min, 5–50% B; 1.0–9.0 min, 50–100% B; 9.0–10.0 min, 100% B; 10.0–11.0 min, 100–5% B; and 11.0–12.0 min, 5% B. The flow rate was 0.35 mL min^{-1} and the injection volume was 5 µL. Meanwhile, the TSQ VantageTM (Thermo Fisher Scientific, San Jose, CA, USA) triple stage quadrupole mass spectrometer was applied for further multi-mycotoxin determination based on multiple reaction monitoring (MRM) using positive and negative electrospray ionization (ESI$^{+/-}$) mode. Based on the selected optimal parent ions, the product ion and their optimized collision energies with argon for each mycotoxin were obtained and summarized in the Table S4. Other optimized parameters were set as follows: Positive spray voltage at +3.0 kV, negative spray voltage at −2.5 kV, capillary temperature at 300 °C, vaporizer temperature at 250 °C, aux gas pressure is 5 psi, and sheath gas pressure is 40 psi. Their final chromatograms is shown in the Figure S3. Related parameters of the UPLC-MS/MS method were verified with the guidelines of the document [39]. Linearity was determined by analyzing each mycotoxin standard solutions. The values (LOD and LOQ) for mycotoxins were determined by the signal-to-noise (S/N) ratios of 3:1 and 10:1, respectively, according to the lowest detectable level for quantitative ion [1,35]. Recovery analysis was conducted using three different concentrations of 13 mycotoxins to fortify simultaneously wheat flour matrices. The spiked concentration ranged from 0.13 to 1200 µg kg^{-1} with six replicates per concentration level (Table 1). The precision and accuracy of the proposed strategy were checked through intra- and inter-day analysis, as described in our previous study, and RSD r or RSD$_R$ less than 20% was evaluated as acceptable [39,40].

4.4. Dietary Risk Assessment and Characterization

The risk of exposure for ingested mycotoxins through wheat flour consumption was assessed by the deterministic and probabilistic approach. The probable daily intake (*PDI n*, µg kg^{-1} bw day^{-1}) of each mycotoxin was calculated by their contamination levels obtained from the analyzed samples combined with the relevant wheat food consumption data, as indicated in the following equation:

$$PDI\ n = (Cn \times \mathrm{CA})/\mathrm{BW} \tag{2}$$

For the point evaluation, where *Cn* is the average content of each mycotoxin *n*. In this study, if a contamination value was not detected, which refers to values lower than LOD values, the substitution values of 0 (lower bound), 1/2 LOD (middle bound) and LOD (upper bound) were used for mycotoxin exposure assessment. CA is the average consumption amount of the commodity (g person^{-1} day^{-1}) and BW is the average body weight of participants (kg). Wheat and wheat-based products consumption data and the demographic information were derived from a 2012~2014 Shanghai Food Consumption Survey (SHFCS) by Fudan University regarding to Shanghai inhabitants (7~60 years old). The average consumption of participants in 7~10 years old groups was the highest in the published study [13]. Therefore, the population groups considered in this study were consistent with a previous study [16]: Total population, adult men, adult women, as well as typical boys and girls (Table S5).

A more accurate or applicable Monte Carlo simulation was also performed using @Risk Industrial 7.5 (Palisade, New York, NY, USA) software for probability assessment, in combination with Microsoft Excel 2016. Mycotoxin contamination data of all analyzed samples at the above three bounds were input into @RISK software and *Cn* could be obtained from the best-fitting distribution for these data. The Anderson-Darling and Kolmogorov-Smirnov tests were selected to evaluate the goodness-of-fit for each distribution by @Risk software. Similarly, *Cn*, CA, and BW were input into @RISK software according to the above formula for Monte Carlo simulation. The exposure distribution of *PDI n* with a confidence interval > 90% were obtained using 10,000 iteration runs. The health risk characterization of each mycotoxin (*%TDI n*) was performed by dividing *PDI n* with their health-based guidance values (*HBGV n*) based on the equation shown below:

$$\%TDI\ n = (PDI\ n/HBGV\ n) \times 100 \tag{3}$$

where *HBGV n* represents the available *TDI n* or *TTC n*. Values of *%TDI n* higher than a hundred indicate a possible health risk scenario. Otherwise, there is no significant risk was observed and a population is not at risk from that exposure. An approximation of exposure assessment was also performed to evaluate consumer's exposure in analyzed samples contaminated with multiple mycotoxins [41]. The *Cn, min* and *Cn, max* derived from multi-mycotoxin contaminated samples. Then, we summed them, and a combined health risk characterization was proposed as follows:

$$\sum_{m=1}^{i} \%TDI\,n, min = \sum_{m=1}^{i} (Cn, min \times CA)/Bw/HBGVn \qquad (4)$$

$$\sum_{m=1}^{i} \%TDI\,n, max = \sum_{m=1}^{i} (Cn, max \times CA)/Bw/HBGVn \qquad (5)$$

Conventionally, values of $\sum_{m=1}^{i} \%TDI$ less than a hundred indicate that the combined exposure level were considered to be acceptable, and people are unlikely to be exposed at a toxic level with possible consequences for health.

4.5. Data Analysis

The UPLC-MS raw data were recognized by Thermo Xcalibur Qual Browser 4.0. Nonparametric statistics were used after the normality and lognormality testing for each group. Distribution characteristics and differences of mycotoxins in various types of wheat flour samples were compared and evaluated using the Kruskal-Wallis or Mann-Whitney test at a significance level of 0.05. All statistical analyses and drawings were performed using GraphPad Prism 9.0 software.

Supplementary Materials: The following supporting information can be downloaded at: https://www.mdpi.com/article/10.3390/toxins14110748/s1, Figure S1. The effect of the extractant types containing 1% formic acid on the recoveries of mycotoxins: (a) 45% Acetonitrile, (b) 55% Acetonitrile, (c) 65% Acetonitrile, (d) 75% Acetonitrile, (e) 85% Acetonitrile, (f) 95% Acetonitrile. The two dashed lines represent recoveries of 60% and 120%, respectively; Figure S2. The distribution characteristics and differences of contamination levels for mycotoxins in various types of wheat flour samples: (A) occurrence of FUS-X in different wheat flour samples; (a) low-gluten wheat flours, (b) medium-gluten wheat flours, (c) high-gluten wheat flours, (d) whole wheat flours; (B) occurrence of TEN in different wheat flour samples; (a) low-gluten wheat flours, (b) medium-gluten wheat flours, (c) high-gluten wheat flours, (d) whole wheat flours; (C) occurrence of AOH in different wheat flour samples; (f) organic wheat flours, g) conventional wheat flours; (D) occurrence of TEN in different wheat flour samples; (f) organic wheat flours, (g) conventional wheat flours; Figure S3. The chromatograms of 13 mycotoxins at middle concentration under optimized chromatographic and mass spectrometry conditions; Table S1. Different combinations of co-occurrence of mycotoxins in wheat flours marketed in China; Table S2. The nonneoplastic and neoplastic effects through wheat flours consumption based on various margin of exposure estimation values of OTA; Table S3. The approximate cumulative exposure risks of four main contamination patterns of mycotoxins co-occurring in real samples for the total population; Table S4. Retention time and MS parameters for the analysis of mycotoxins; Table S5. The average daily intake of wheat and wheat-based products in Shanghai and participants for average body weight.

Author Contributions: Conceptualization, H.Z., N.L. and A.W.; methodology, H.Z. and A.X.; software, Z.Y. and H.Z.; validation, Z.Y. and H.Z.; formal Analysis, H.Z., A.X. and M.L.; investigation A.X. and M.L.; resources, L.Q., H.L. and A.W.; data curation, H.Z. and A.X.; writing—original draft preparation, H.Z.; writing—review and editing, N.L. and A.W.; visualization,; supervision, N.L. and A.W.; project administration, L.Q., H.L., N.L. and A.W.; funding acquisition, L.Q., H.L., N.L. and A.W. All authors have read and agreed to the published version of the manuscript.

Funding: This work was supported by the Shanghai Agriculture Applied Technology Development Program, China (Grant No. X2019-02-08-00-02-F01146).

Institutional Review Board Statement: Not applicable.

Informed Consent Statement: Not applicable.

Data Availability Statement: The data that support the findings of this study are available from the corresponding author upon reasonable request.

Conflicts of Interest: The authors declare that there is no known competing financial interest or personal relationships that could have appeared to influence the work reported in this paper.

References

1. Zhang, Y.; Pei, F.; Fang, Y.; Li, P.; Zhao, Y.; Shen, F.; Zou, Y.; Hu, Q. Comparison of concentration and health risks of 9 Fusarium mycotoxins in commercial whole wheat flour and refined wheat flour by multi-IAC-HPLC. *Food Chem.* **2019**, *275*, 763–769. [CrossRef] [PubMed]
2. Wang, J.; Hasanalieva, G.; Wood, L.; Markellou, E.; Iversen, P.O.; Bernhoft, A.; Seal, C.; Baranski, M.; Vigar, V.; Ernst, L.; et al. Effect of wheat species (*Triticum aestivum* vs. *T. spelta*), farming system (organic vs. conventional) and flour type (wholegrain vs. white) on composition of wheat flour; results of a retail survey in the UK and Germany-1. Mycotoxin content. *Food Chem.* **2020**, *327*, 127011. [CrossRef] [PubMed]
3. Yan, Z.; Chen, W.; van der Lee, T.; Waalwijk, C.; van Diepeningen, A.D.; Feng, J.; Zhang, H.; Liu, T. Evaluation of Fusarium Head Blight Resistance in 410 Chinese Wheat Cultivars Selected for Their Climate Conditions and Ecological Niche Using Natural Infection Across Three Distinct Experimental Sites. *Front Plant Sci.* **2022**, *13*, 916282. [CrossRef] [PubMed]
4. Miraglia, M.; Marvin, H.J.; Kleter, G.A.; Battilani, P.; Brera, C.; Coni, E.; Cubadda, F.; Croci, L.; De Santis, B.; Dekkers, S.; et al. Climate change and food safety: An emerging issue with special focus on Europe. *Food Chem. Toxicol.* **2009**, *47*, 1009–1021. [CrossRef]
5. Szulc, J.; Ruman, T. Laser Ablation Remote-Electrospray Ionisation Mass Spectrometry (LARESI MSI) Imaging-New Method for Detection and Spatial Localization of Metabolites and Mycotoxins Produced by Moulds. *Toxins* **2020**, *12*, 720. [CrossRef]
6. Wan, J.; Chen, B.; Rao, J. Occurrence and preventive strategies to control mycotoxins in cereal-based food. *Compr. Rev. Food Sci. Food Saf.* **2020**, *19*, 928–953. [CrossRef]
7. Schaarschmidt, S.; Fauhl-Hassek, C. The Fate of Mycotoxins During the Processing of Wheat for Human Consumption. *Compr. Rev. Food Sci. Food Saf.* **2018**, *17*, 556–593. [CrossRef]
8. Khaneghah, A.M.; Martins, L.M.; von Hertwig, A.M.; Bertoldo, R.; Sant'Ana, A.S. Deoxynivalenol and its masked forms: Characteristics, incidence, control and fate during wheat and wheat based products processing—A review. *Trends Food Sci. Technol.* **2018**, *71*, 13–24. [CrossRef]
9. Annunziata, L.; Schirone, M.; Visciano, P.; Campana, G.; De Massis, M.R.; Migliorati, G. Determination of aflatoxins, deoxynivalenol, ochratoxin A and zearalenone in organic wheat flour under different storage conditions. *Int. J. Food Sci. Technol.* **2021**, *56*, 4139–4148. [CrossRef]
10. Farhadi, A.; Fakhri, Y.; Kachuei, R.; Vasseghian, Y.; Huseyn, E.; Mousavi Khaneghah, A. Prevalence and concentration of fumonisins in cereal-based foods: A global systematic review and meta-analysis study. *Environ. Sci. Pollut. Res. Int.* **2021**, *28*, 20998–21008. [CrossRef]
11. Zhang, K.; Schaab, M.R.; Southwood, G.; Tor, E.R.; Aston, L.S.; Song, W.; Eitzer, B.; Majumdar, S.; Lapainis, T.; Mai, H.; et al. A Collaborative Study: Determination of Mycotoxins in Corn, Peanut Butter, and Wheat Flour Using Stable Isotope Dilution Assay (SIDA) and Liquid Chromatography-Tandem Mass Spectrometry (LC-MS/MS). *J. Agric Food Chem.* **2017**, *65*, 7138–7152. [CrossRef] [PubMed]
12. Meerpoel, C.; Vidal, A.; Andjelkovic, M.; De Boevre, M.; Tangni, E.K.; Huybrechts, B.; Devreese, M.; Croubels, S.; De Saeger, S. Dietary exposure assessment and risk characterization of citrinin and ochratoxin A in Belgium. *Food Chem. Toxicol.* **2021**, *147*, 111914. [CrossRef] [PubMed]
13. Han, Z.; Nie, D.; Ediage, E.N.; Yang, X.; Wang, J.; Chen, B.; Li, S.; On, S.L.; De Saeger, S.; Wu, A. Cumulative health risk assessment of co-occurring mycotoxins of deoxynivalenol and its acetyl derivatives in wheat and maize: Case study, Shanghai, China. *Food Chem. Toxicol.* **2014**, *74*, 334–342. [CrossRef] [PubMed]
14. Li, F.; Jiang, D.; Zhou, J.; Chen, J.; Li, W.; Zheng, F. Mycotoxins in wheat flour and intake assessment in Shandong province of China. *Food Addit. Contam. Part B Surveill* **2016**, *9*, 170–175. [CrossRef] [PubMed]
15. Sun, J.; Wu, Y. Evaluation of dietary exposure to deoxynivalenol (DON) and its derivatives from cereals in China. *Food Control.* **2016**, *69*, 90–99. [CrossRef]
16. Yang, X.; Zhao, Z.; Tan, Y.; Chen, B.; Zhou, C.; Wu, A. Risk profiling of exposuRes to multiclass contaminants through cereals and cereal-based products consumption: A case study for the inhabitants in Shanghai, China. *Food Control.* **2020**, *109*, 106964. [CrossRef]
17. Varga, E.; Fodor, P.; Soros, C. Multi-mycotoxin LC-MS/MS method validation and its application to fifty-four wheat flours in Hungary. *Food Addit. Contam. Part A Chem. Anal. Control Expo. Risk Assess.* **2021**, *38*, 670–680. [CrossRef]
18. Elaridi, J.; Yamani, O.; Al Matari, A.; Dakroub, S.; Attieh, Z. Determination of Ochratoxin A (OTA), Ochratoxin B (OTB), T-2, and HT-2 Toxins in Wheat Grains, Wheat Flour, and Bread in Lebanon by LC-MS/MS. *Toxins* **2019**, *11*, 471. [CrossRef]
19. Hajok, I.; Kowalska, A.; Piekut, A.; Cwielag-Drabek, M. A risk assessment of dietary exposure to ochratoxin A for the Polish population. *Food Chem.* **2019**, *284*, 264–269. [CrossRef]

20. Puntscher, H.; Kutt, M.L.; Skrinjar, P.; Mikula, H.; Podlech, J.; Frohlich, J.; Marko, D.; Warth, B. Tracking emerging mycotoxins in food: Development of an LC-MS/MS method for free and modified Alternaria toxins. *Anal. Bioanal. Chem.* **2018**, *410*, 4481–4494. [CrossRef]

21. Puntscher, H.; Cobankovic, I.; Marko, D.; Warth, B. Quantitation of free and modified Alternaria mycotoxins in European food products by LC-MS/MS. *Food Control* **2019**, *102*, 157–165. [CrossRef]

22. Ji, X.; Xiao, Y.; Wang, W.; Lyu, W.; Wang, X.; Li, Y.; Deng, T.; Yang, H. Mycotoxins in cereal-based infant foods marketed in China: Occurrence and risk assessment. *Food Control* **2022**, *138*, 108998. [CrossRef]

23. Commission Regulation (EC). Setting Maximum Levels for Certain Contaminants in Foodstuffs (Document No. EC/1881/2006). Available online: https://www.legislation.gov.uk/eur/2006/1881 (accessed on 10 August 2021).

24. Walker, R.; Meyland, I.; Tritscher, A.; JoInt. FAO/WHO Expert Committee on Food Additives, Seventy-Second Meeting (JECFA/72/SC). 2010. Available online: https://www.researchgate.net/publication/242567110 (accessed on 16 November 2021).

25. JoInt. *FAO/WHO Expert Committee on Food Additives (JECFA). 56 St Report of JECFA-Safty Evaluation of Certain Mycotoxins in Food*; World Health Organization: Geneva, Switzerland, 2002.

26. EFSA Panel on Contaminants in the Food Chain (CONTAM). Appropriateness to set a group health-based guidance value for zearalenone and its modified forms. *EFSA J.* **2016**, *14*, e04425. [CrossRef]

27. SCF-Scientific Committee on Food. Opinion of the Scientific Committee on Food on Fusarium Toxins. Part 6: Group Evaluation of T-2 Toxin, HT-2 Toxin, Nivalenol and Deoxynivalenol. SCF/CS/CNTM/MYC/27. 2022, p. 1.e12. Available online: https://food.ec.europa.eu/system/files/2016-10/cs_contaminants_catalogue_fusarium_out123_en.pdf (accessed on 18 February 2022).

28. Arcella, D.; Eskola, M.; Gómez Ruiz, J.A. Dietary exposure assessment to Alternaria toxins in the European population. *EFSA J.* **2016**, *14*, 4654. [CrossRef]

29. Andrade, P.D.; Dias, J.V.; Souza, D.M.; Brito, A.P.; van Donkersgoed, G.; Pizzutti, I.R.; Caldas, E.D. Mycotoxins in cereals and cereal-based products: Incidence and probabilistic dietary risk assessment for the Brazilian population. *Food Chem. Toxicol.* **2020**, *143*, 111572. [CrossRef]

30. Wang, X.; Yang, D.; Qin, M.; Xu, H.; Zhang, L.; Zhang, L. Risk assessment and spatial analysis of deoxynivalenol exposure in Chinese population. *Mycotoxin Res.* **2020**, *36*, 419–427. [CrossRef]

31. Oueslati, S.; Berrada, H.; Juan-García, A.; Mañes, J.; Juan, C. Multiple Mycotoxin Determination on Tunisian Cereals-Based Food and Evaluation of the Population Exposure. *Food Anal. Methods* **2020**, *13*, 1271–1281. [CrossRef]

32. Juan, C.; Ritieni, A.; Manes, J. Occurrence of Fusarium mycotoxins in Italian cereal and cereal products from organic farming. *Food Chem.* **2013**, *141*, 1747–1755. [CrossRef]

33. Sakuma, H.; Watanabe, Y.; Furusawa, H.; Yoshinari, T.; Akashi, H.; Kawakami, H.; Saito, S.; Sugita-Konishi, Y. Estimated dietary exposure to mycotoxins after taking into account the cooking of staple foods in Japan. *Toxins* **2013**, *5*, 1032–1042. [CrossRef]

34. Vidal, A.; Ambrosio, A.; Sanchis, V.; Ramos, A.J.; Marin, S. Enzyme bread improvers affect the stability of deoxynivalenol and deoxynivalenol-3-glucoside during breadmaking. *Food Chem.* **2016**, *208*, 288–296. [CrossRef]

35. Llorens, P.; Pietrzak-Fiecko, R.; Molto, J.C.; Manes, J.; Juan, C. Development of an Extraction Method of Aflatoxins and Ochratoxin A from Oral, Gastric and Intestinal Phases of Digested Bread by In Vitro Model. *Toxins* **2022**, *14*, 38. [CrossRef] [PubMed]

36. Escriva, L.; Agahi, F.; Vila-Donat, P.; Manes, J.; Meca, G.; Manyes, L. Bioaccessibility Study of Aflatoxin B1 and Ochratoxin A in Bread Enriched with Fermented Milk Whey and/or Pumpkin. *Toxins* **2021**, *14*, 6. [CrossRef] [PubMed]

37. Frangiamone, M.; Alonso-Garrido, M.; Font, G.; Cimbalo, A.; Manyes, L. Pumpkin extract and fermented whey individually and in combination alleviated AFB1- and OTA-induced alterations on neuronal differentiation invitro. *Food Chem. Toxicol.* **2022**, *164*, 113011. [CrossRef] [PubMed]

38. Soleimany, F.; Jinap, S.; Abas, F. Determination of mycotoxins in cereals by liquid chromatography tandem mass spectrometry. *Food Chem.* **2012**, *130*, 1055–1060. [CrossRef]

39. European Commission. Guidance Document on Method Validation and Quality Control Procedures for Pesticide Residues Analysis in Food and Feed (Document no. SANTE/12682/2019). Available online: https://www.eurl-pesticides.eu/userfiles/file/EurlALL/AqcGuidance_SANTE_2019_12682.pdf (accessed on 8 October 2021).

40. Zhou, H.; Liu, N.; Yan, Z.; Yu, D.; Wang, L.; Wang, K.; Wei, X.; Wu, A. Development and validation of the one-step purification method coupled to LC-MS/MS for simultaneous determination of four aflatoxins in fermented tea. *Food Chem.* **2021**, *354*, 129497. [CrossRef] [PubMed]

41. Rodriguez-Carrasco, Y.; Ruiz, M.J.; Font, G.; Berrada, H. Exposure estimates to Fusarium mycotoxins through cereals intake. *Chemosphere* **2013**, *93*, 2297–2303. [CrossRef]